NATIONAL AERONAUTICS AND SPACE ADMINISTRATION

Technical Memorandum 33-736

Volume II

Mission Design Data for Venus, Mars, and Jupiter Through 1990

Andrey B. Sergeyevsky

JET PROPULSION LABORATORY
CALIFORNIA INSTITUTE OF TECHNOLOGY
PASADENA, CALIFORNIA
September 1, 1975

PREFACE

This document is divided into three volumes. Volume I comprises the mission design data for Venus, Volume II the data for Mars, and Volume III the data for Jupiter.

TABLE OF CONTENTS

LIST OF FIGURES

MISSION DESIGN DATA FOR VENUS, MARS AND JUPITER THROUGH 1990

Andrey B. Sergeyevsky

A. INTRODUCTION

This document presents mission design data for direct transfer trajectories from Earth to three planets – Venus, Mars and Jupiter, extending previously published information (see Refs. 1, 2, 3, 4 and 5) through the 1990 departure opportunity.

The primary purpose of this effort is to provide the mission analyst with graphical information, sufficient for preliminary mission design and evaluation. The data follows closely the format of Reference 4 and reflects methods of Reference 2. A specially modified version of the Space Research Conic Program (SPARC) (see Ref. 6) was used to generate the trajectory information presented. The data were automatically contour-plotted on the SC4020 plotter using the General Plot Program (GPP) (see Ref. 7), then hand retouched and labeled. A special program (VIEWPE) was constructed to provide planetary positional data in graphical form, plotted on the SC4020, and presented in original format.

The data are arranged in three sections by arrival planet, in natural sequence. Each section consists of two parts – the trajectory characteristics for all available opportunities to the particular planet, in chronological order, followed by that planet's positional data for every calendar year, from 1975 to 1995.

The persevering and encouraging insistence of management, especially that of Mr. Willard E. Bollman to carry this effort through to completion, as well as the graphic and editorial support of Mr. Richard W. Rackus are gratefully acknowledged.

B. DESCRIPTION OF TRAJECTORY CHARACTERISTICS DATA

1. General

The data represent trajectory performance information plotted in the departure date/arrival date space, thus defining all possible transfer trajectories between the two bodies, within the time-span considered. Fourteen individual parameters are contour-plotted on the departure energy (C_3) background contour chart, for each opportunity. The following opportunities are presented:

To Venus:	1975, 1976/7, 1978, 1980, 1981, 1983, 1984/5, 1986, 1988, 1989/90.
To Mars:	1979, 1981/2, 1983/4, 1985/6, 1988, 1990.
To Jupiter:	1977, 1978, 1979, 1980/81, 1981/82, 1983, 1984, 1985, 1986, 1987, 1988, 1989, 1990.

2. Definition of Terms

The following parameters are displayed on the contour plots:

C_3 Earth departure energy (km^2/sec^2); same as the square of departure hyperbolic excess velocity $V_\infty^2 = C_3 = V_I^2 - 2GM/R_I$, where

V_I = conic injection velocity (km/sec)

GM = gravitational constant times mass of the attracting body, from Reference 8:

GM_{VENUS} = 0.32486010E6 (km^3/sec^2)

GM_{EARTH} = 0.39860115E6

GM_{MARS} = 0.42828444E5

$GM_{JUPITER}$ = 0.12670772E9

R_I = $R_S + h_I$, Injection radius (km), sum of surface radius $R_{S_{PLANET}}$ and injection altitude h_I, where (see Ref. 8):

$R_{S_{VENUS}}$ = 6052 (km)

$R_{S_{EARTH}}$ = 6378.16

$R_{S_{MARS}}$ = 3393.4

$R_{S_{JUPITER}}$ = 71372

TF Time of flight (Days)

CD Earth to planet communication distance at arrival (km)

VHP Arrival hyperbolic excess velocity

$$V_\infty = \sqrt{V^2 - \frac{2GM}{R}}, \text{ (km/sec)},$$

where

V = Heliocentric conic arrival velocity at heliocentric radius R (km).

Arrival Planet Orbit insertion velocity increment ΔV, at periapse, may be computed from V_∞:

$$\Delta V = \sqrt{V_\infty^2 + \frac{GM}{R_P}} - \sqrt{\frac{2GM\,R_A}{R_p(R_A + R_p)}}$$

where R_p and R_A are planetocentric periapse and apoapse radii (km), respectively. Similarly, if specific capture orbit period P(sec) and periapse radius R_p are desired:

$$\Delta V = \sqrt{V^2 + \frac{2GM}{R_p}} - \sqrt{\frac{2GM}{R_p} - \sqrt[3]{\left(\frac{2GM\pi}{P}\right)^2}}$$

B-PLANE A plane normal to the incoming V_∞ - vector and passing through the center of planet.

T-AXIS Axis in B-plane, parallel to ecliptic (Earth mean orbital) plane (see Figure 1).

DLA Geocentric declination (vs. mean Earth equator of launch date) of the departure V_∞ - vector. May impose launch constraints. (deg)

ZAL Angle between departure V_∞ vector and Sun-Earth vector. Equivalent to Earth-probe-Sun angle, several days out. (deg)

INC Heliocentric inclination of transfer trajectory with mean ecliptic (Earth orbital) plane of launch date. (deg)

ZAP Angle between arrival V_∞ vector and the arrival planet-to-Sun vector. Equivalent to planet-probe-Sun angle at far encounter; for subsolar impact would be equal to 180°(deg).

ETS Angle in arrival B-plane, measured from T-axis, clockwise, to projection of Sun-to-planet vector. Equivalent to solar occultation region center-line. (deg)

LVI Planetocentric latitude of vertical impact vs arrival planet equator. Note that Venusian north is below ecliptic, while Mars' and Jupiter's is above. Equivalent to declination of the incoming asymptote (i.e., the negative of incoming V_∞ vector) in planetary equator system.

ZAE Angle between arrival V_∞ vector and the planet-to-Earth vector. Equivalent to planet-probe-Earth angle at far encounter. (Deg.)

ETE Angle in arrival B-plane, measured from T-axis, clockwise, to projection of Earth-to-planet vector. Equivalent to Earth occultation region centerline. (deg)

THA Angle in arrival B-plane, from T-axis, clockwise, to major axis of error dispersion ellipse (0 - 180 deg).

SG1 Semi-major axis magnitude of B-plane dispersion ellipse, resulting from a spherically distributed V_∞ velocity vector error of 0.1 m/sec on departure asymptote (km).

SG2 Semi-minor axis of above dispersion ecliptse (km).

SG3	Arrival time dispersion, normal to B-plane, for above error model (sec).

C. DESCRIPTION OF PLANETARY POSITIONAL DATA

1. General

The data represent planetary geometry-related information plotted versus calendar arrival date at the target planet. Each set of seven plots represents the annual time history of 19 parameters, and may be used for flyby and orbiter missions.

2. Description of Curve Labels

P	Target planet, equivalent to probe approaching or in orbit about target planet.
E	Earth
S	Sun
CA	Cone Angle, i.e., Sun-probe-object (Earth or Canopus, etc.) angle. (See Figure 2.)
KA	Clock Angle, i.e., angle between projections of the Probe-Canopus and probe-object vectors into the plane normal to the sun-line (for which CA = 90°). (See Figure 2.)
RISEXX	Rise time (GMT) of planet through 6° horizon mask at DSN Station No. XX. (e.g., XX = 14 = GOLDSTONE, 43 = CANBERRA, 63 = MADRID.)
SETXX	Set time (GMT) of planet through 6° horizon mask at DSN Station No. XX.
YR/M/D	Year, Month, Date.

3. Description of Plots

Plot	Y-axis label	
a)	DECLIN	Geocentric Earth equatorial declination of planet (P), planetocentric planetary equatorial declination of Earth (E) and Sun (S). Note that Venusian north is below ecliptic.
b)	EC.LON	Heliocentric ecliptic longitude of planet.
c)	CA,KA	Cone (ECA) and Clock (EKA) angle of Earth and cone angle of Canopus (CCA) as seen from a Sun-Canopus oriented spacecraft near target planet, P (see Figure 2).
d)	DISTANCE	Sun-Planet distance (SP) and Earth-Planet communication distance (EP) in mill. km.
e)	SUN-EARTH-PLANET	Sun-Earth-Planet angle (SEP), indicating times of superior (SEP $\simeq$ 0) and inferior (SEP $\simeq$ 180°) conjunction; SEP > 5° is a communications constraint.
f)	STATION RISE/SET	Rise and Set times (GMT) of planet at 3 DSN Stations on Earth, 6° mask.

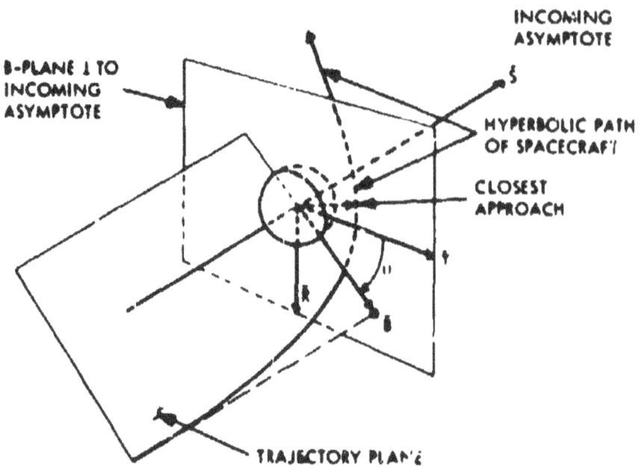

$\bar{B}$ = TARGET PARAMETER
θ = ORIENTATION OF $\bar{B}$
$\bar{S}$ = INCOMING ASYMPTOTE
$\bar{T}$ = PARALLEL TO ECLIPTIC PLANE AND ⊥ TO $\bar{S}$
$\bar{R}$ = $\bar{S} \times \bar{T}$

Figure 1. Definition of B-Plane

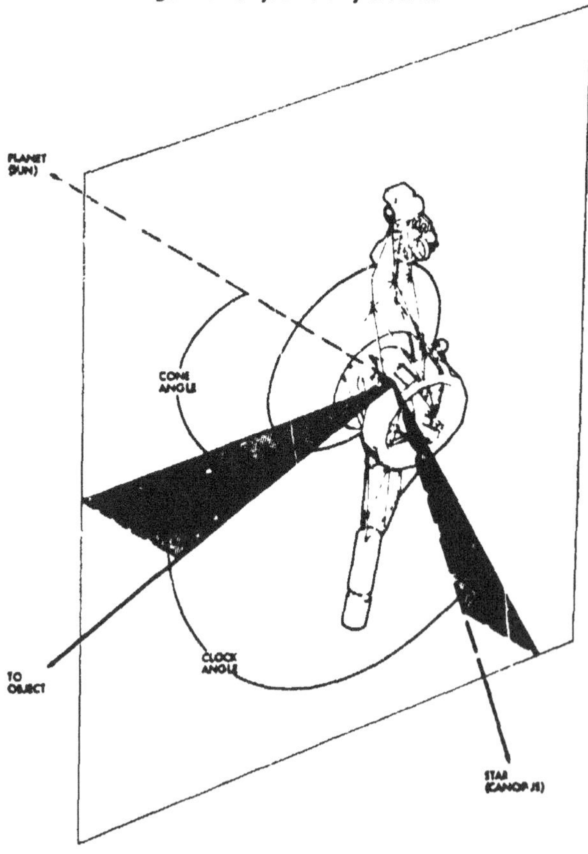

Figure 2. Definition of Cone and Clock Angle

REFERENCES

1. Clarke, V. C., Jr., Bollman, W. E., Roth, R. T., Scholey, W. J., "Design Parameters for Ballistic Interplanetary Trajectories Part I. One-way Transfers to Mars and Venus," JPL TR 32-77, January 1963.

2. Clarke, V. C., Jr., Bollman, W. E., Feritis, P. H., Roth, R. Y., "Design Parameters for Ballistic Interplanetary Trajectories Part II. One-way Transfers to Mercury and Jupiter," JPL TR 32-77, January 1966.

3. Richards, R. J., Roth, R. Y., "Earth-Mars Trajectories," JPL TM 33-100, June 1965.

4. Kohlhase, C. E., Bollman, W. E., "Trajectory Selection Considerations for Voyager Missions to Mars During the 1971-1977 Time Period," JPL TM 33-210, September 1965.

5. Wallace, R. A., "Trajectory Considerations for a Mission to Jupiter in 1972," JPL TM 33-375, March 1968.

6. Roth, R., Zorian, M. D., "Space Research Conic Program, Phase III," JPL 900-130, Rev. A, May 1969.*

7. "General Plot Program," JPL 900-341, Anon., May 1970.*

8. Melbourne, W. G., Mulholland, T. D., Sjogren, W. L., Sturms, F. M., Jr., "Constants and Related Information for Astrodynamic Calculations, 1968," JPL TR 32-1306, July 1968.

*JPL Internal Document

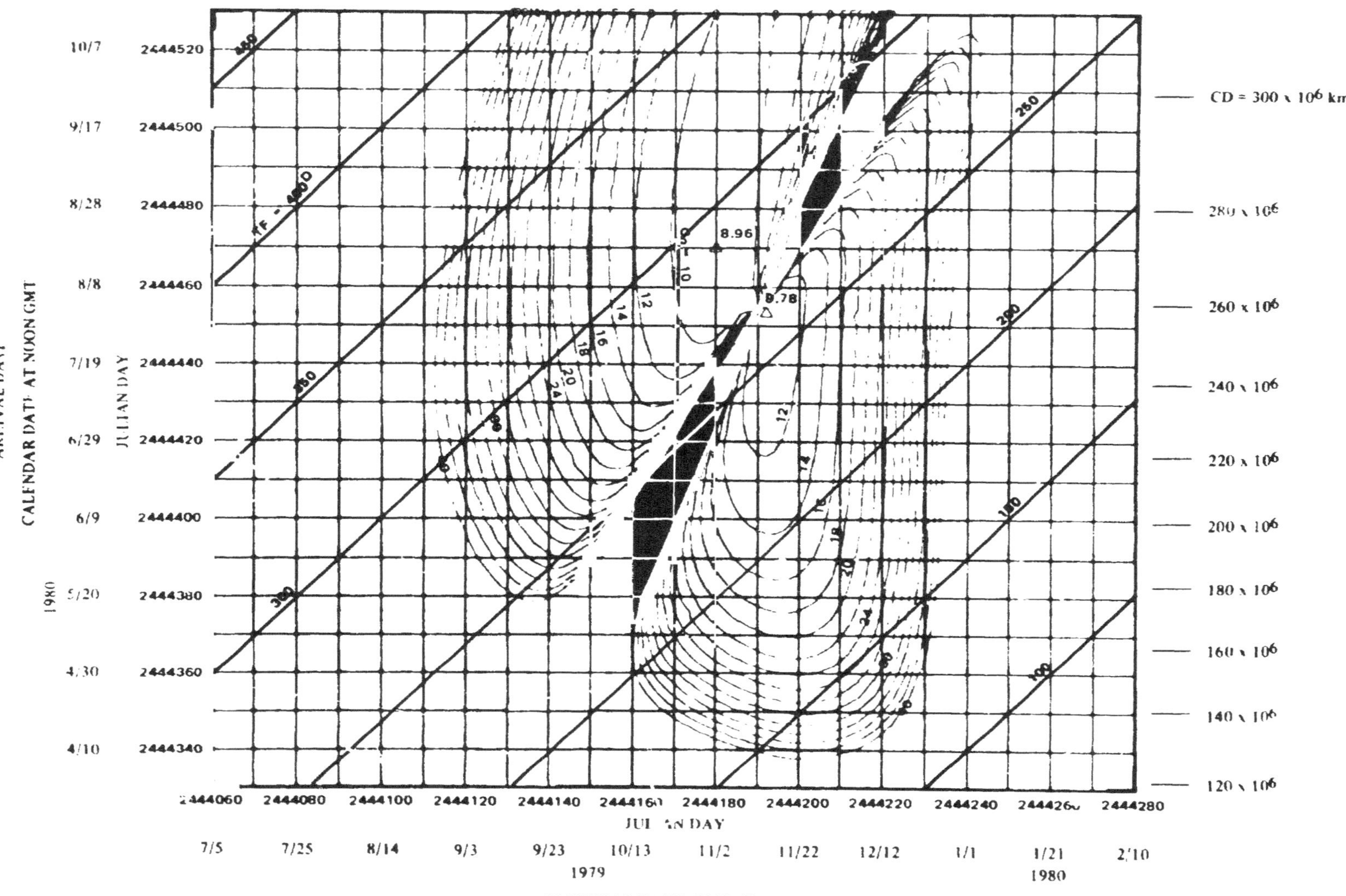

CONTOURS OF C_3 AND FLIGHT TIMES EARTH TO MARS 1979-80

C_3
♂
1979

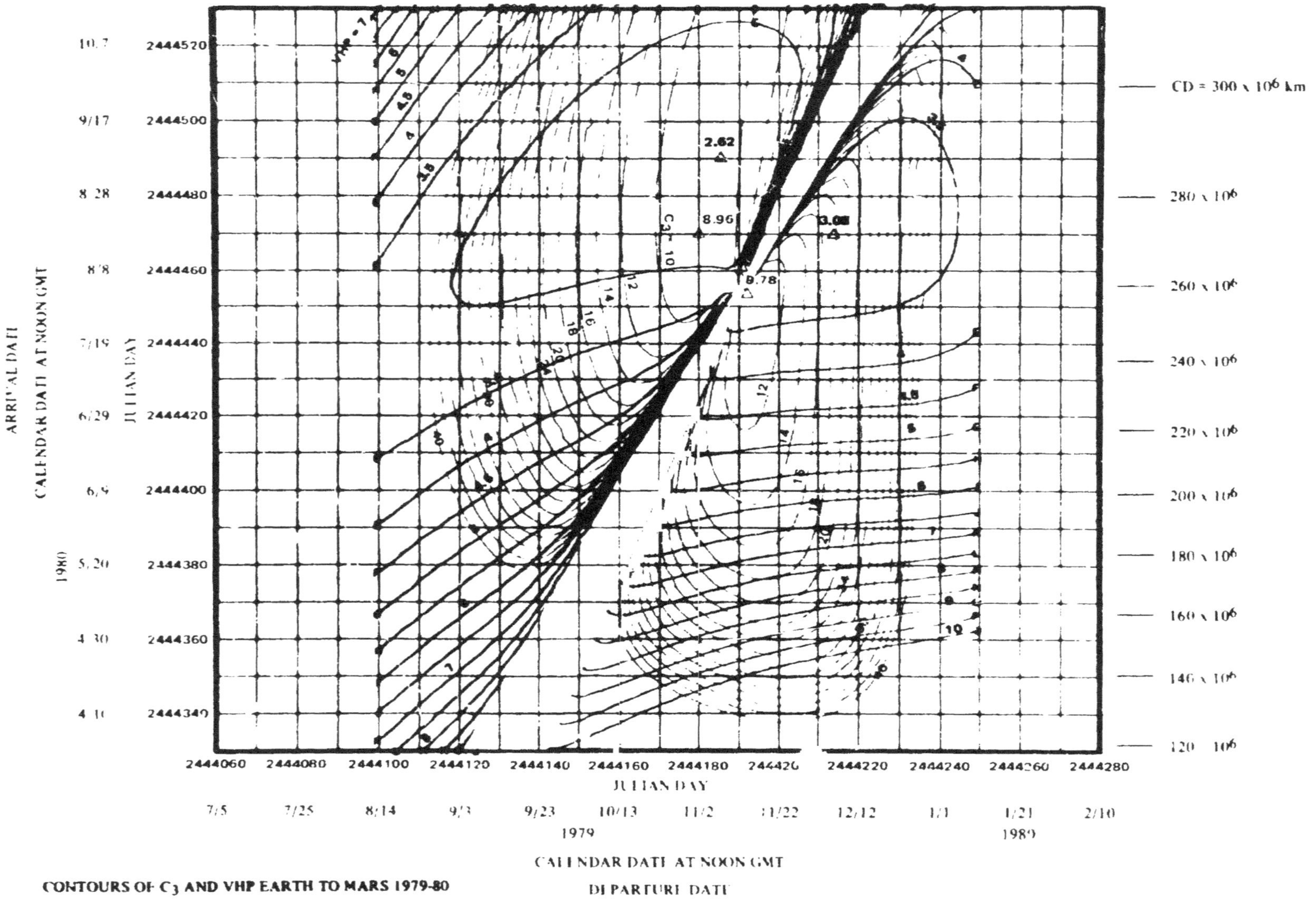

ARRIVAL DATE
CALENDAR DATE AT NOON GMT
1980

Calendar date	Julian day
10/7	2444520
9/17	2444500
8/28	2444480
8/8	2444460
7/19	2444440
6/29	2444420
6/9	2444400
5/20	2444380
4/30	2444360
4/10	2444340

CD = 300 x 10^6 km
280 x 10^6
260 x 10^6
240 x 10^6
220 x 10^6
200 x 10^6
180 x 10^6
160 x 10^6
140 x 10^6
120 x 10^6

JULIAN DAY
2444060 2444080 2444100 2444120 2444140 2444160 2444180 2444200 2444220 2444240 2444260 2444280

7/5 7/25 8/14 9/3 9/23 10/13 11/2 11/22 12/12 1/1 1/21 2/10
1979 1980

CALENDAR DATE AT NOON GMT
DEPARTURE DATE

VHP = 7 6 5 4.5 4 3.5
C_3 = 10 12 14 16 18 20 40
2.62 8.96 8.78 3.08

CONTOURS OF C_3 AND VHP EARTH TO MARS 1979-80

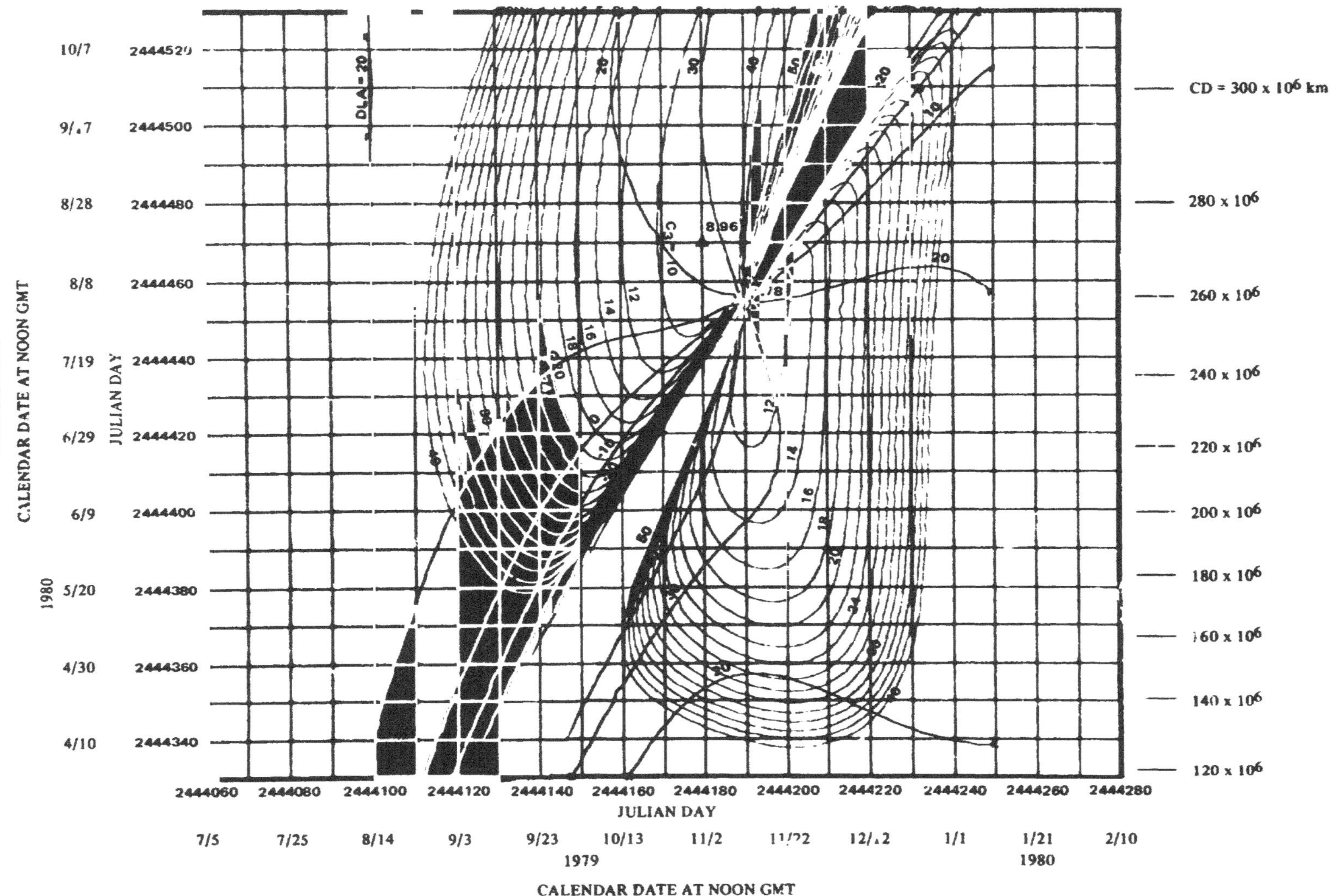

CONTOURS OF C_3 AND DLA EARTH TO MARS 1979-80

DLA
♂
1979

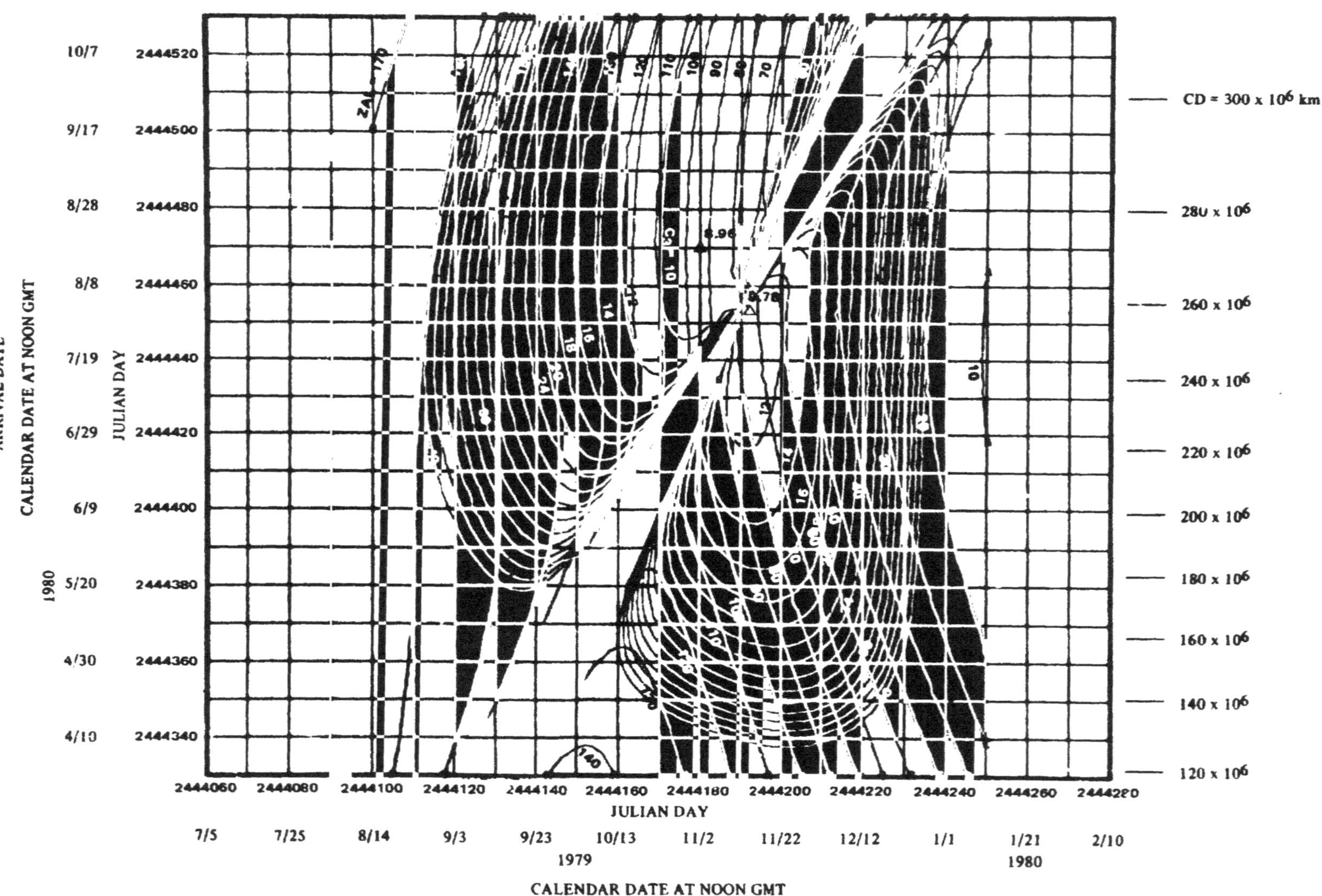

CONTOURS OF C_3 AND ZAL EARTH TO MARS 1979-80

ARRIVAL DATE
CALENDAR DATE AT NOON GMT

JULIAN DAY

Arrival (1980)	Julian Day
10/7	2444520
9/17	2444500
8/28	2444480
8/8	2444460
7/19	2444440
6/29	2444420
6/9	2444400
5/20	2444380
4/30	2444360
4/10	2444340

CD = 300 x 10^6 km
280 x 10^6
260 x 10^6
240 x 10^6
220 x 10^6
200 x 10^6
180 x 10^6
160 x 10^6
140 x 10^6
120 x 10^6

INC = 1.0
C_3 = 10

Julian Day	2444060	2444080	2444100	2444120	2444140	2444160	2444180	2444200	2444220	2444240	2444260	2444280
Calendar date	7/5	7/25	8/14	9/3	9/23	10/13	11/2	11/22	12/12	1/1	1/21	2/10

1979 1980

CALENDAR DATE AT NOON GMT

DEPARTURE DATE

CONTOURS OF C_3 AND INC EARTH TO MARS 1979-80

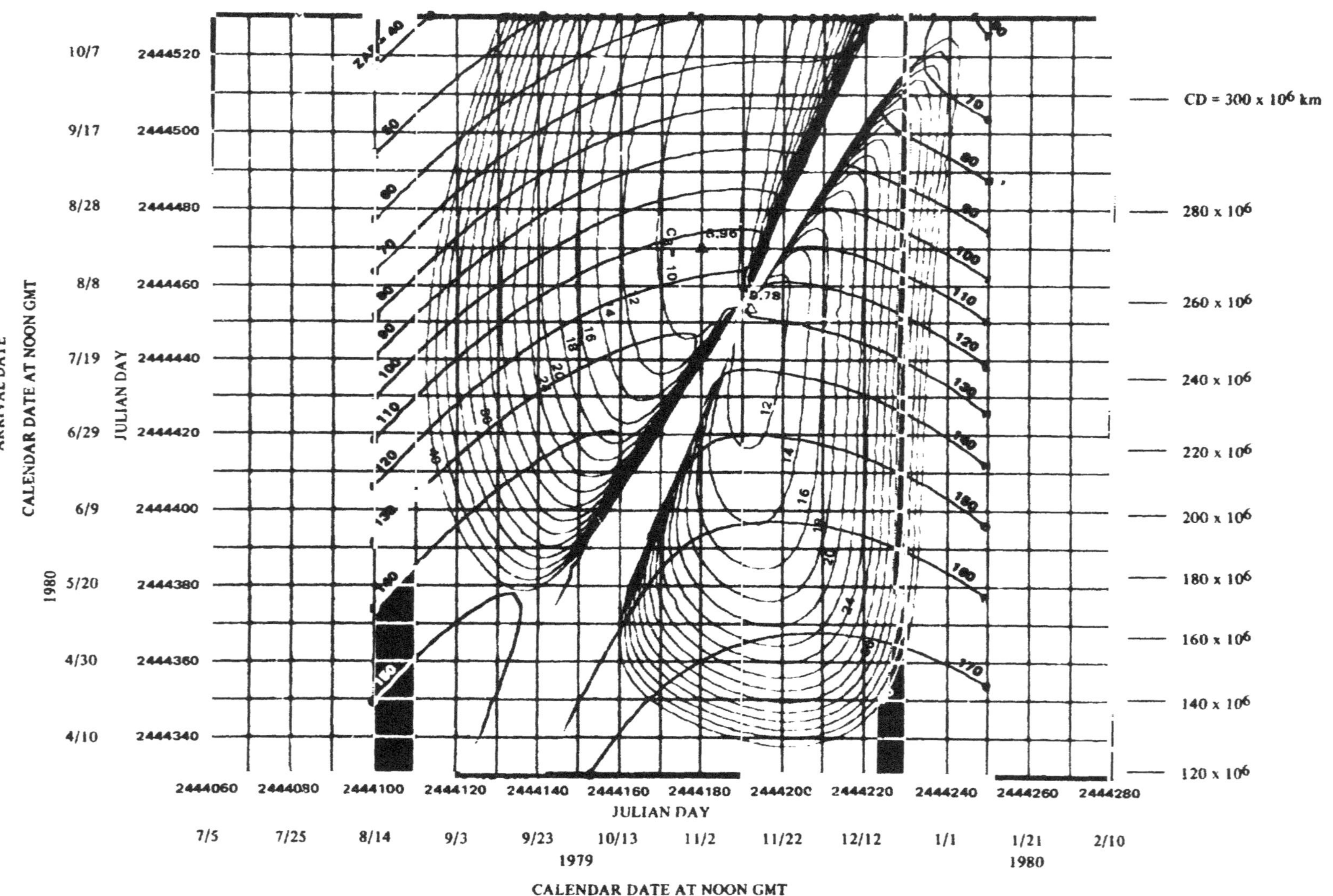

CONTOURS OF C_3 AND ZAP EARTH TO MARS 1979-80

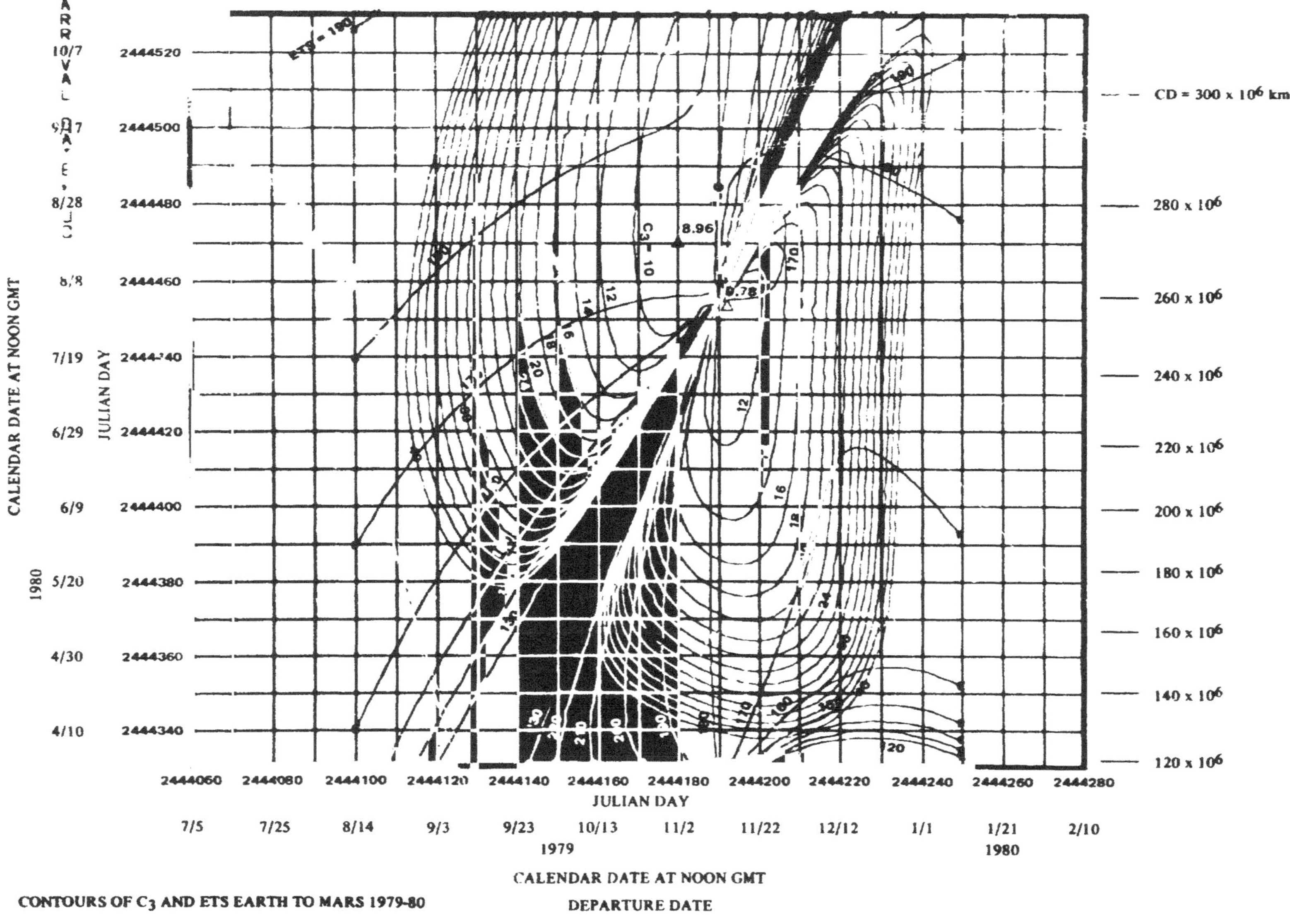

CONTOURS OF C_3 AND ETS EARTH TO MARS 1979-80

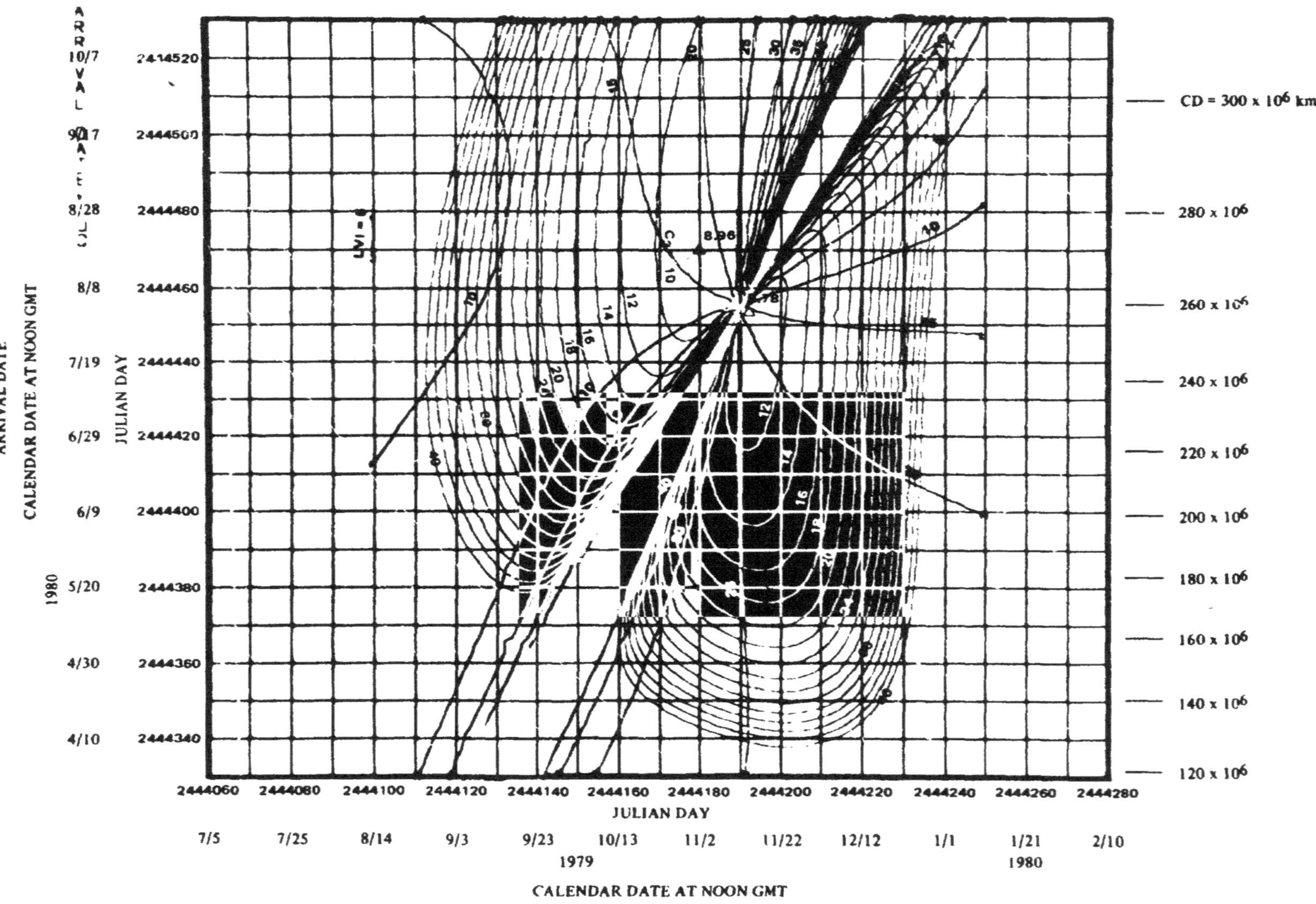

CONTOURS OF C_3 AND LVI EARTH TO MARS 1979-80

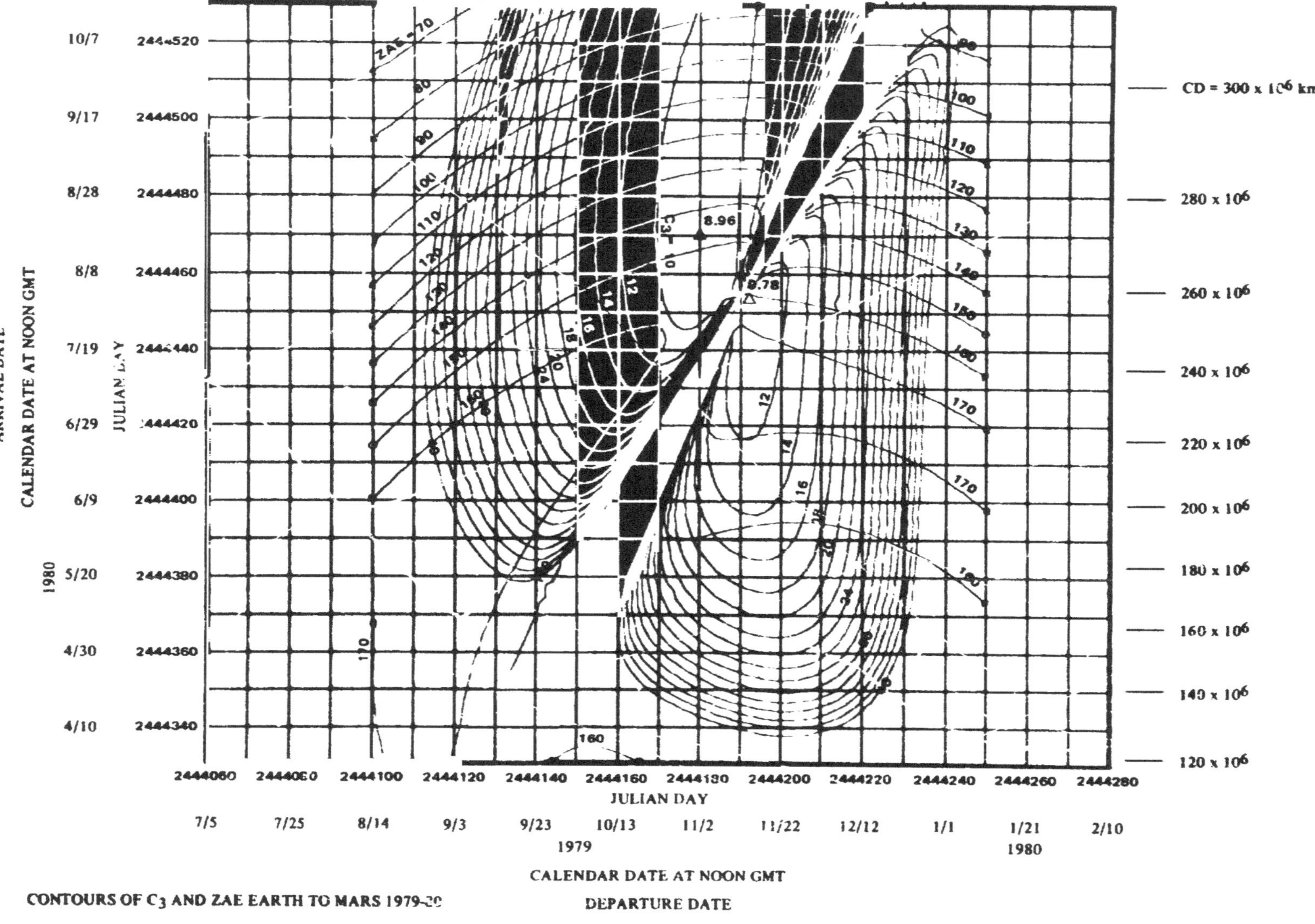

CONTOURS OF C_3 AND ZAE EARTH TO MARS 1979-80

ZAE
♂
1979

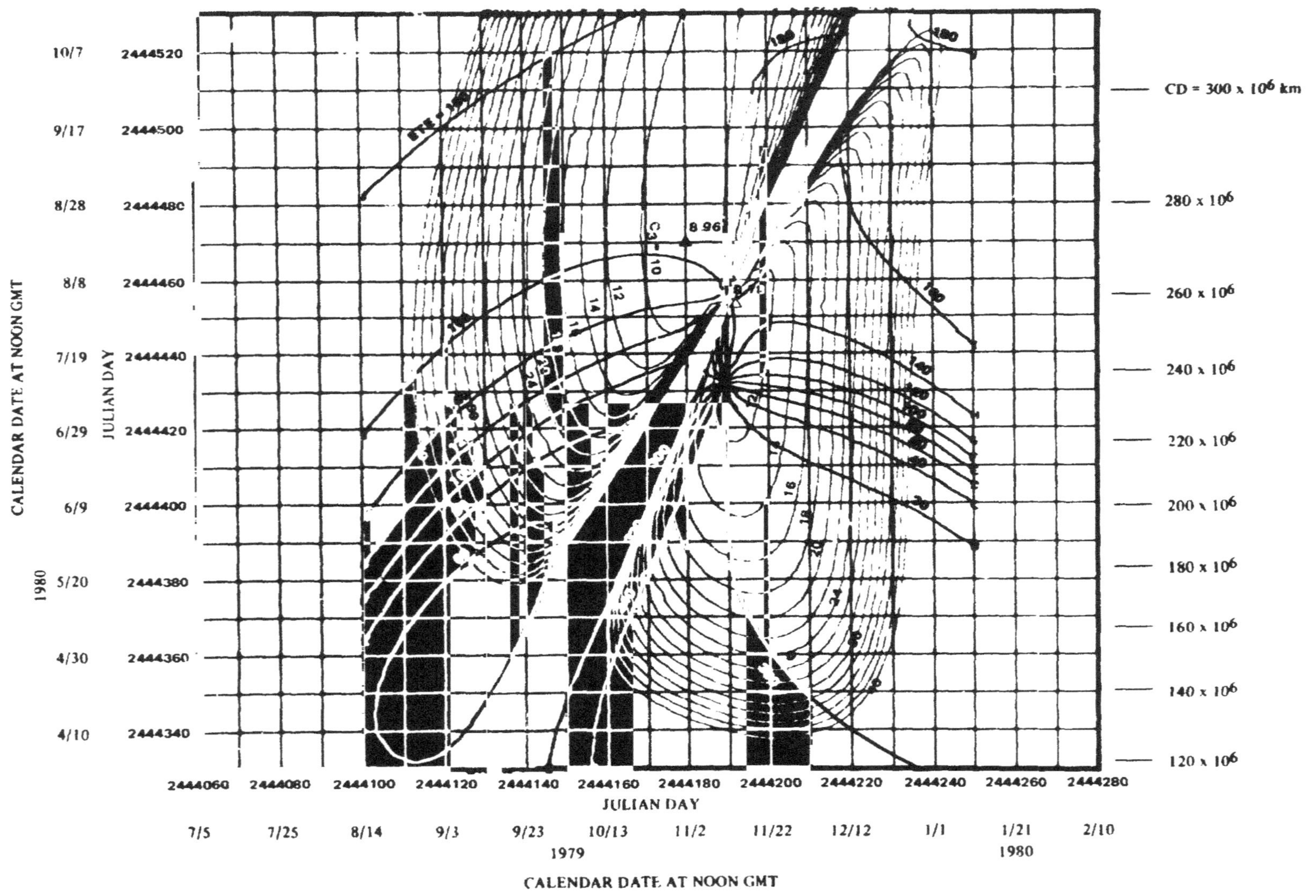

CONTOURS OF C_3 AND ETE EARTH TO MARS 1979-80

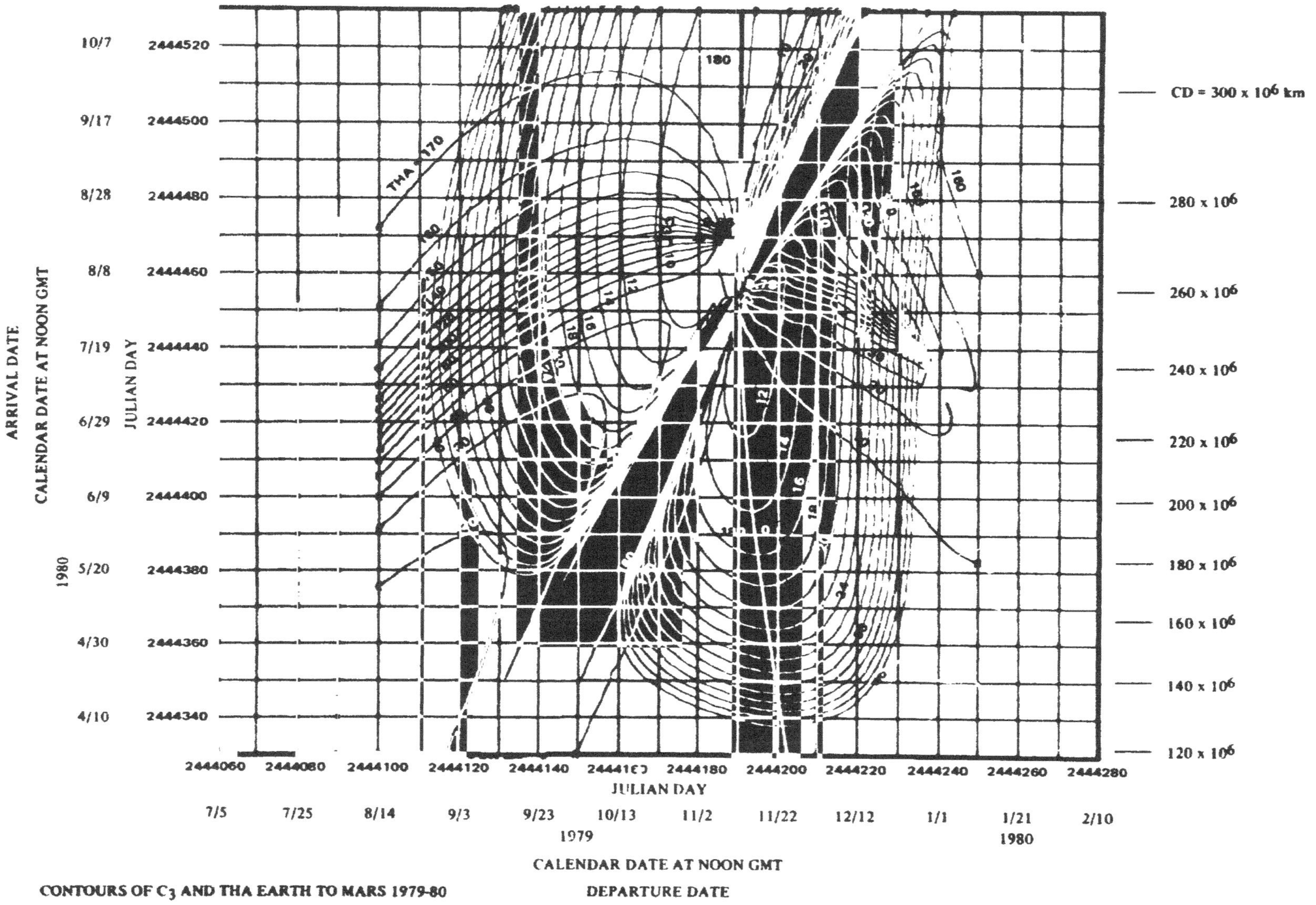

ARRIVAL DATE
CALENDAR DATE AT NOON GMT

JULIAN DAY

Arrival (1980)	Julian day
10/7	2444520
9/17	2444500
8/28	2444480
8/8	2444460
7/19	2444440
6/29	2444420
6/9	2444400
5/20	2444380
4/30	2444360
4/10	2444340

CD = 300 x 10^6 km
280 x 10^6
260 x 10^6
240 x 10^6
220 x 10^6
200 x 10^6
180 x 10^6
160 x 10^6
140 x 10^6
120 x 10^6

THA = 170

Julian day	Departure
2444060	7/5
2444080	7/25
2444100	8/14
2444120	9/3
2444140	9/23
2444160	10/13
2444180	11/2
2444200	11/22
2444220	12/12
2444240	1/1
2444260	1/21
2444280	2/10

JULIAN DAY

1979 1980

CALENDAR DATE AT NOON GMT

DEPARTURE DATE

CONTOURS OF C_3 AND THA EARTH TO MARS 1979-80

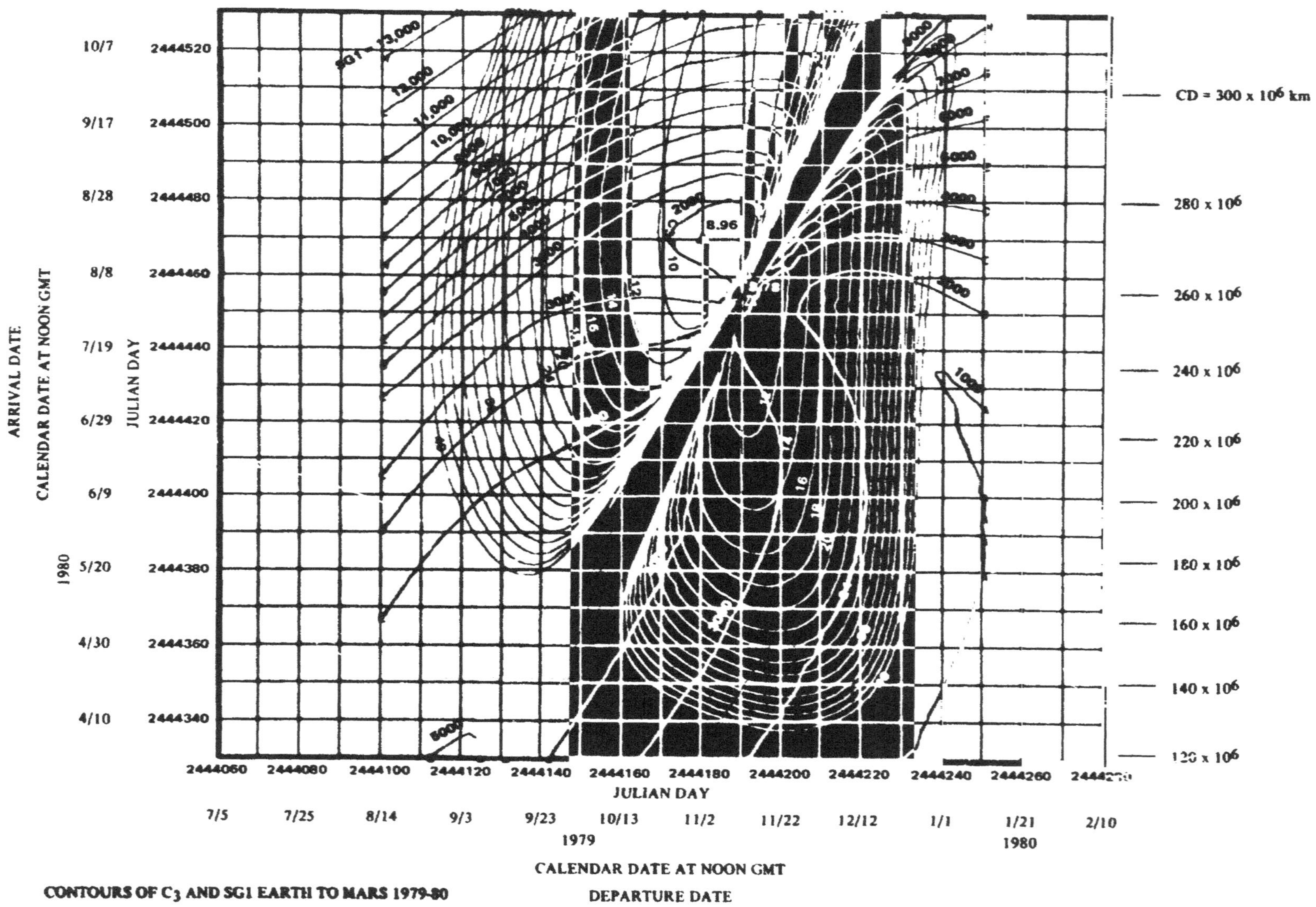

SG1

1979

ARRIVAL DATE

CALENDAR DATE AT NOON GMT

Calendar date	Julian day
10/7	2444520
9/17	2444500
8/28	2444480
8/8	2444460
7/19	2444440
6/29	2444420
6/9	2444400
5/20 1980	2444380
4/30	2444360
4/10	2444340

SG1 = 13,000, 12,000, 11,000, 10,000, 9000, 8000, 7000, 6000, 5000, 4000, 3000, 2000, 1000

C_3 = 8.96, 10, 12, 14, 16, 18, 20

CD = 300 x 10^6 km
280 x 10^6
260 x 10^6
240 x 10^6
220 x 10^6
200 x 10^6
180 x 10^6
160 x 10^6
140 x 10^6
120 x 10^6

Julian day	Calendar date
2444060	7/5
2444080	7/25
2444100	8/14
2444120	9/3
2444140	9/23
2444160	10/13 1979
2444180	11/2
2444200	11/22
2444220	12/12
2444240	1/1
2444260	1/21 1980
2444280	2/10

JULIAN DAY

CALENDAR DATE AT NOON GMT

DEPARTURE DATE

CONTOURS OF C_3 AND SG1 EARTH TO MARS 1979-80

SG2 ♂ 1979

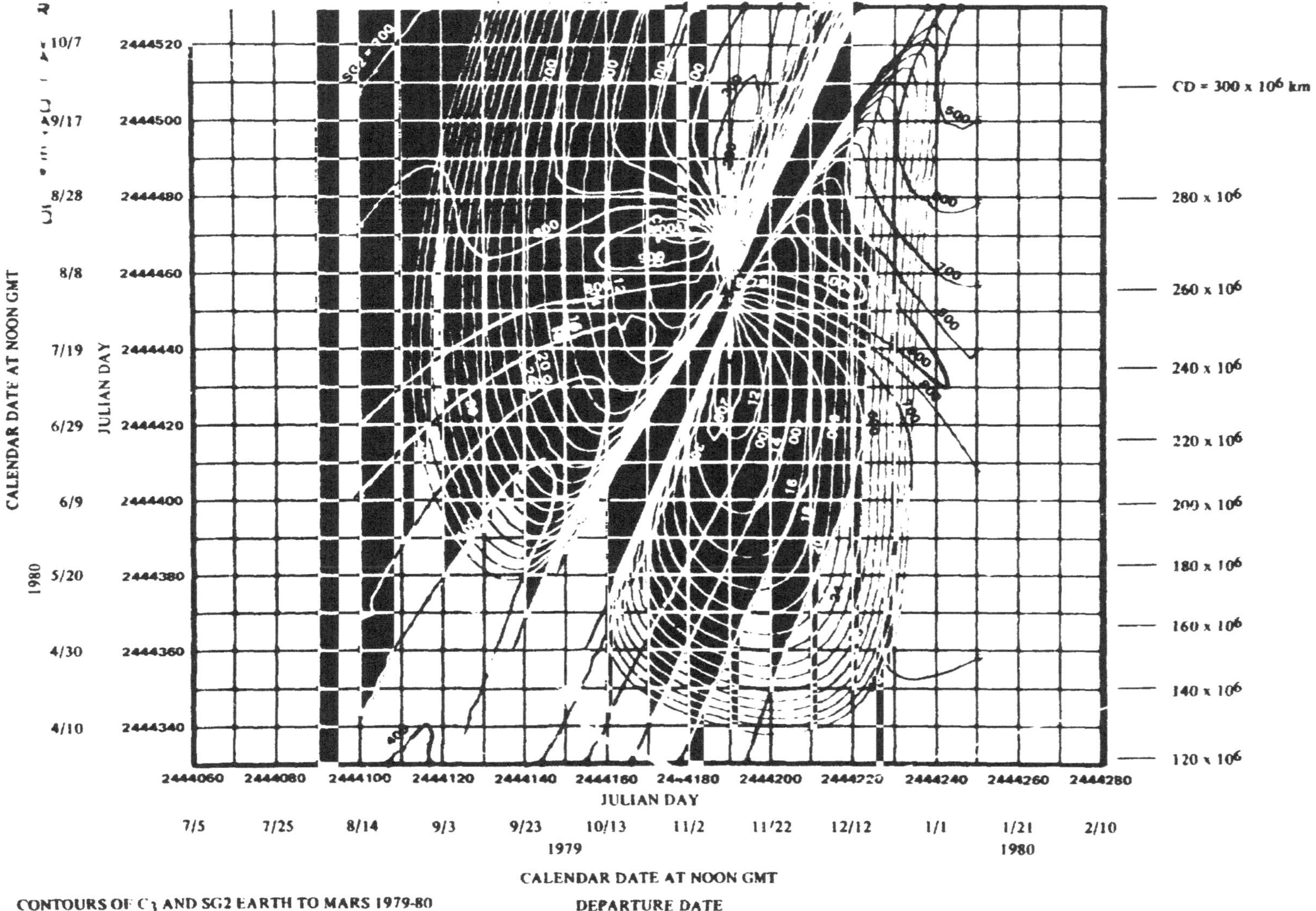

CONTOURS OF C_3 AND SG2 EARTH TO MARS 1979-80

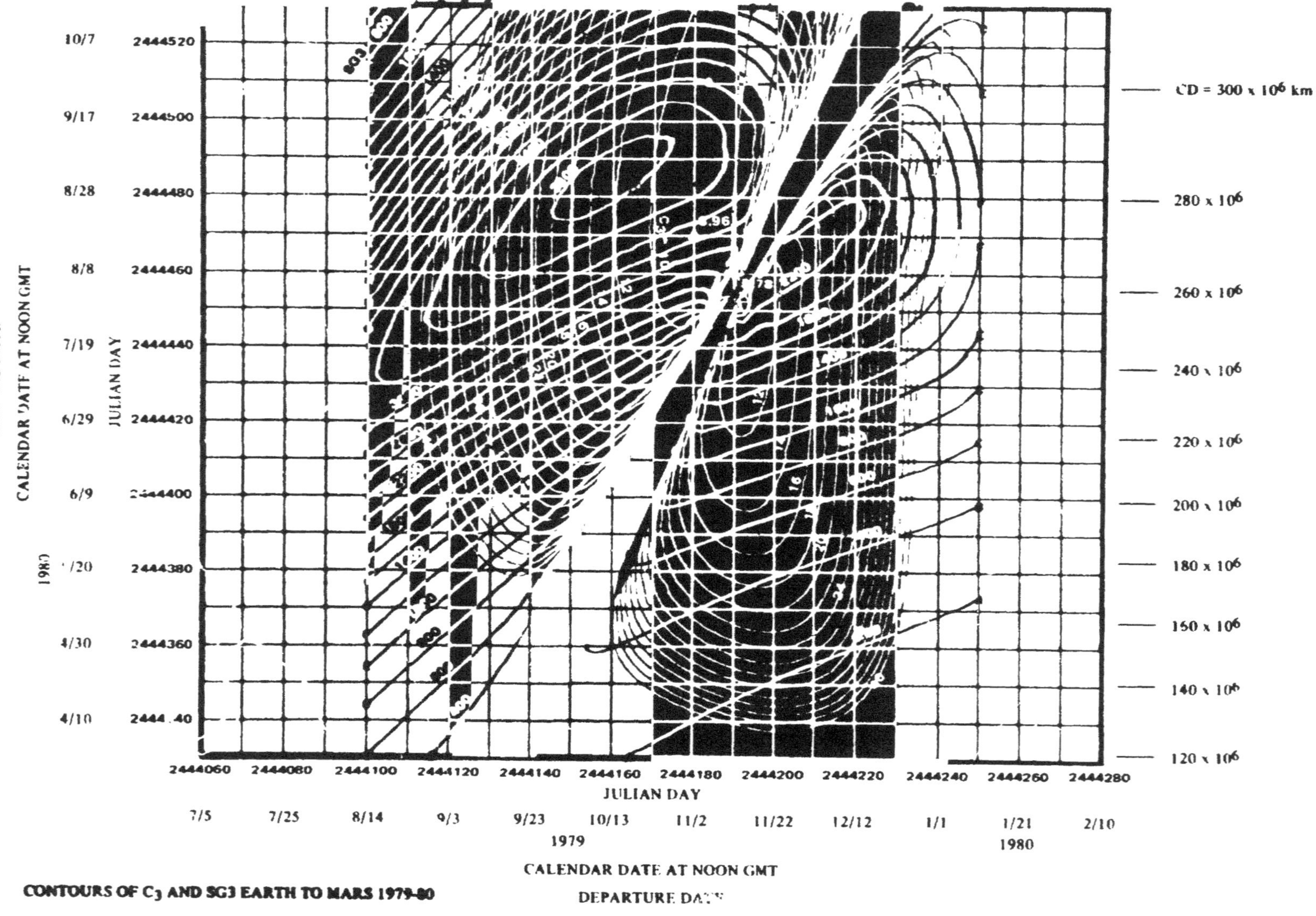

CONTOURS OF C_3 AND SG3 EARTH TO MARS 1979-80

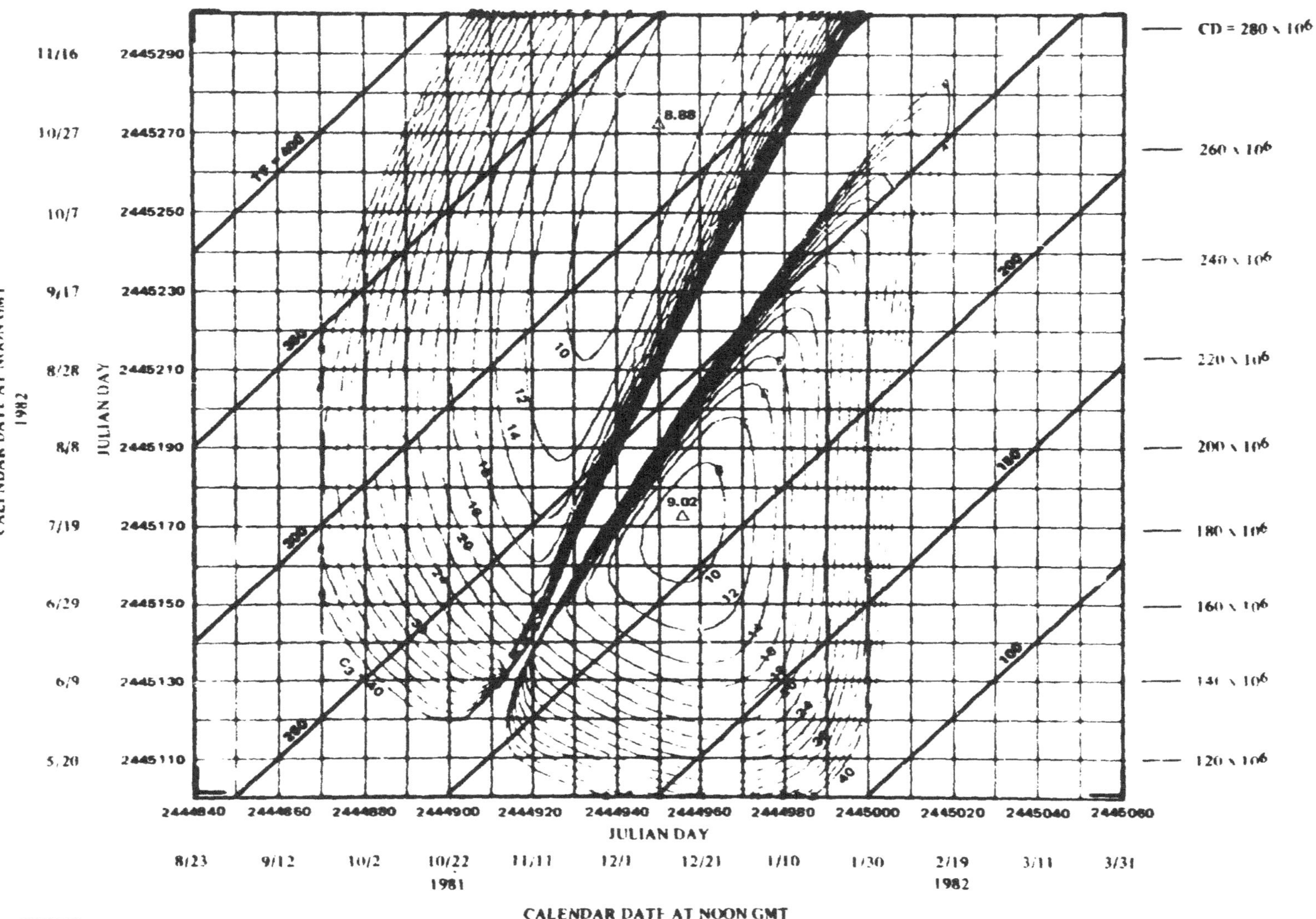

CONTOURS OF C_3 AND FLIGHT TIMES EARTH TO MARS 1981-82

♂ C_3 1981

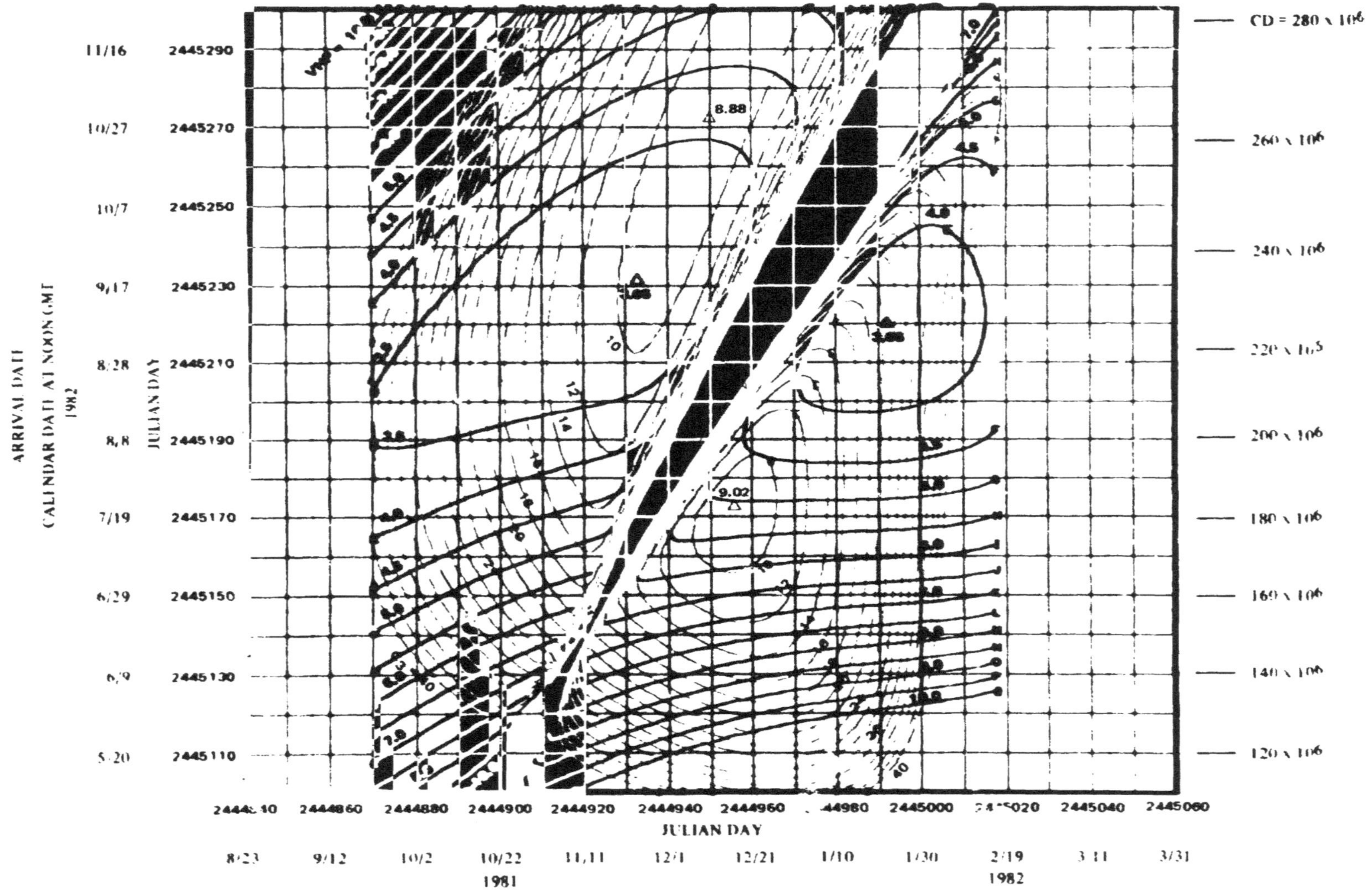

CD = 280 x 10⁶
260 x 10⁶
240 x 10⁶
220 x 10⁶
200 x 10⁶
180 x 10⁶
160 x 10⁶
140 x 10⁶
120 x 10⁶

ARRIVAL DATE
CALENDAR DATE AT NOON GMT
1982

CALENDAR DATE	JULIAN DAY
11/16	2445290
10/27	2445270
10/7	2445250
9/17	2445230
8/28	2445210
8/8	2445190
7/19	2445170
6/29	2445150
6/9	2445130
5/20	2445110

8.88
9.02

JULIAN DAY	2444840	2444860	2444880	2444900	2444920	2444940	2444960	2444980	2445000	2445020	2445040	2445060
CALENDAR DATE	8/23	9/12	10/2	10/22 (1981)	11/11	12/1	12/21	1/10	1/30	2/19 (1982)	3/11	3/31

CALENDAR DATE AT NOON GMT

DEPARTURE DATE

CONTOURS OF C_3 AND VHP EARTH TO MARS 1981-82

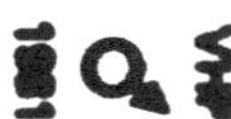

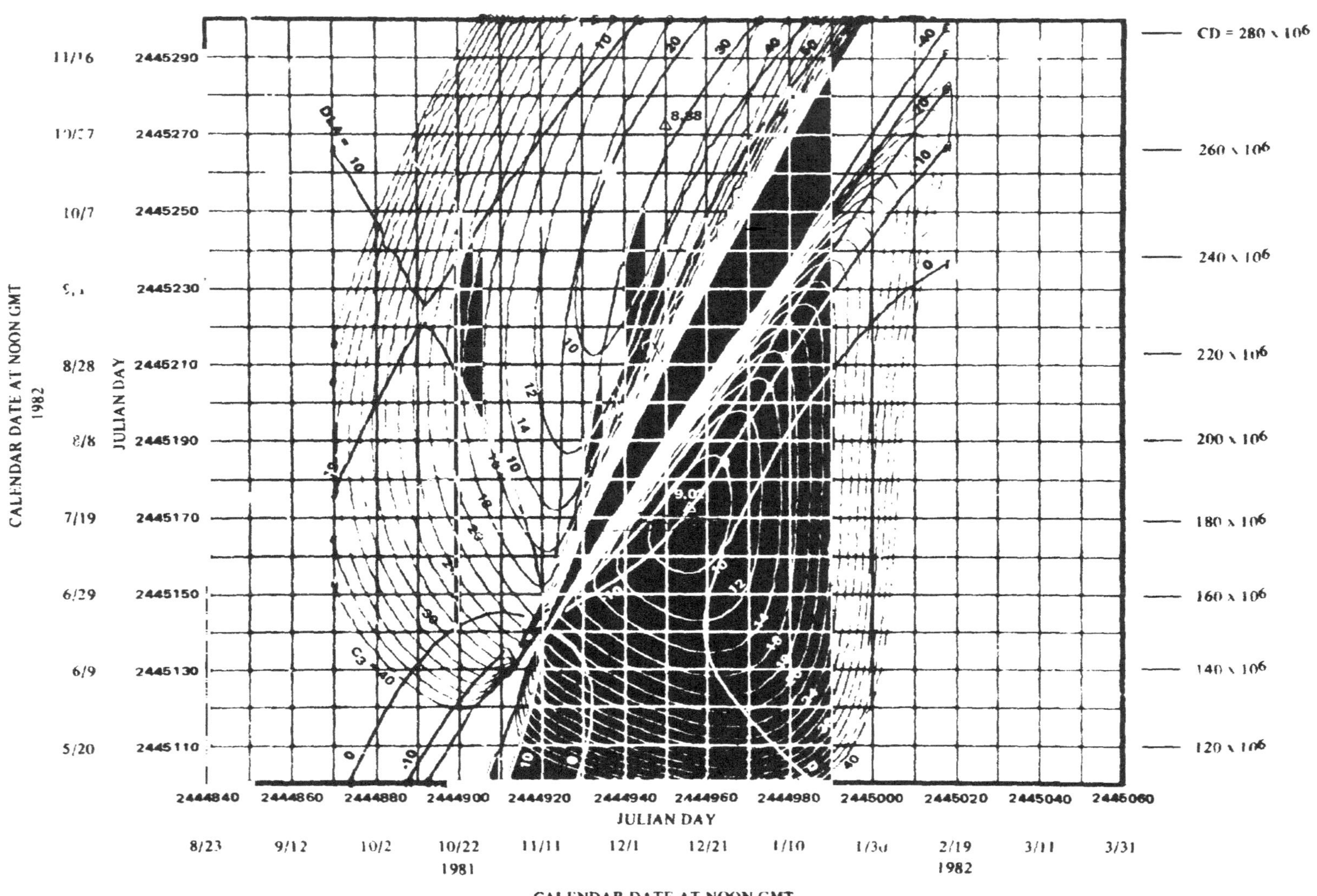

CONTOURS OF C_3 AND DLA EARTH TO MARS 1981-82

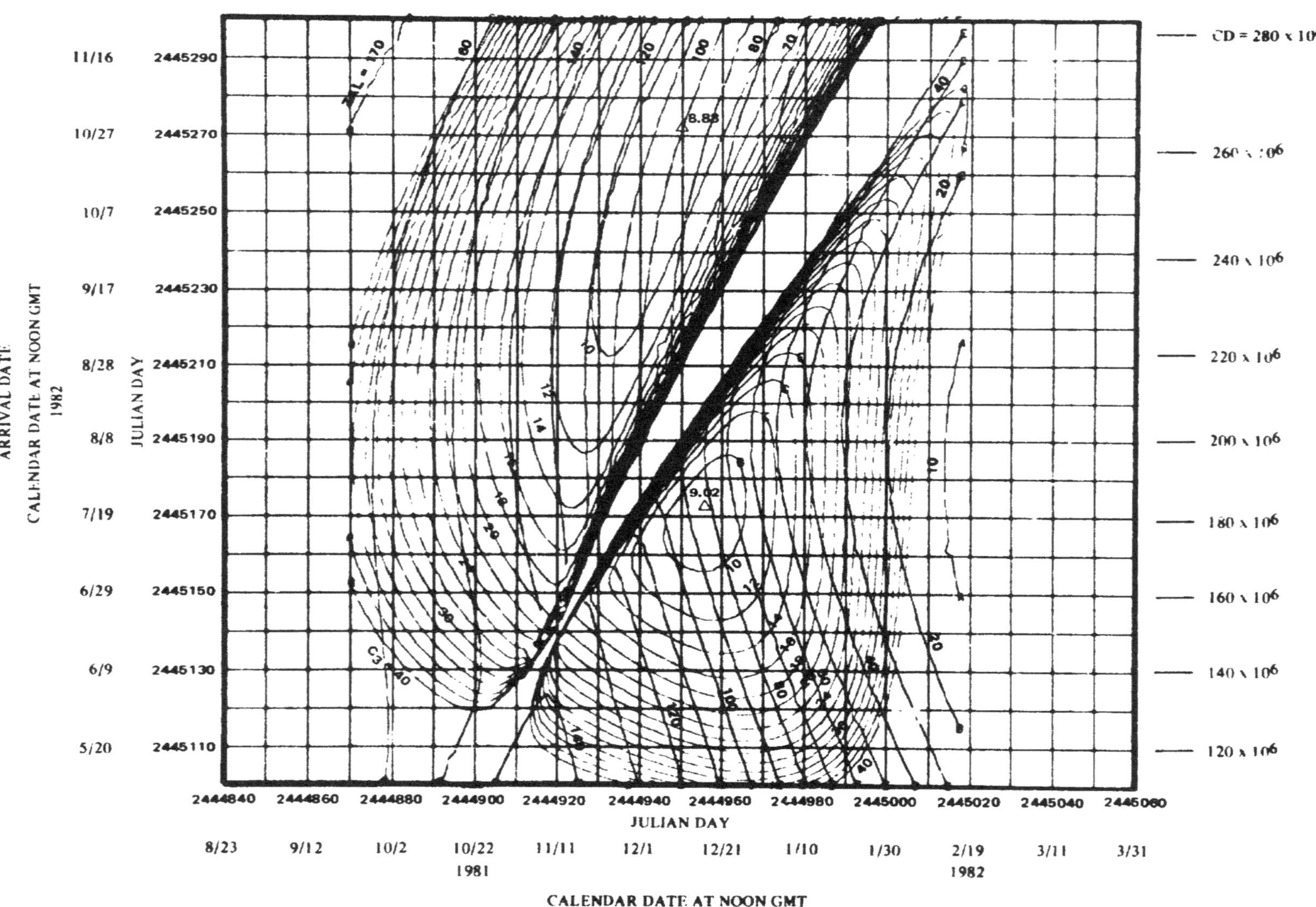

CONTOURS OF C_3 AND ZAL EARTH TO MARS 1981-82

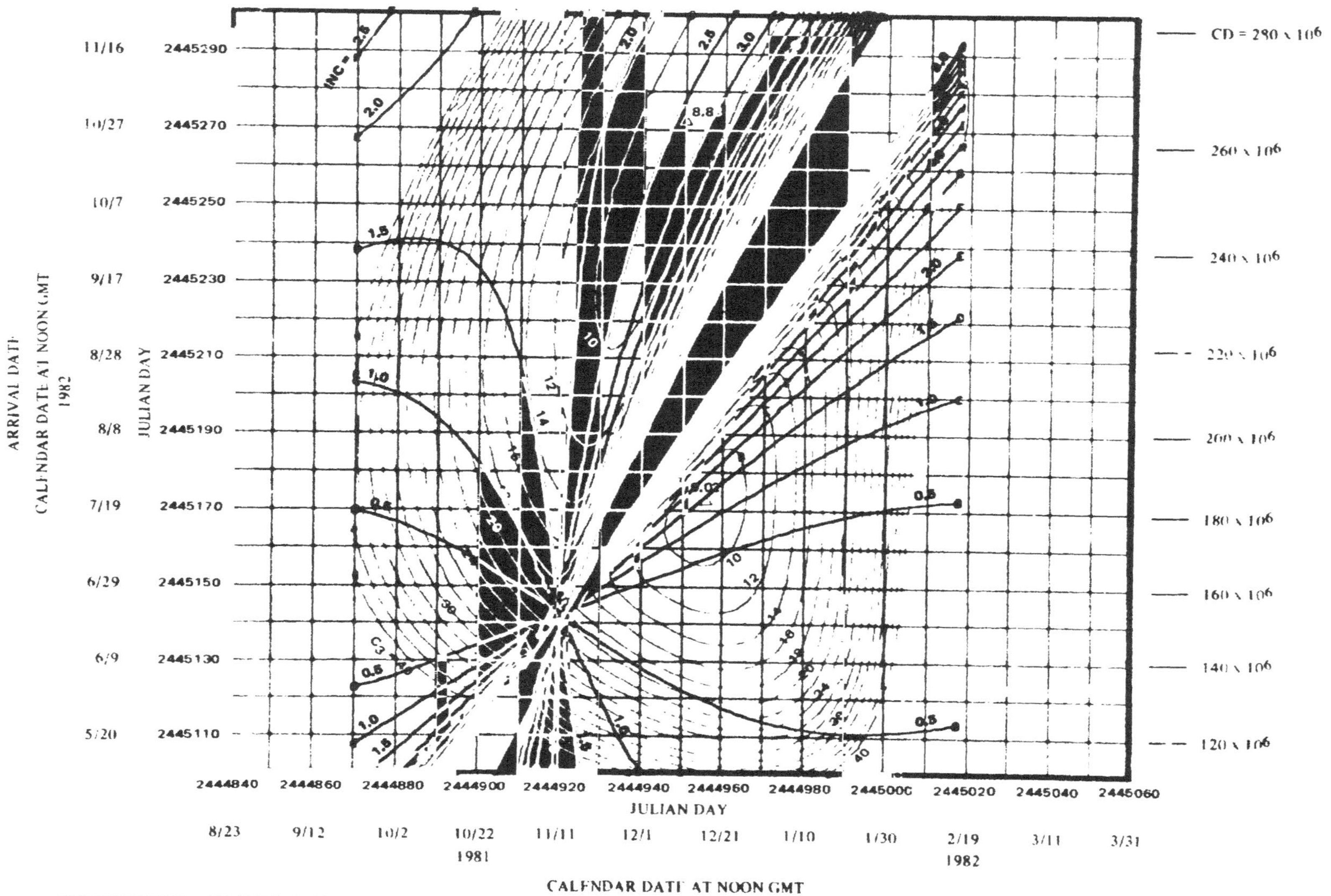

CD = 280 x 10^6

260 x 10^6

240 x 10^6

220 x 10^6

200 x 10^6

180 x 10^6

160 x 10^6

140 x 10^6

120 x 10^6

ARRIVAL DATE
CALENDAR DATE AT NOON GMT
1982

Calendar date	Julian day
11/16	2445290
10/27	2445270
10/7	2445250
9/17	2445230
8/28	2445210
8/8	2445190
7/19	2445170
6/29	2445150
6/9	2445130
5/20	2445110

Julian day	Calendar date
2444840	8/23
2444860	9/12
2444880	10/2
2444900	10/22 1981
2444920	11/11
2444940	12/1
2444960	12/21
2444980	1/10
2445000	1/30
2445020	2/19 1982
2445040	3/11
2445060	3/31

JULIAN DAY

CALENDAR DATE AT NOON GMT

DEPARTURE DATE

INC = 2.5 · 2.0 · 1.5 · 1.0 · 0.5 · 2.5 · 3.0 · 8.8 · 10 · 12 · 14 · 16 · 20 · 30 · 40 · C_3

CONTOURS OF C_3 AND INC EARTH TO MARS 1981-82

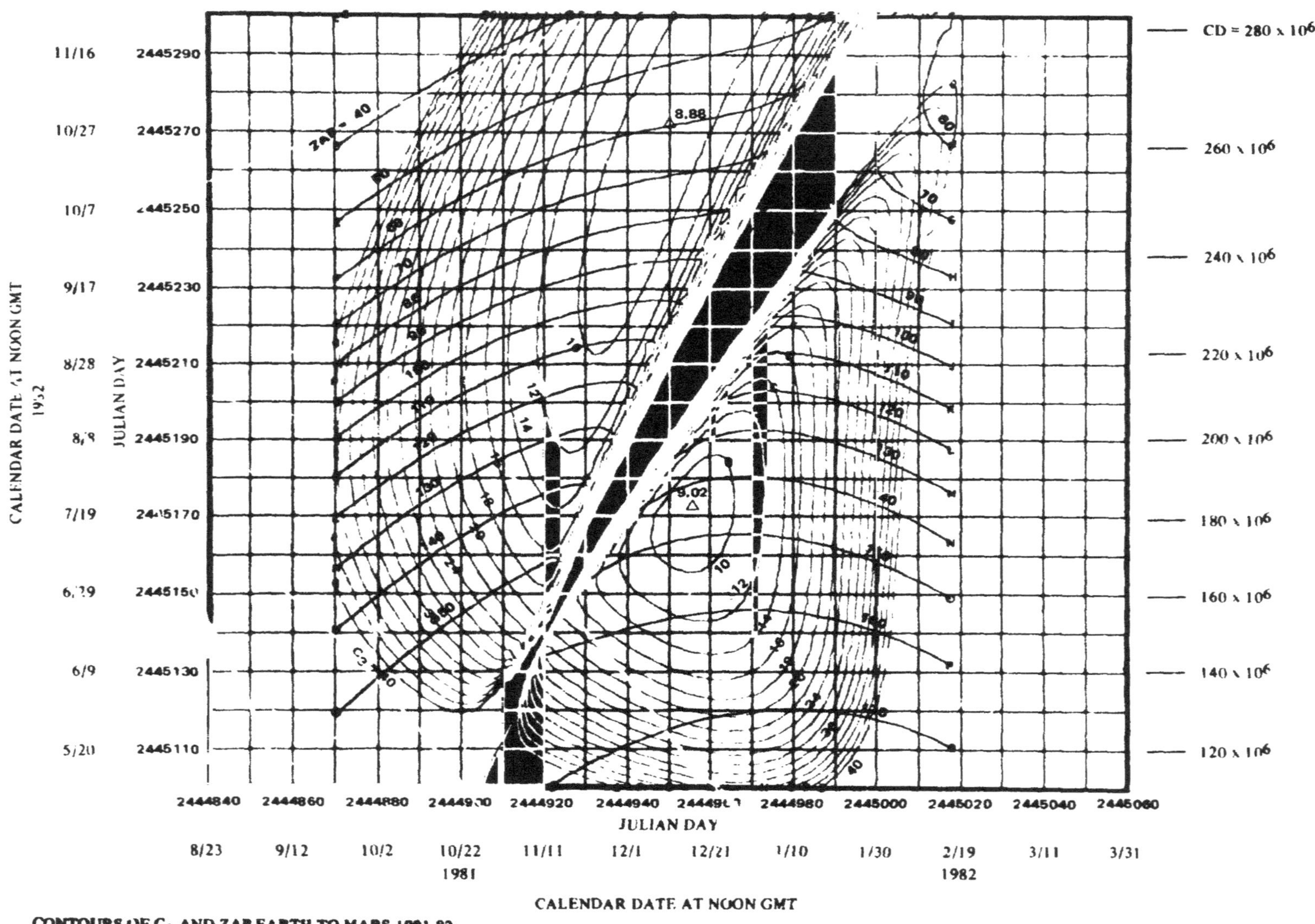

CONTOURS OF C_3 AND ZAP EARTH TO MARS 1981-82

ZAP ♂ 1981

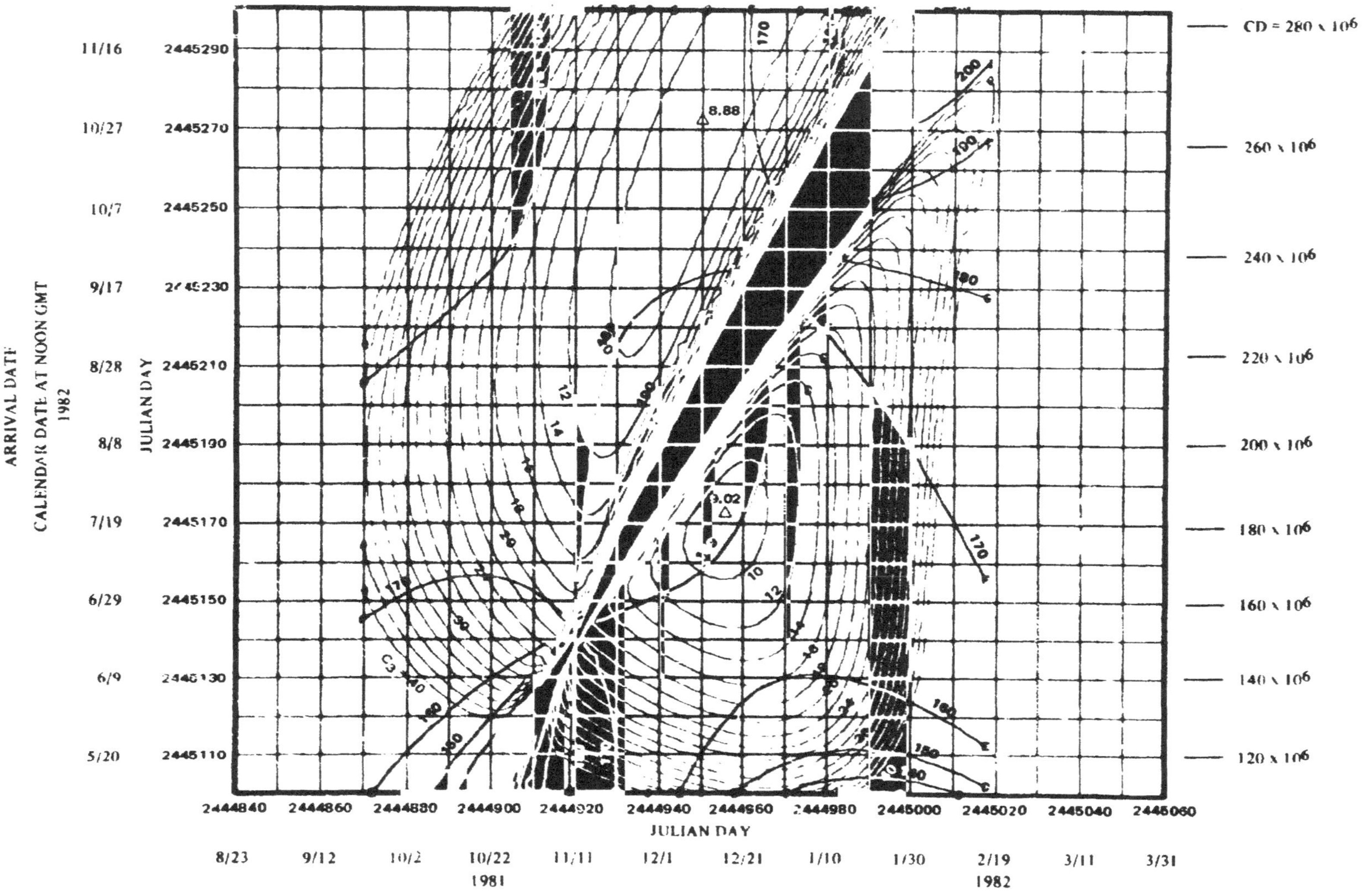

ARRIVAL DATE
CALENDAR DATE AT NOON GMT
1982

JULIAN DAY

Arrival (1982)	Julian Day
11/16	2445290
10/27	2445270
10/7	2445250
9/17	2445230
8/28	2445210
8/8	2445190
7/19	2445170
6/29	2445150
6/9	2445130
5/20	2445110

CD = 280 x 10^6
260 x 10^6
240 x 10^6
220 x 10^6
200 x 10^6
180 x 10^6
160 x 10^6
140 x 10^6
120 x 10^6

8.88
3.02
C_3 = 40

2444840	2444860	2444880	2444900	2444920	2444940	2444960	2444980	2445000	2445020	2445040	2445060
8/23	9/12	10/2	10/22 1981	11/11	12/1	12/21	1/10	1/30	2/19 1982	3/11	3/31

JULIAN DAY

CALENDAR DATE AT NOON GMT
DEPARTURE DATE

CONTOURS OF C_3 AND ETS EARTH TO MARS 1981-82

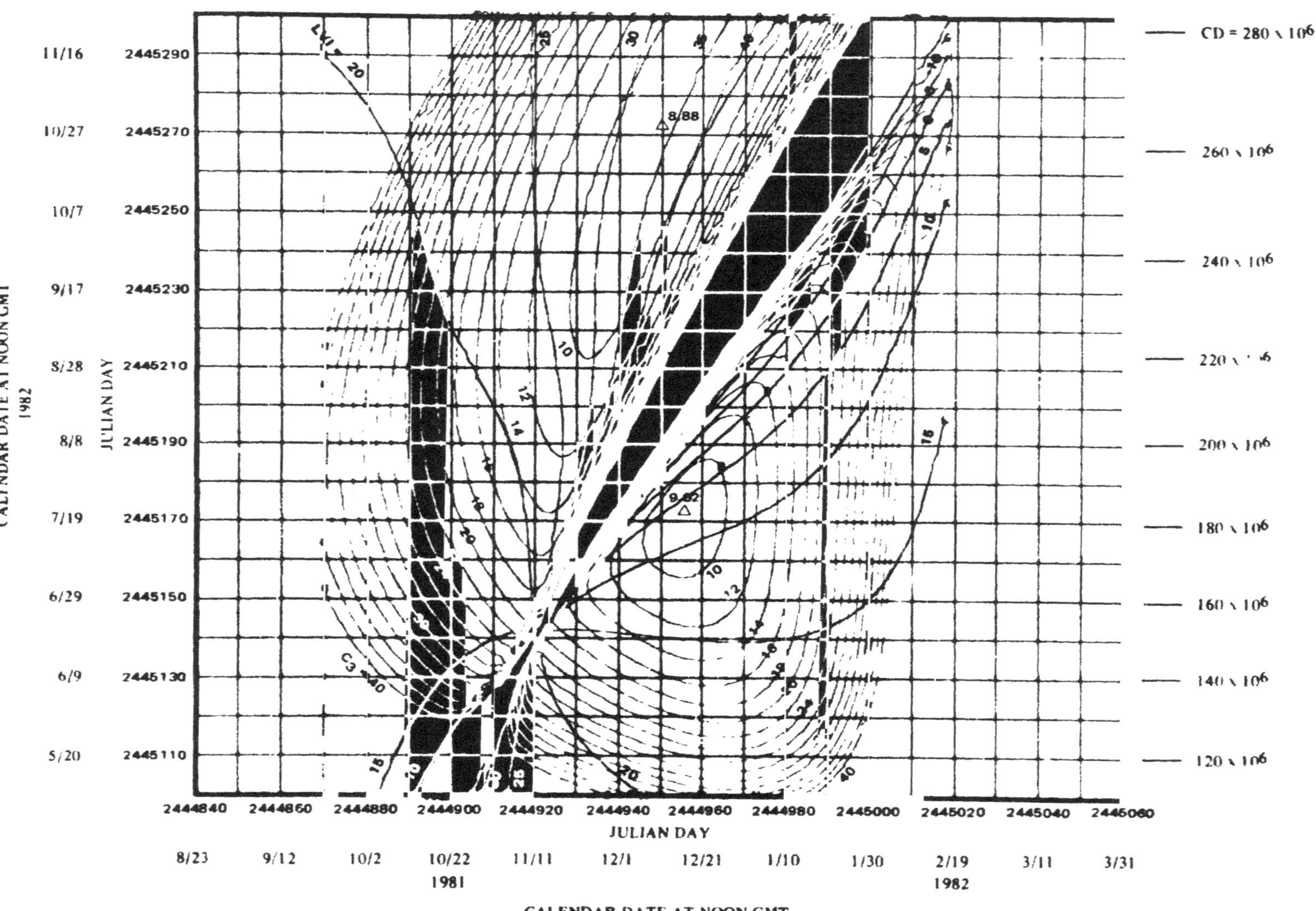

CONTOURS OF C_3 AND LVI EARTH TO MARS 1981-82

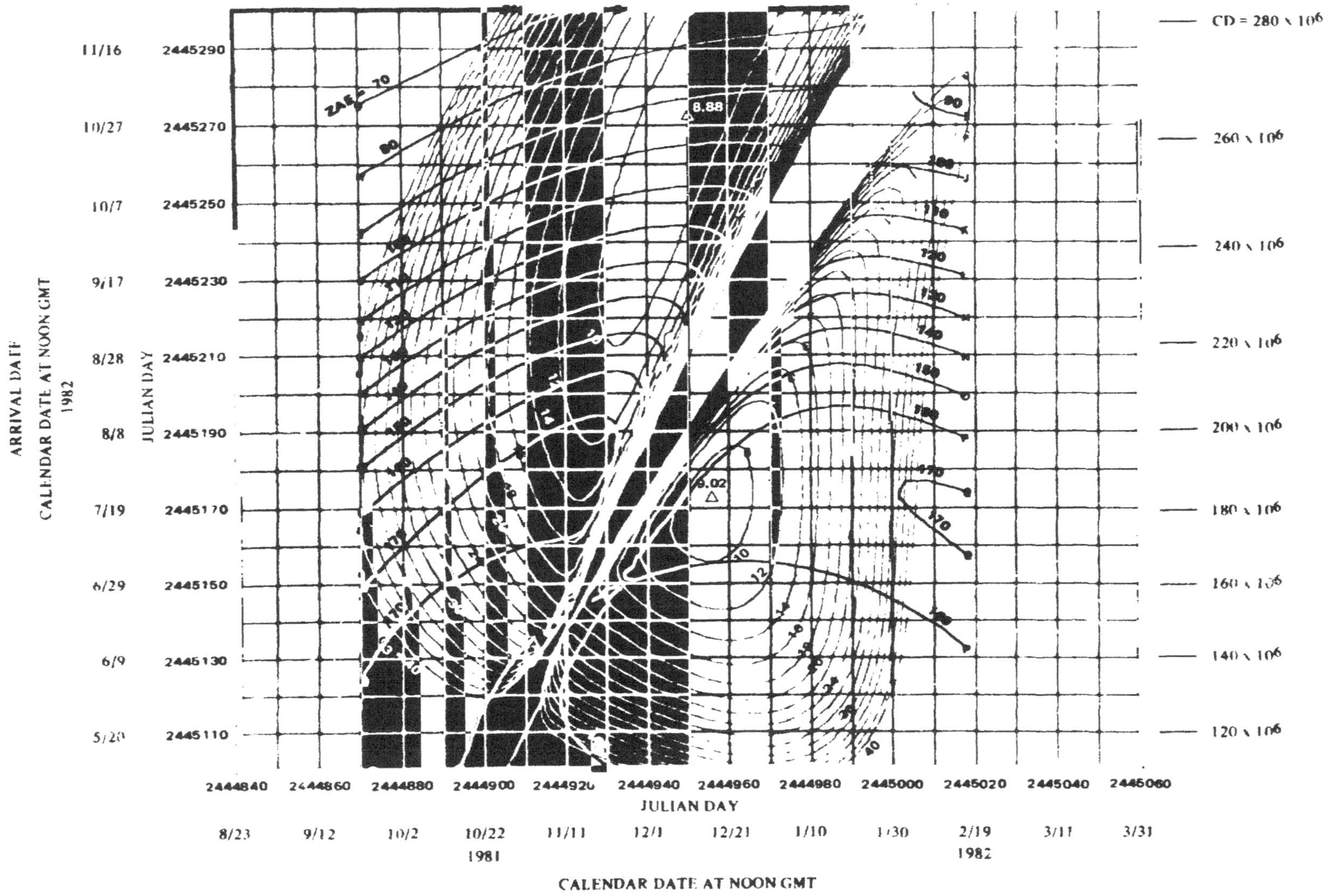

CD = 280 x 10^6
260 x 10^6
240 x 10^6
220 x 10^6
200 x 10^6
180 x 10^6
160 x 10^6
140 x 10^6
120 x 10^6

ARRIVAL DATE
CALENDAR DATE AT NOON GMT
1982

JULIAN DAY

Arrival date	Julian day
11/16	2445290
10/27	2445270
10/7	2445250
9/17	2445230
8/28	2445210
8/8	2445190
7/19	2445170
6/29	2445150
6/9	2445130
5/20	2445110

ZAE 70
80
8.88
9.02

Julian day	2444840	2444860	2444880	2444900	2444920	2444940	2444960	2444980	2445000	2445020	2445040	2445060
Calendar date	8/23	9/12	10/2	10/22 1981	11/11	12/1	12/21	1/10	1/30	2/19 1982	3/11	3/31

JULIAN DAY

CALENDAR DATE AT NOON GMT
DEPARTURE DATE

CONTOURS OF C_3 AND ZAE EARTH TO MARS 1981-82

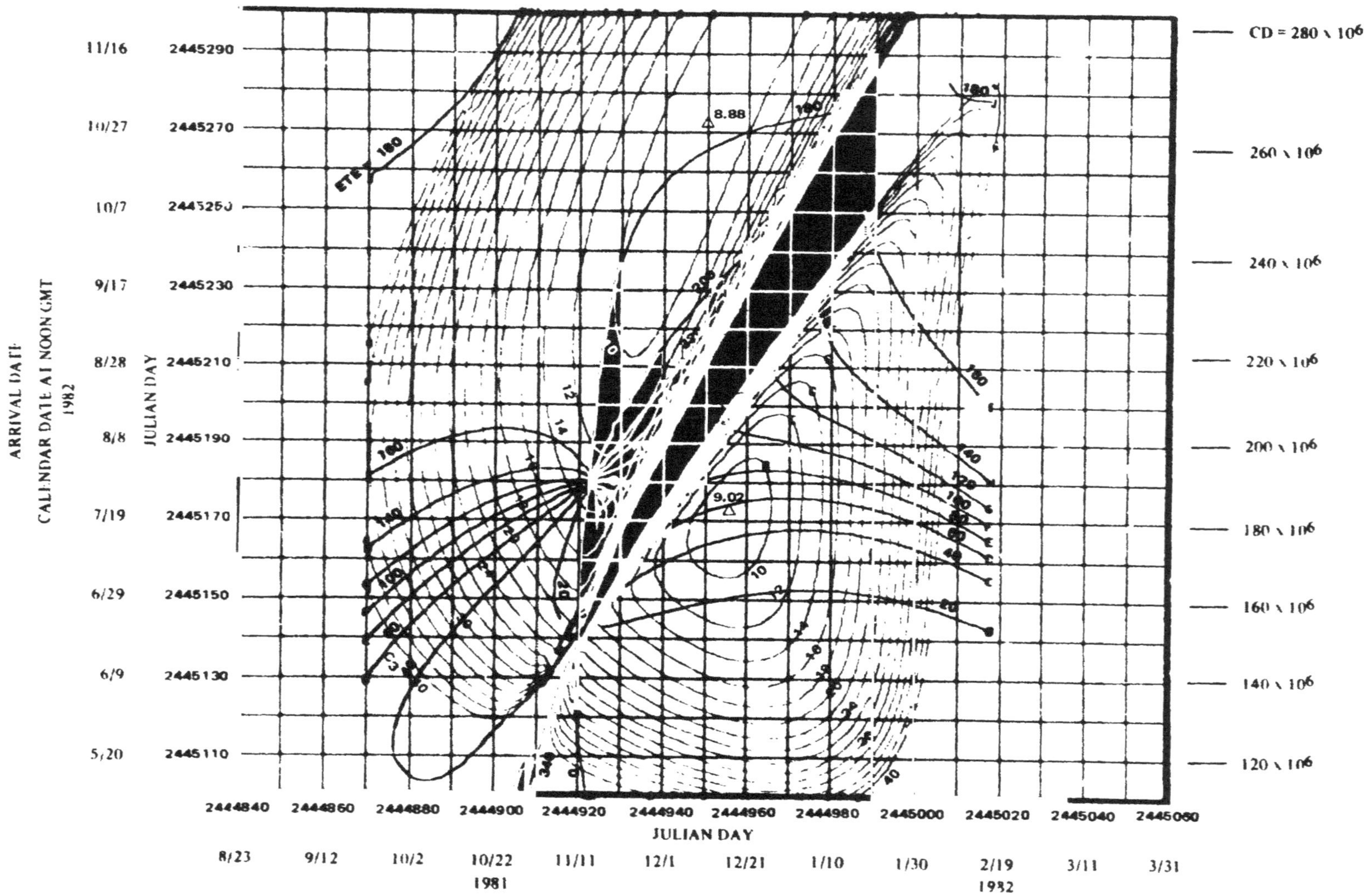

ARRIVAL DATE
CALENDAR DATE AT NOON GMT
1982

Calendar date	Julian day	CD
11/16	2445290	CD = 280 x 10^6
10/27	2445270	260 x 10^6
10/7	2445250	240 x 10^6
9/17	2445230	220 x 10^6
8/28	2445210	200 x 10^6
8/8	2445190	180 x 10^6
7/19	2445170	160 x 10^6
6/29	2445150	140 x 10^6
6/9	2445130	120 x 10^6
5/20	2445110	

JULIAN DAY	2444840	2444860	2444880	2444900	2444920	2444940	2444960	2444980	2445000	2445020	2445040	2445060
CALENDAR DATE AT NOON GMT	8/23	9/12	10/2	10/22 1981	11/11	12/1	12/21	1/10	1/30	2/19 1982	3/11	3/31

DEPARTURE DATE

ETE 180; 8.88; 9.02; C_3

CONTOURS OF C_3 AND ETE EARTH TO MARS 1981-82

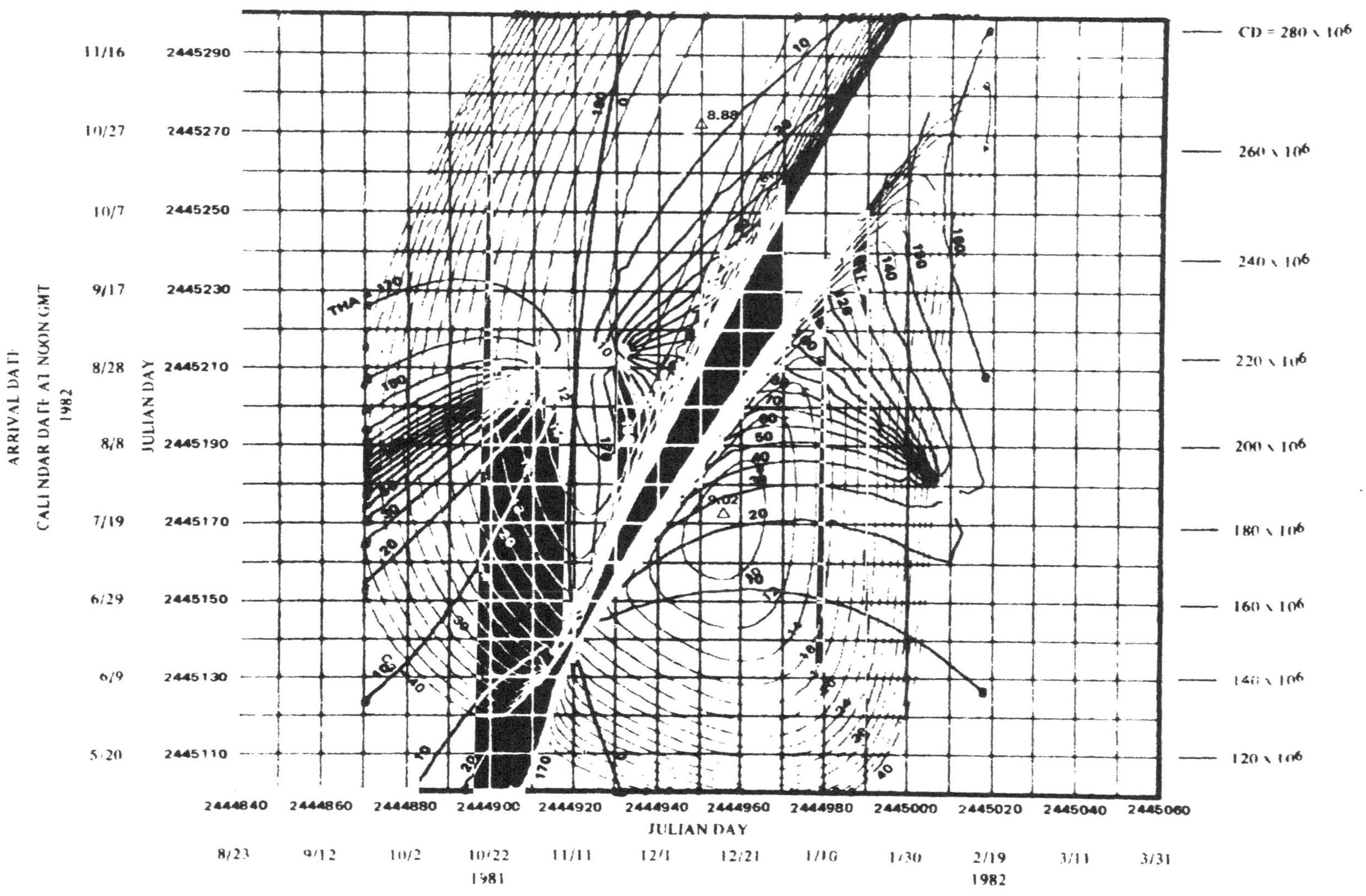

ARRIVAL DATE

CALENDAR DATE AT NOON GMT

1982

JULIAN DAY

11/16	2445290	CD = 280 x 10^6
10/27	2445270	260 x 10^6
10/7	2445250	240 x 10^6
9/17	2445230	220 x 10^6
8/28	2445210	200 x 10^6
8/8	2445190	180 x 10^6
7/19	2445170	160 x 10^6
6/29	2445150	140 x 10^6
6/9	2445130	120 x 10^6
5/20	2445110	

THA = 170

8.88

9.02

2444840	2444860	2444880	2444900	2444920	2444940	2444960	2444980	2445000	2445020	2445040	2445060
8/23	9/12	10/2	10/22 1981	11/11	12/1	12/21	1/10	1/30	2/19 1982	3/11	3/31

JULIAN DAY

CALENDAR DATE AT NOON GMT

DEPARTURE DATE

CONTOURS OF C_3 AND THA EARTH TO MARS 1981-82

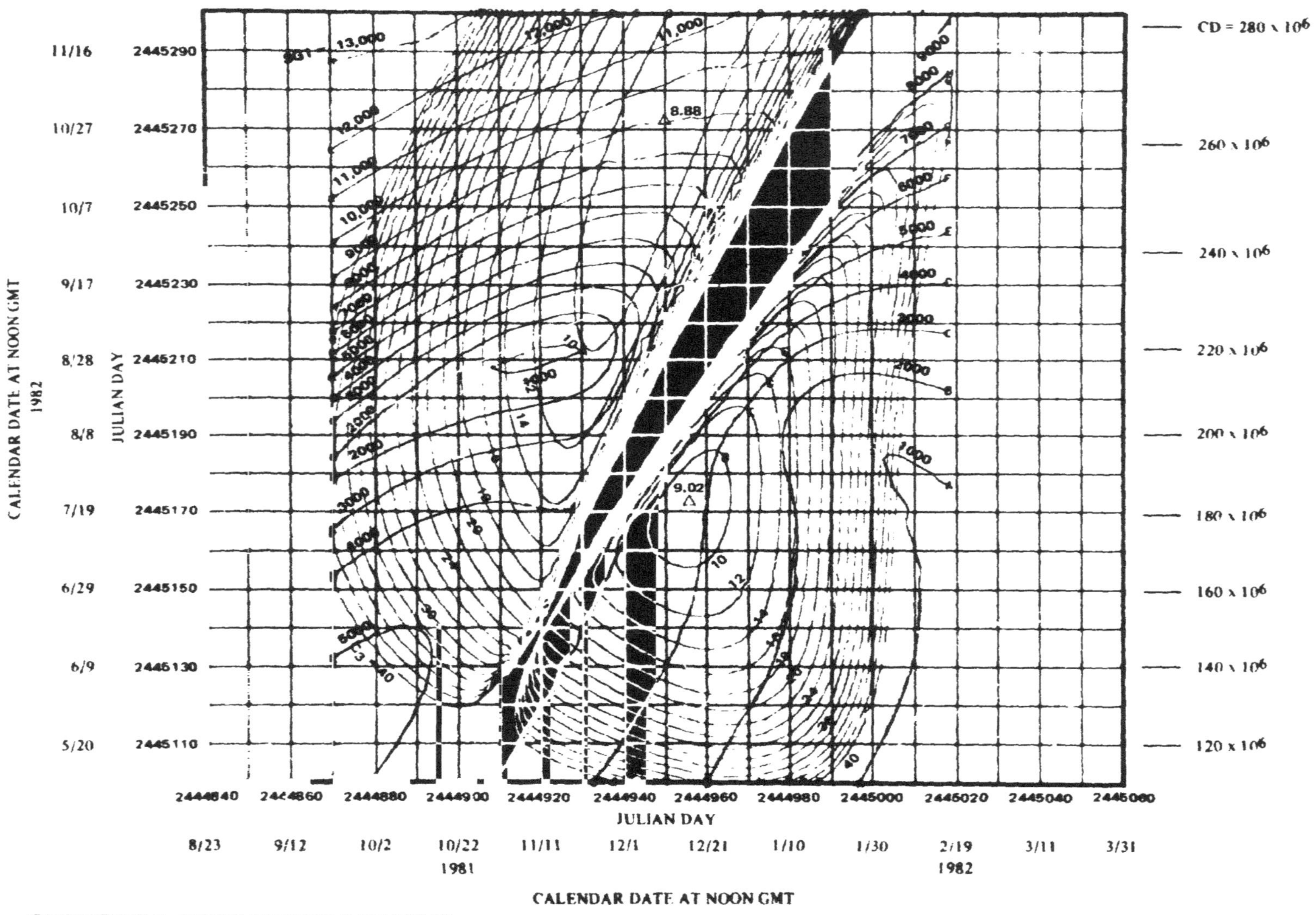

CONTOURS OF C_3 AND SG1 EARTH TO MARS 1981-82

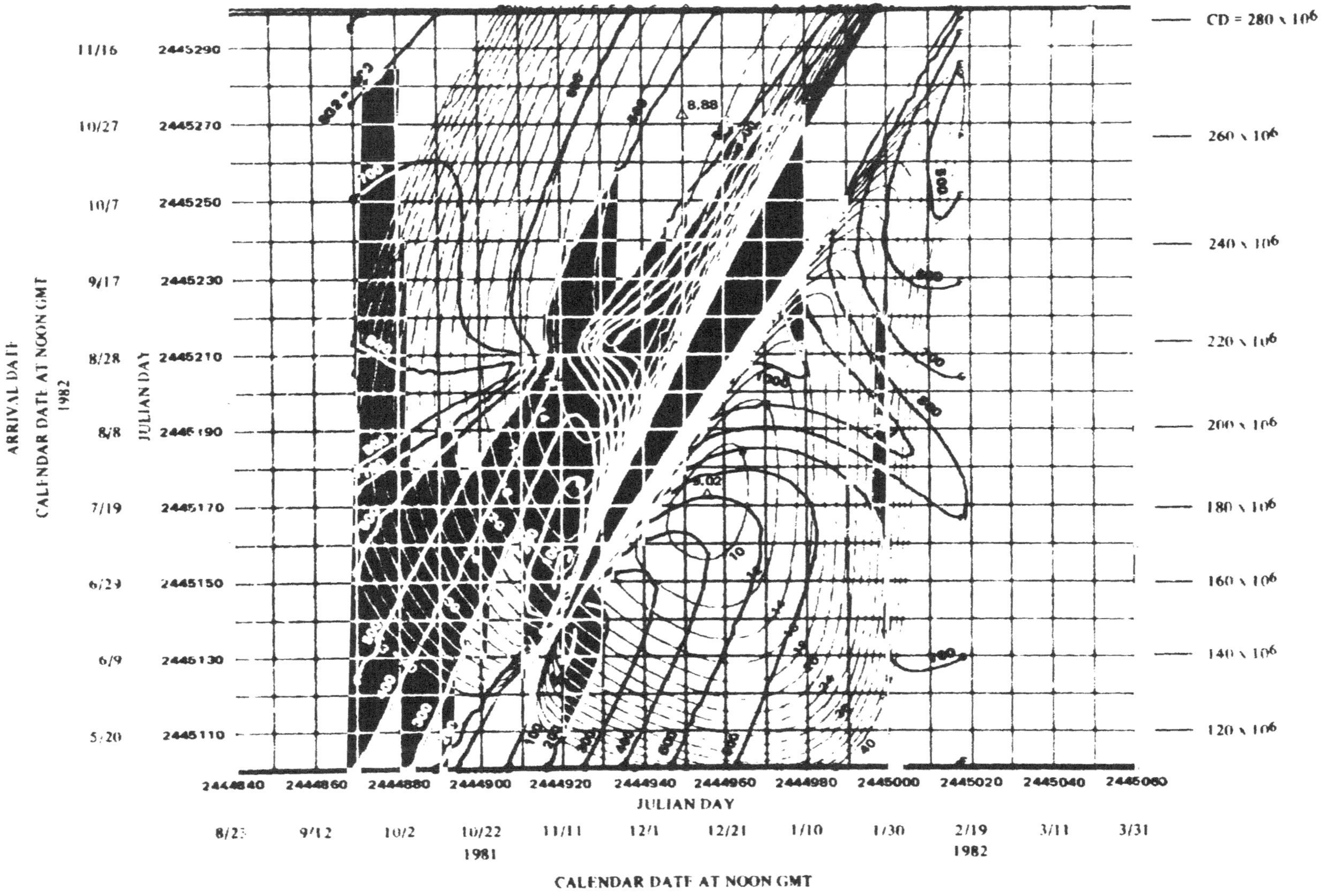

ARRIVAL DATE
CALENDAR DATE AT NOON GMT
1982

JULIAN DAY

Calendar date	Julian day
11/16	2445290
10/27	2445270
10/7	2445250
9/17	2445230
8/28	2445210
8/8	2445190
7/19	2445170
6/29	2445150
6/9	2445130
5/20	2445110

CD = 280 x 10^6
260 x 10^6
240 x 10^6
220 x 10^6
200 x 10^6
180 x 10^6
160 x 10^6
140 x 10^6
120 x 10^6

8.88
9.02

2444840 2444860 2444880 2444900 2444920 2444940 2444960 2444980 2445000 2445020 2445040 2445060

JULIAN DAY

8/23 9/12 10/2 10/22 11/11 12/1 12/21 1/10 1/30 2/19 3/11 3/31

1981 1982

CALENDAR DATE AT NOON GMT

DEPARTURE DATE

CONTOURS OF C_3 AND SG2 EARTH TO MARS 1981-82

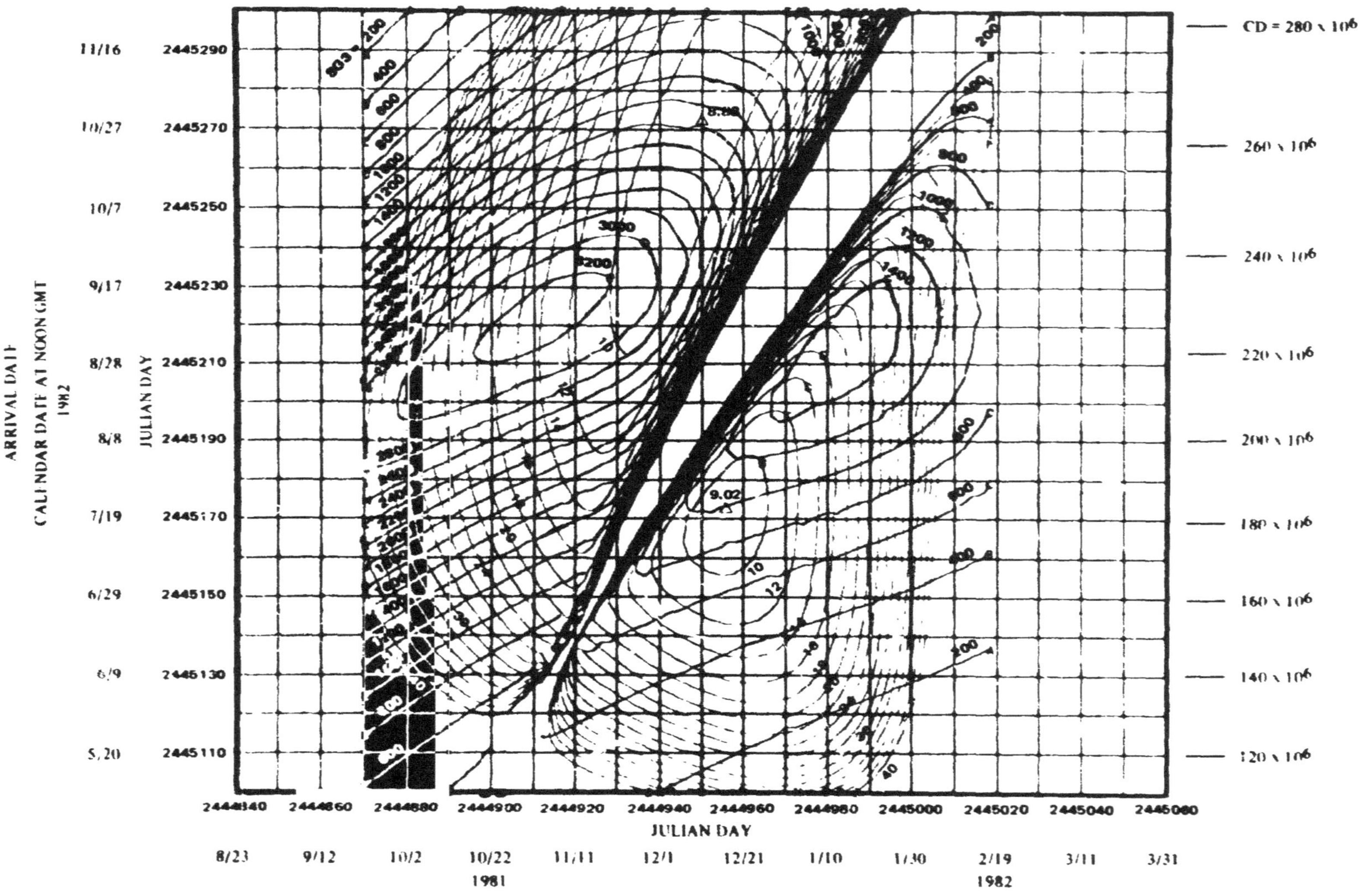

ARRIVAL DATE CALENDAR DATE AT NOON GMT 1982	JULIAN DAY	CD
11/16	2445290	CD = 280 x 10^6
10/27	2445270	260 x 10^6
10/7	2445250	240 x 10^6
9/17	2445230	220 x 10^6
8/28	2445210	200 x 10^6
8/8	2445190	180 x 10^6
7/19	2445170	160 x 10^6
6/29	2445150	140 x 10^6
6/9	2445130	120 x 10^6
5/20	2445110	

JULIAN DAY	2444840	2444860	2444880	2444900	2444920	2444940	2444960	2444980	2445000	2445020	2445040	2445060
CALENDAR DATE AT NOON GMT	8/23	9/12	10/2	10/22 1981	11/11	12/1	12/21	1/10	1/30	2/19 1982	3/11	3/31

DEPARTURE DATE

CONTOURS OF C_3 AND SG3 EARTH TO MARS 1981-82

C_3 ♂ 1984

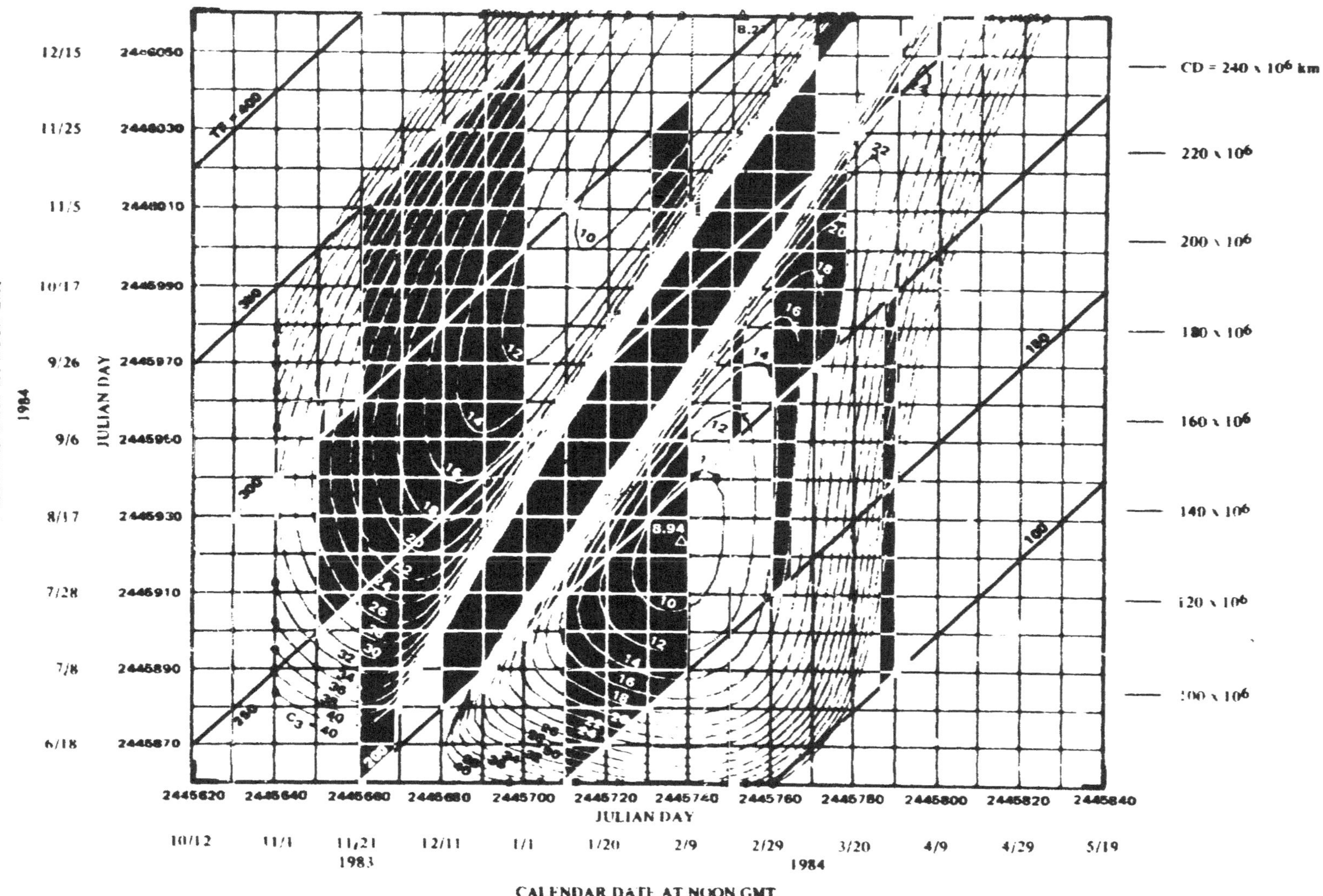

CONTOURS OF C_3 AND FLIGHT TIMES EARTH TO MARS 1984

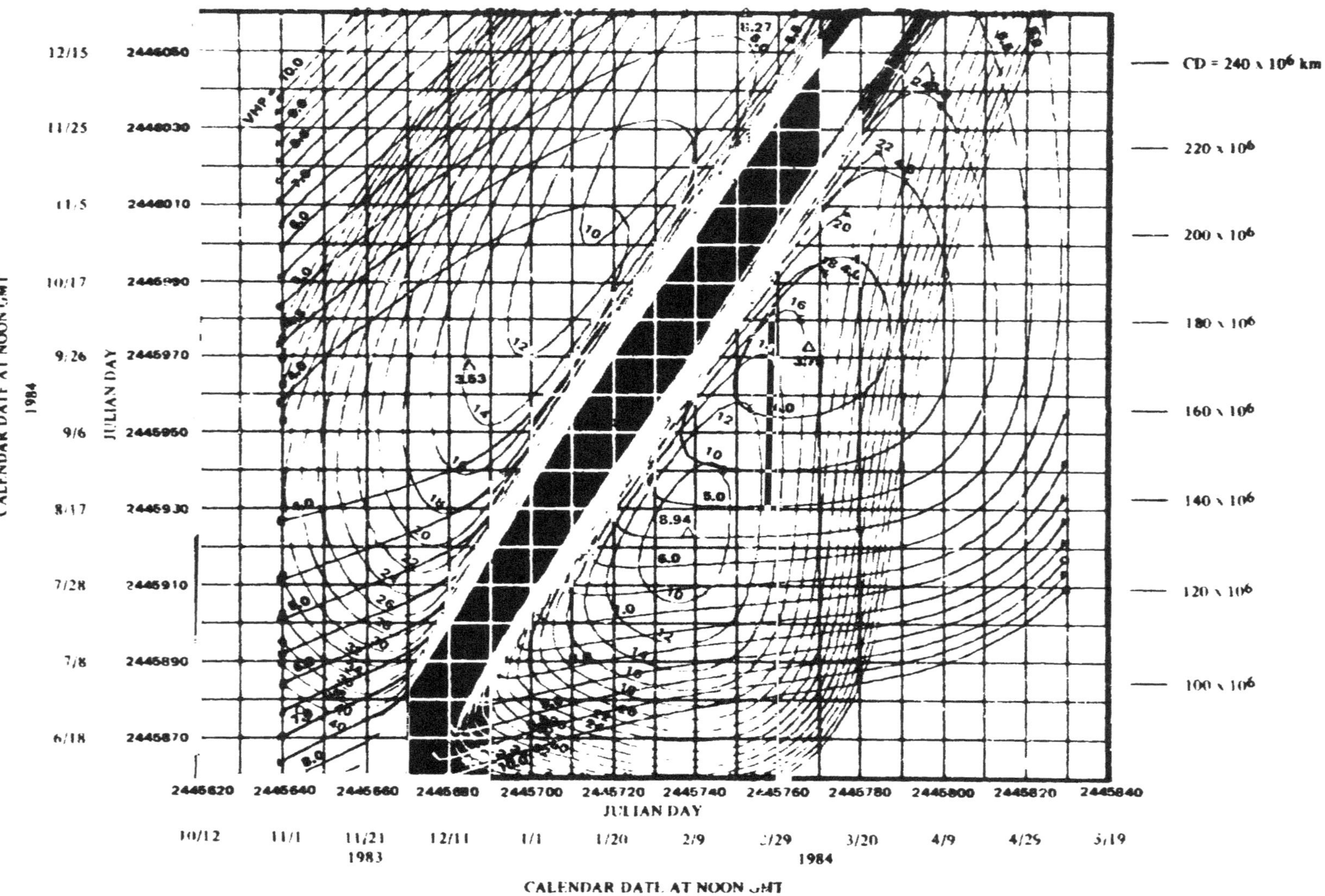

CONTOURS OF C_3 AND VHP EARTH TO MARS 1984

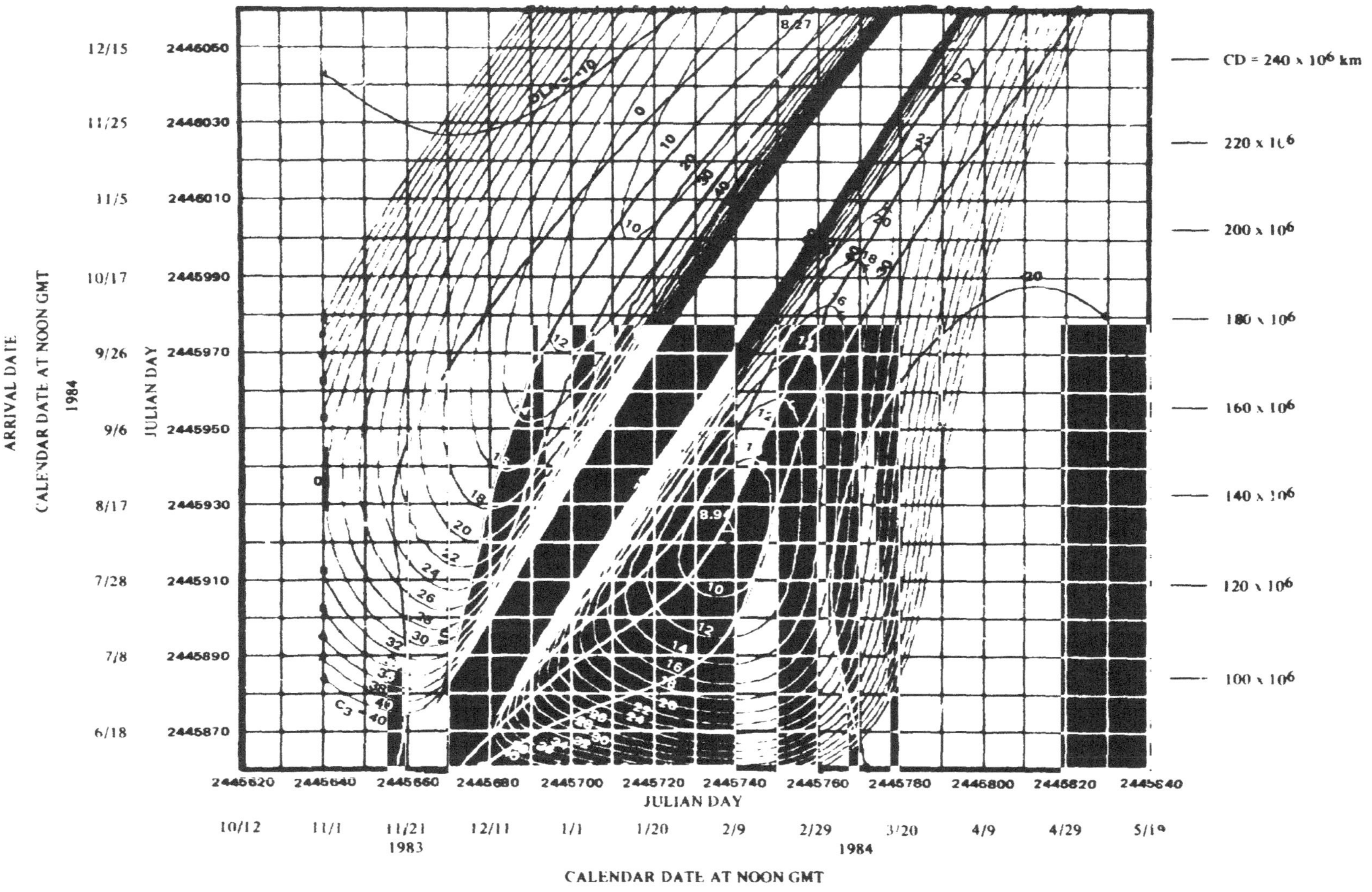

ARRIVAL DATE
CALENDAR DATE AT NOON GMT
1984

JULIAN DAY

Calendar date	Julian day
12/15	2446050
11/25	2446030
11/5	2446010
10/17	2445990
9/26	2445970
9/6	2445950
8/17	2445930
7/28	2445910
7/8	2445890
6/18	2445870

CD = 240 x 10^6 km
220 x 10^6
200 x 10^6
180 x 10^6
160 x 10^6
140 x 10^6
120 x 10^6
100 x 10^6

Julian day	Calendar date
2445620	10/12
2445640	11/1
2445660	11/21 1983
2445680	12/11
2445700	1/1
2445720	1/20
2445740	2/9
2445760	2/29 1984
2445780	3/20
2445800	4/9
2445820	4/29
2445840	5/19

JULIAN DAY

CALENDAR DATE AT NOON GMT
DEPARTURE DATE

CONTOURS OF C_3 AND DLA EARTH TO MARS 1984

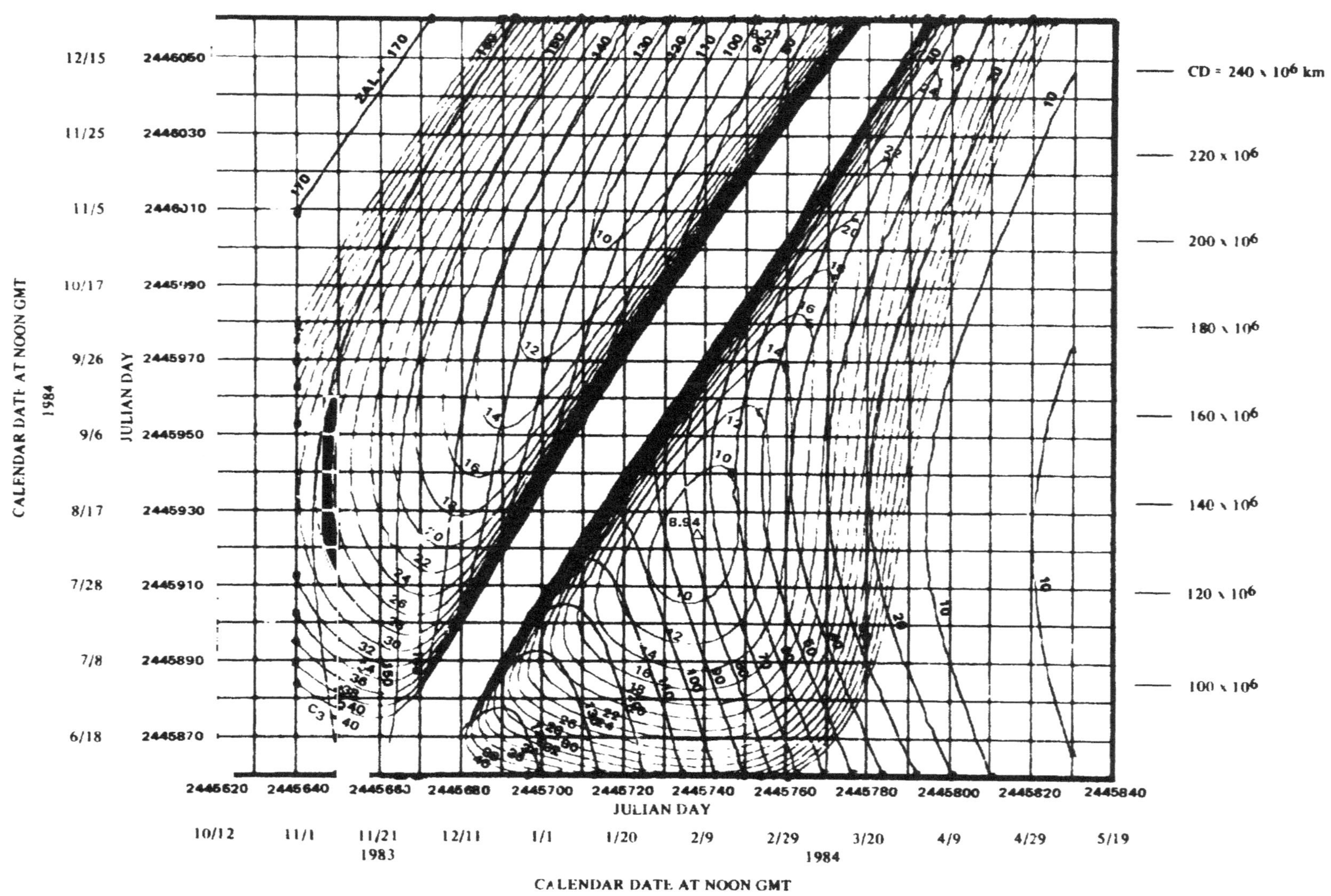

CONTOURS OF C_3 AND ZAL EARTH TO MARS 1984

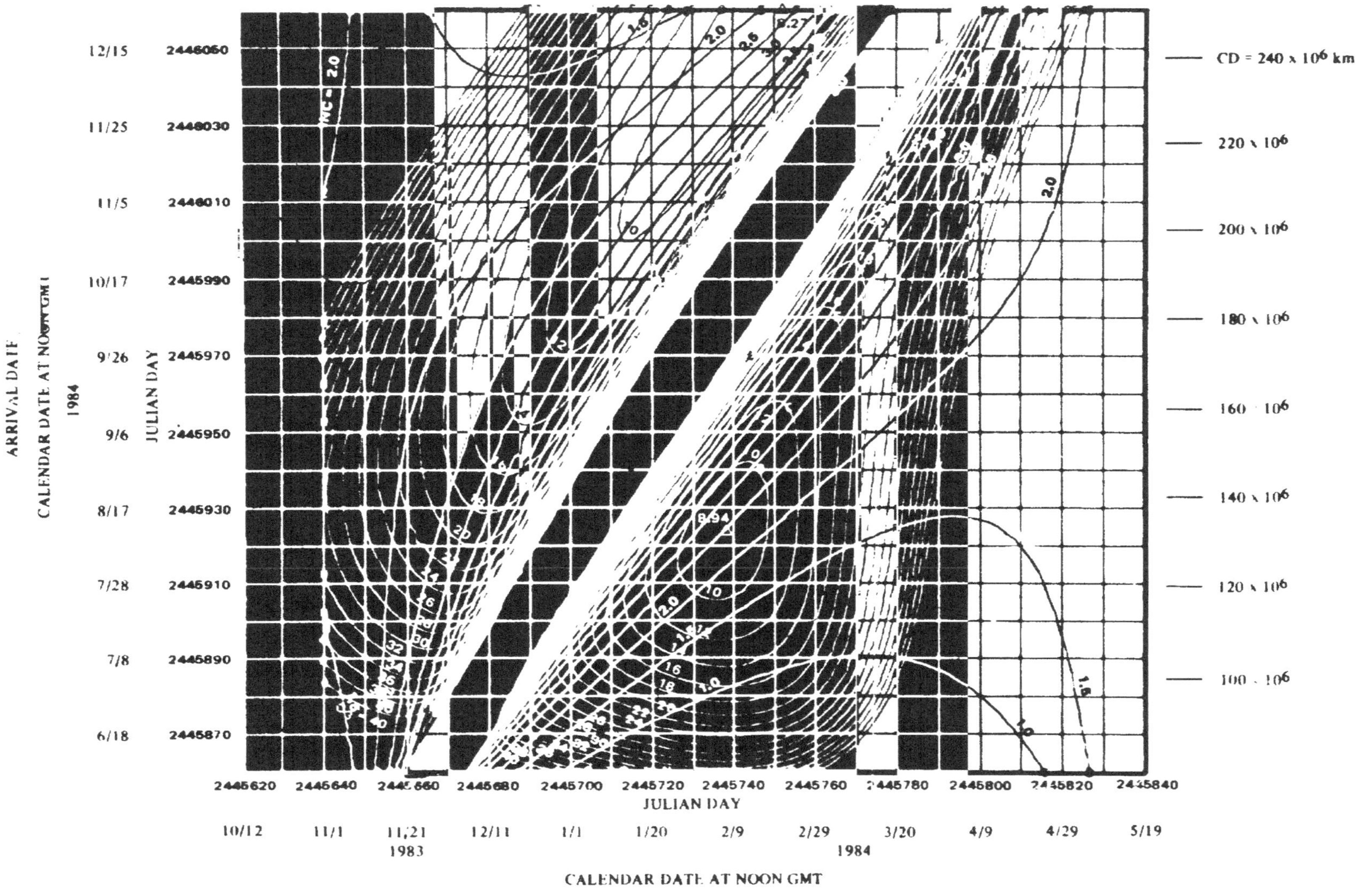

ARRIVAL DATE

CALENDAR DATE AT NOON GMT

1984

JULIAN DAY

Arrival date	Julian day	CD
12/15	2446050	CD = 240 x 10^6 km
11/25	2446030	220 x 10^6
11/5	2446010	200 x 10^6
10/17	2445990	180 x 10^6
9/26	2445970	160 x 10^6
9/6	2445950	140 x 10^6
8/17	2445930	120 x 10^6
7/28	2445910	100 x 10^6
7/8	2445890	
6/18	2445870	

Julian day	2445620	2445640	2445660	2445680	2445700	2445720	2445740	2445760	2445780	2445800	2445820	2445840
Calendar date	10/12	11/1	11/21 1983	12/11	1/1	1/20	2/9	2/29 1984	3/20	4/9	4/29	5/19

JULIAN DAY

CALENDAR DATE AT NOON GMT

DEPARTURE DATE

CONTOURS OF C_3 AND INC EARTH TO MARS 1984

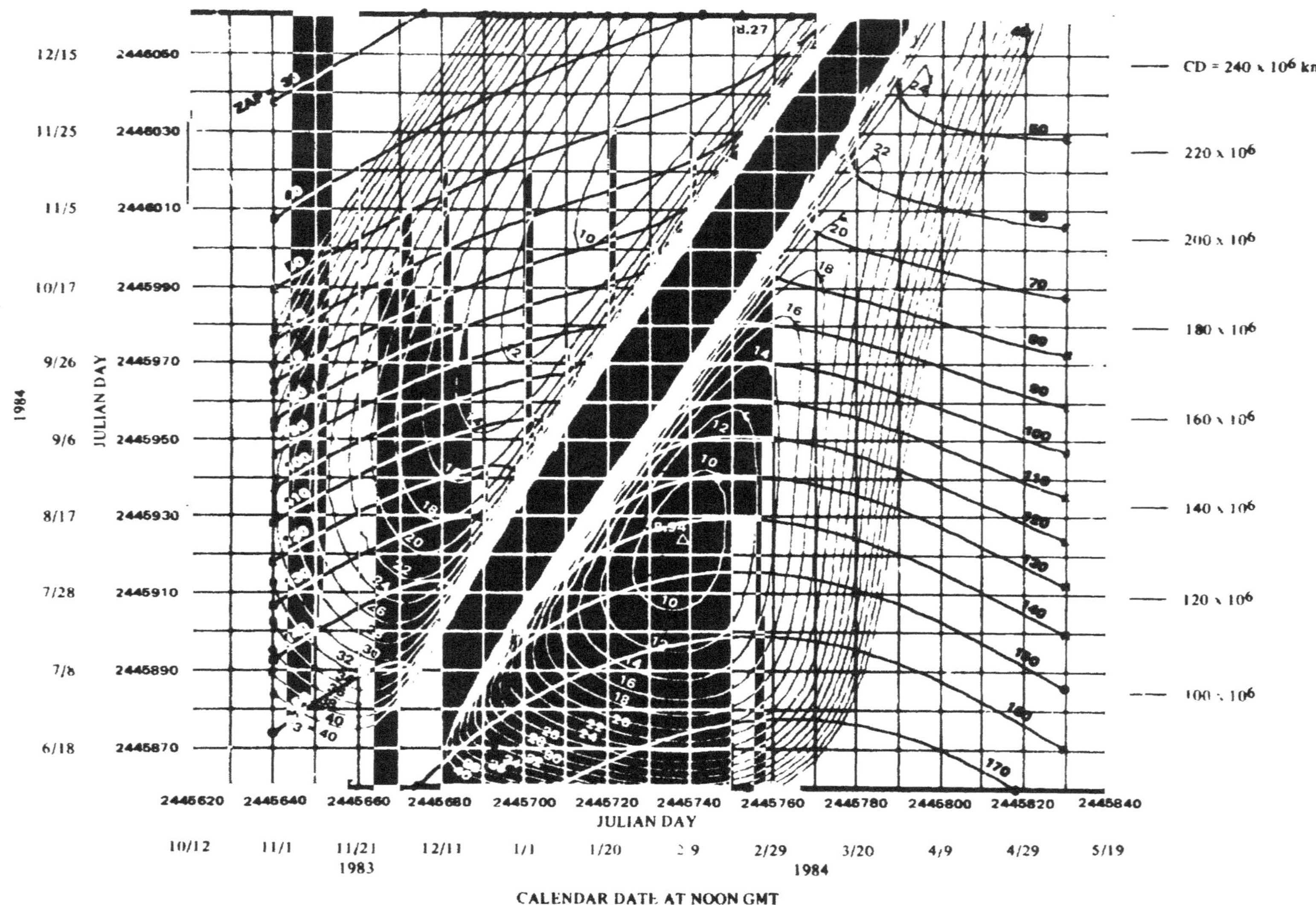

CONTOURS OF C_3 AND ZAP EARTH TO MARS 1984

ZAP ♂ 1984

ETS ♂ 1984

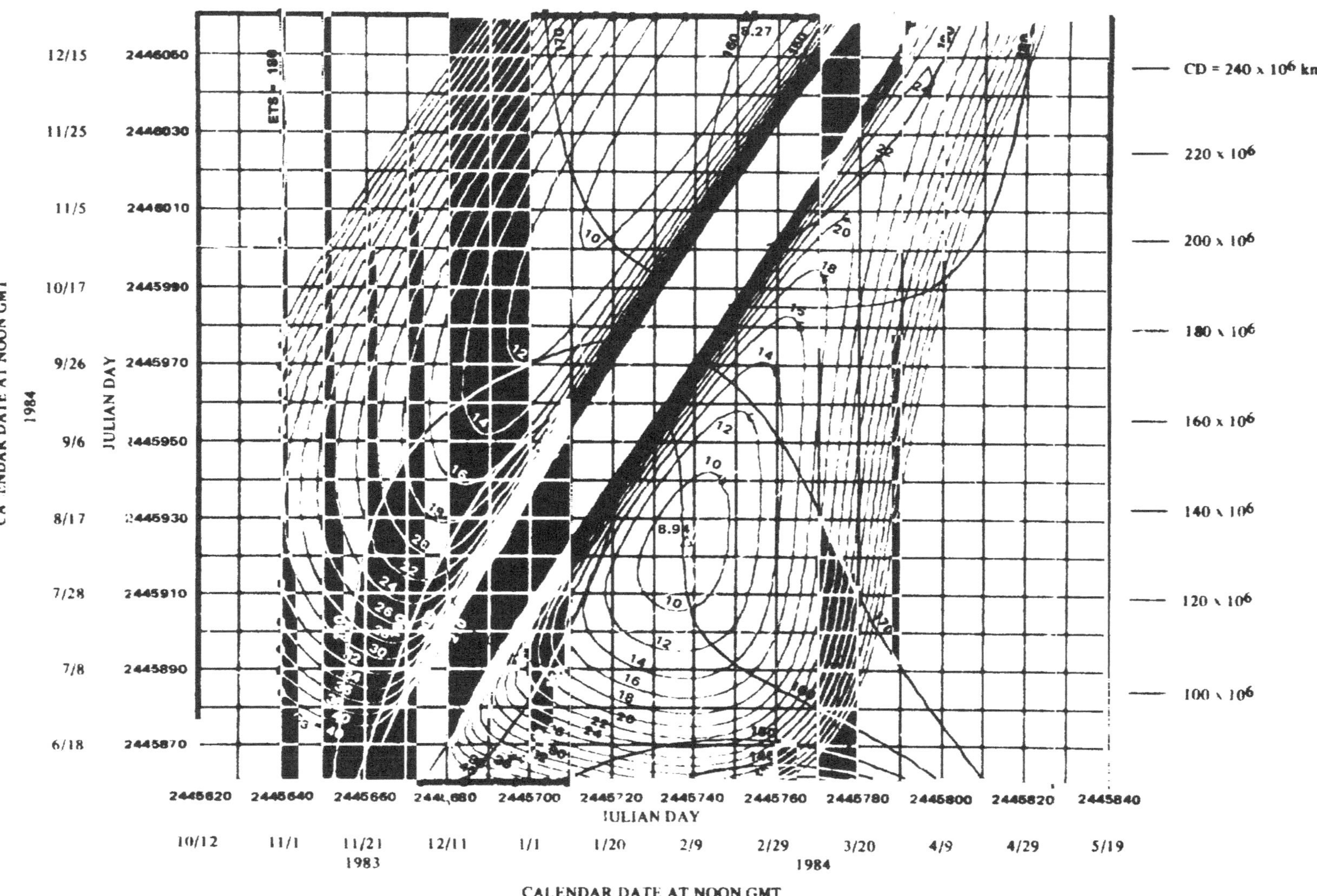

CONTOURS OF C_3 AND ETS EARTH TO MARS 1984

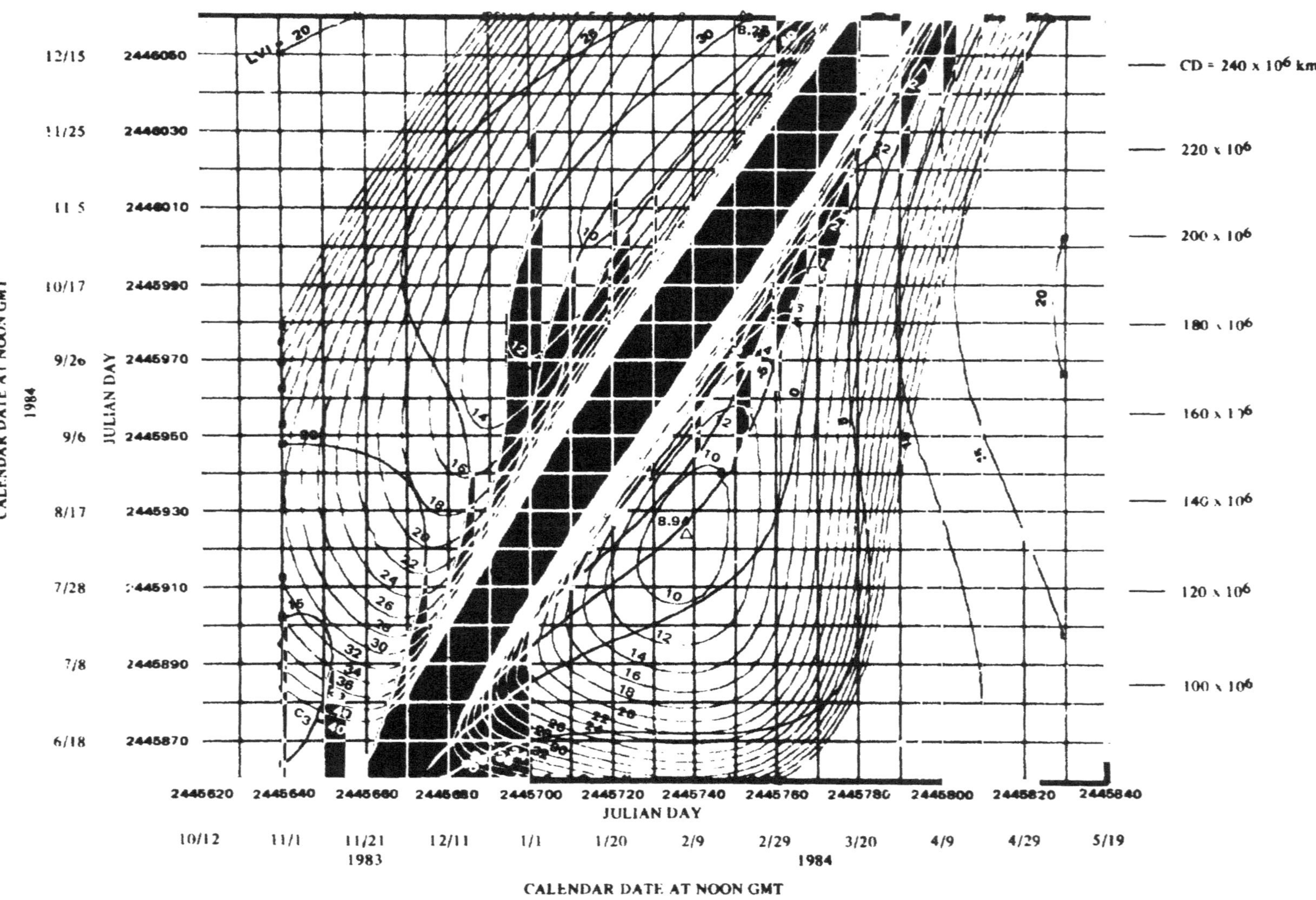

CONTOURS OF C_3 AND LVI EARTH TO MARS 1984

ZAE ♂ 1984

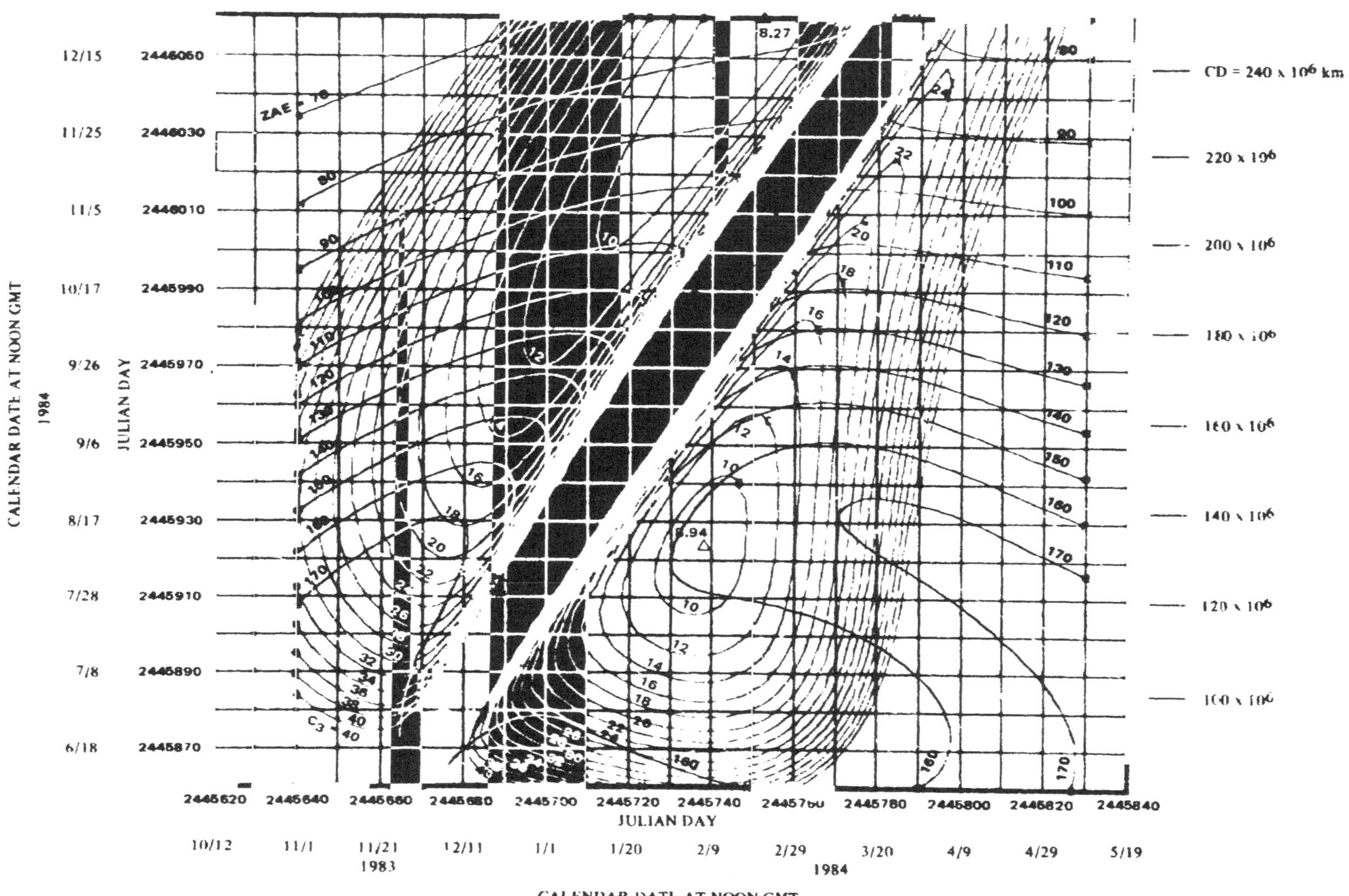

CONTOURS OF C_3 AND ZAE EARTH TO MARS 1984

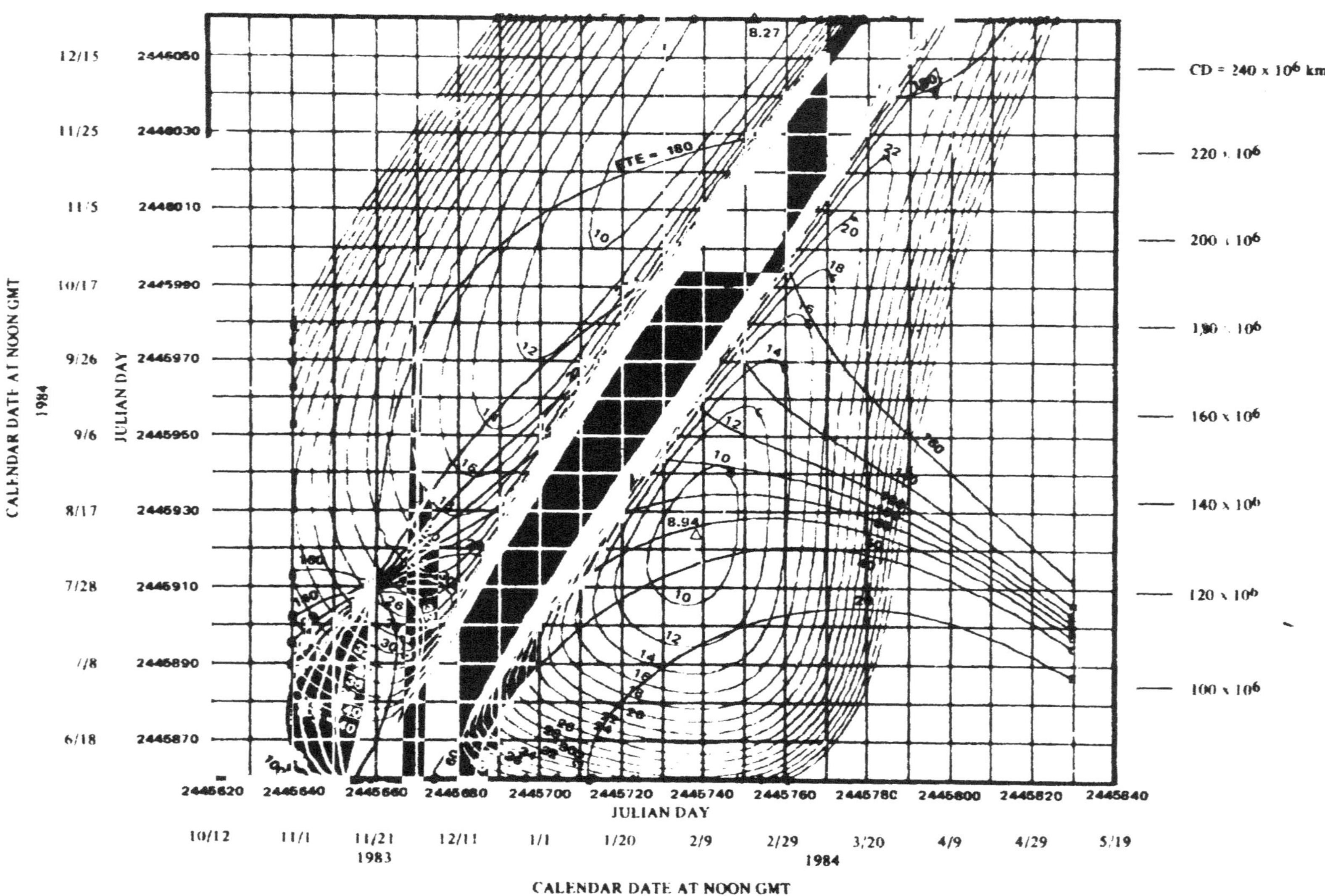

CONTOURS OF C_3 AND ETE EARTH TO MARS 1984

ETE
♂
1984

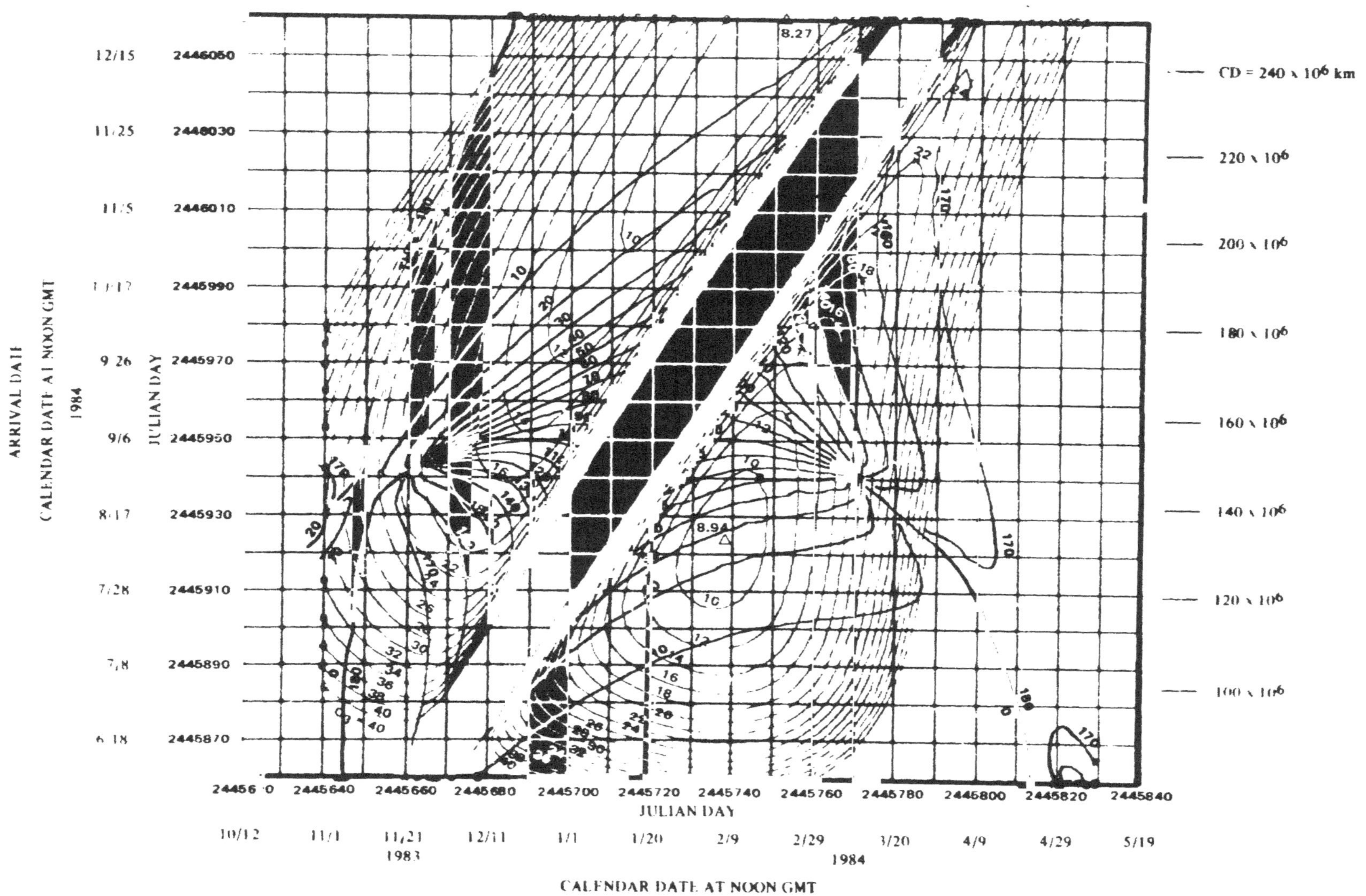

THA

1984

ARRIVAL DATE
CALENDAR DATE AT NOON GMT
1984

JULIAN DAY

12/15	2446050
11/25	2446030
11/5	2446010
10/17	2445990
9/26	2445970
9/6	2445950
8/17	2445930
7/28	2445910
7/8	2445890
6/18	2445870

CD = 240 x 10^6 km
220 x 10^6
200 x 10^6
180 x 10^6
160 x 10^6
140 x 10^6
120 x 10^6
100 x 10^6

2445620	2445640	2445660	2445680	2445700	2445720	2445740	2445760	2445780	2445800	2445820	2445840
10/12	11/1	11/21	12/11	1/1	1/20	2/9	2/29	3/20	4/9	4/29	5/19

JULIAN DAY

1983 1984

CALENDAR DATE AT NOON GMT
DEPARTURE DATE

CONTOURS OF C_3 AND THA EARTH TO MARS 1984

SG1 ♂ 1984

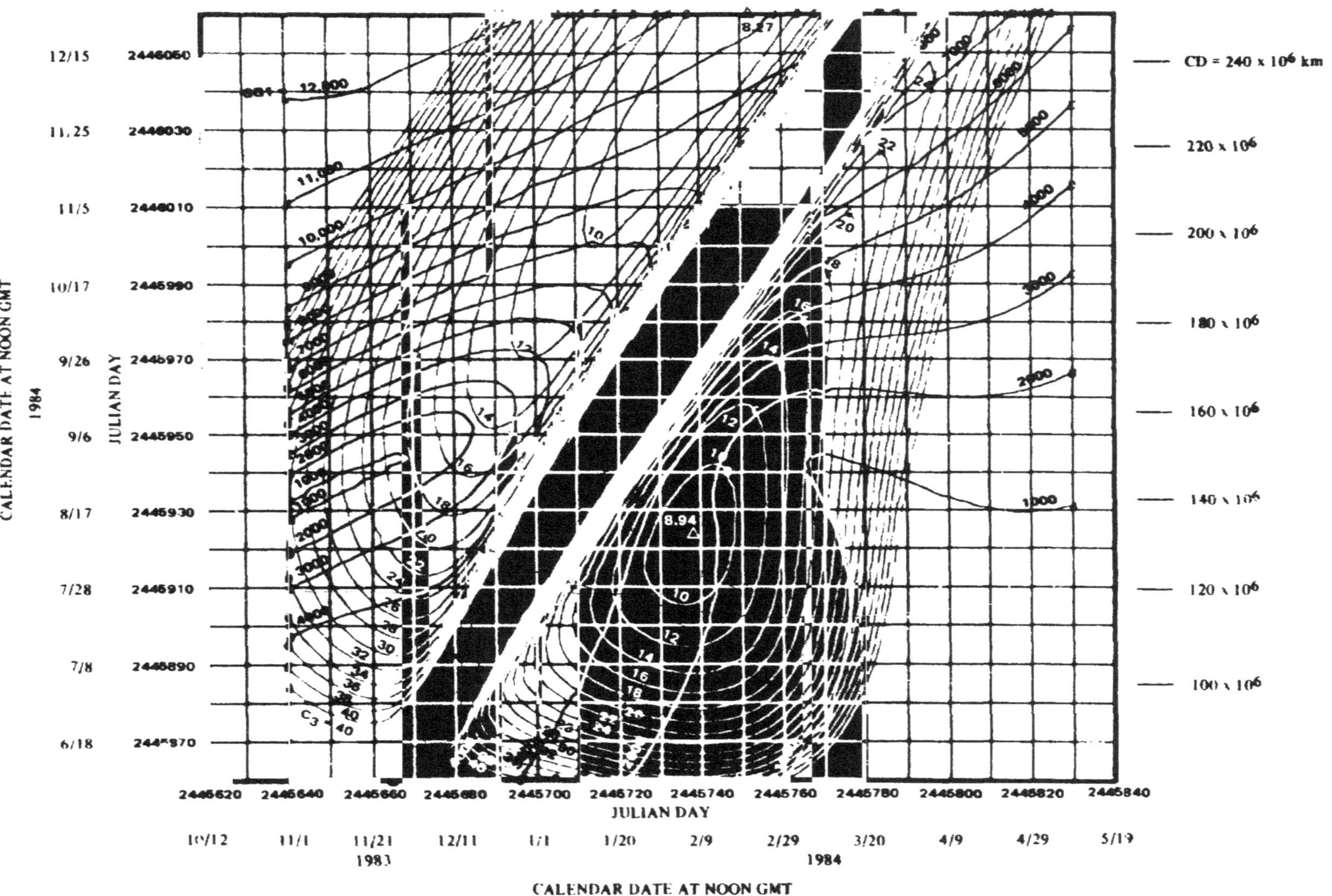

CONTOURS OF C_3 AND SG1 EARTH TO MARS 1984

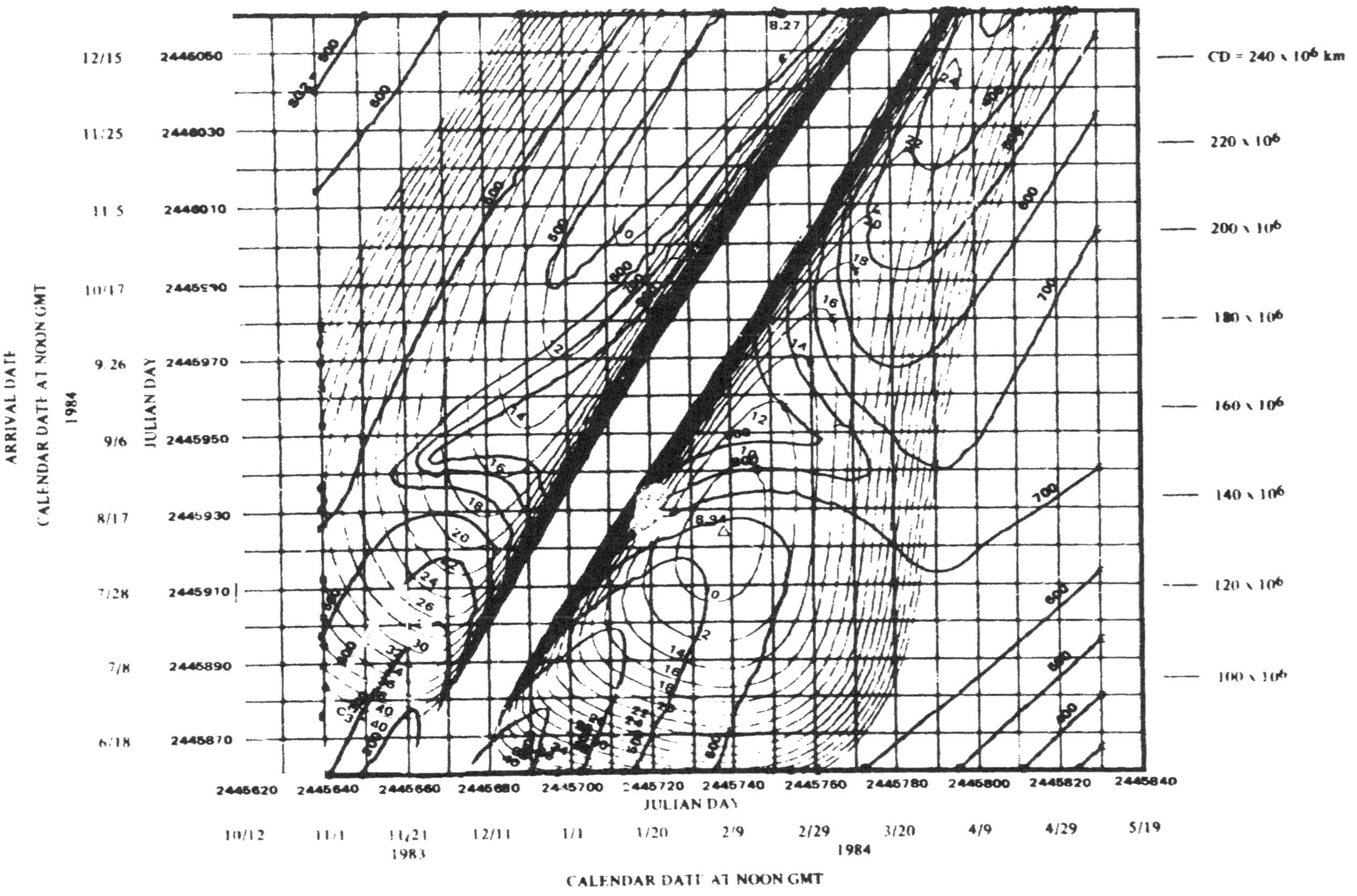

ARRIVAL DATE
CALENDAR DATE AT NOON GMT
1984

Arrival date	Julian day
12/15	2446050
11/25	2446030
11/5	2446010
10/17	2445990
9/26	2445970
9/6	2445950
8/17	2445930
7/28	2445910
7/8	2445890
6/18	2445870

— CD = 240 x 10^6 km
— 220 x 10^6
— 200 x 10^6
— 180 x 10^6
— 160 x 10^6
— 140 x 10^6
— 120 x 10^6
— 100 x 10^6

Julian day	2445620	2445640	2445660	2445680	2445700	2445720	2445740	2445760	2445780	2445800	2445820	2445840
Departure date	10/12	11/1	11/21	12/11	1/1	1/20	2/9	2/29	3/20	4/9	4/29	5/19

1983 … 1984

JULIAN DAY

CALENDAR DATE AT NOON GMT
DEPARTURE DATE

CONTOURS OF C_3 AND SG2 EARTH TO MARS 1984

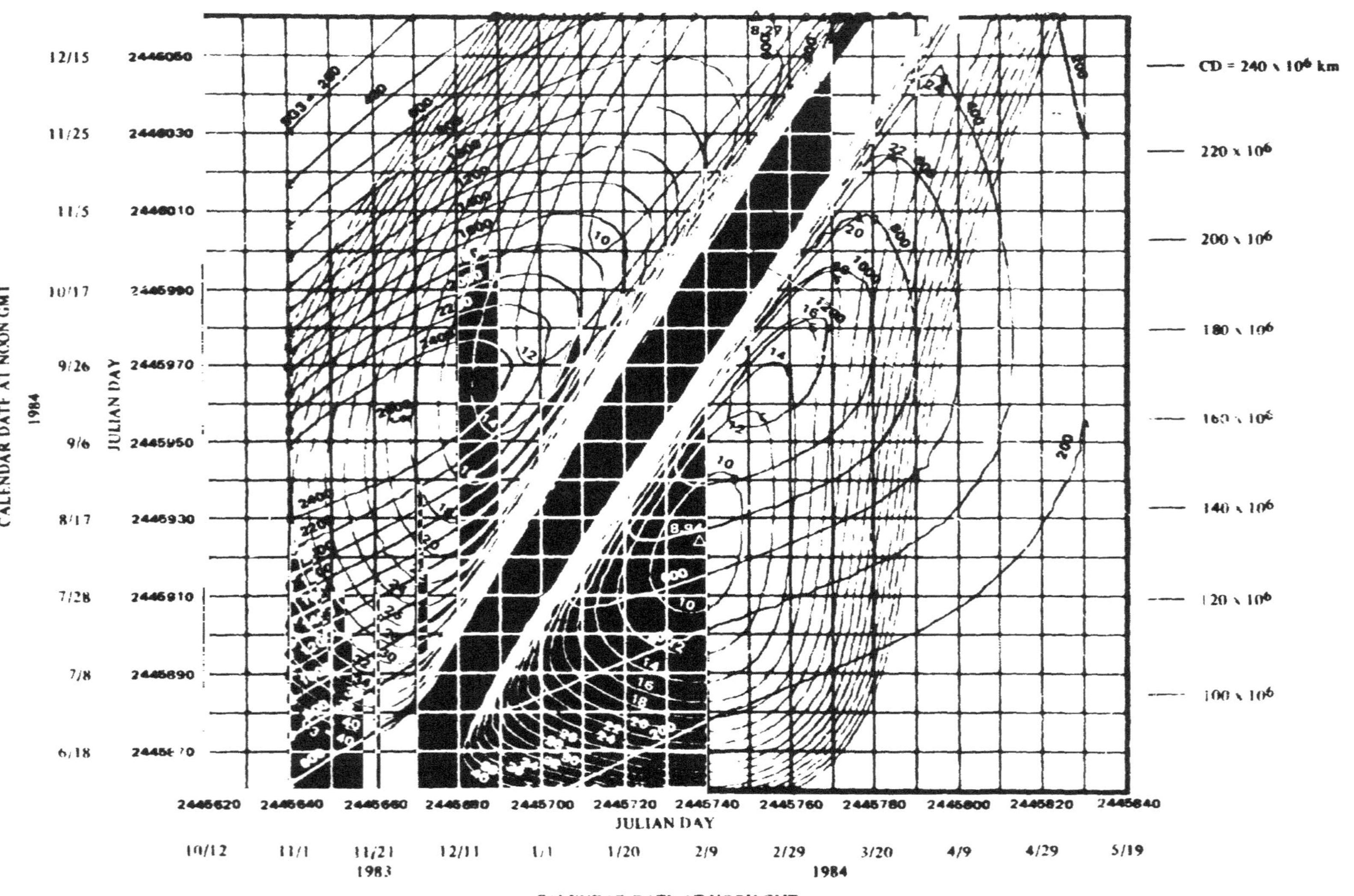

CONTOURS OF C_3 AND SG3 EARTH TO MARS 1984

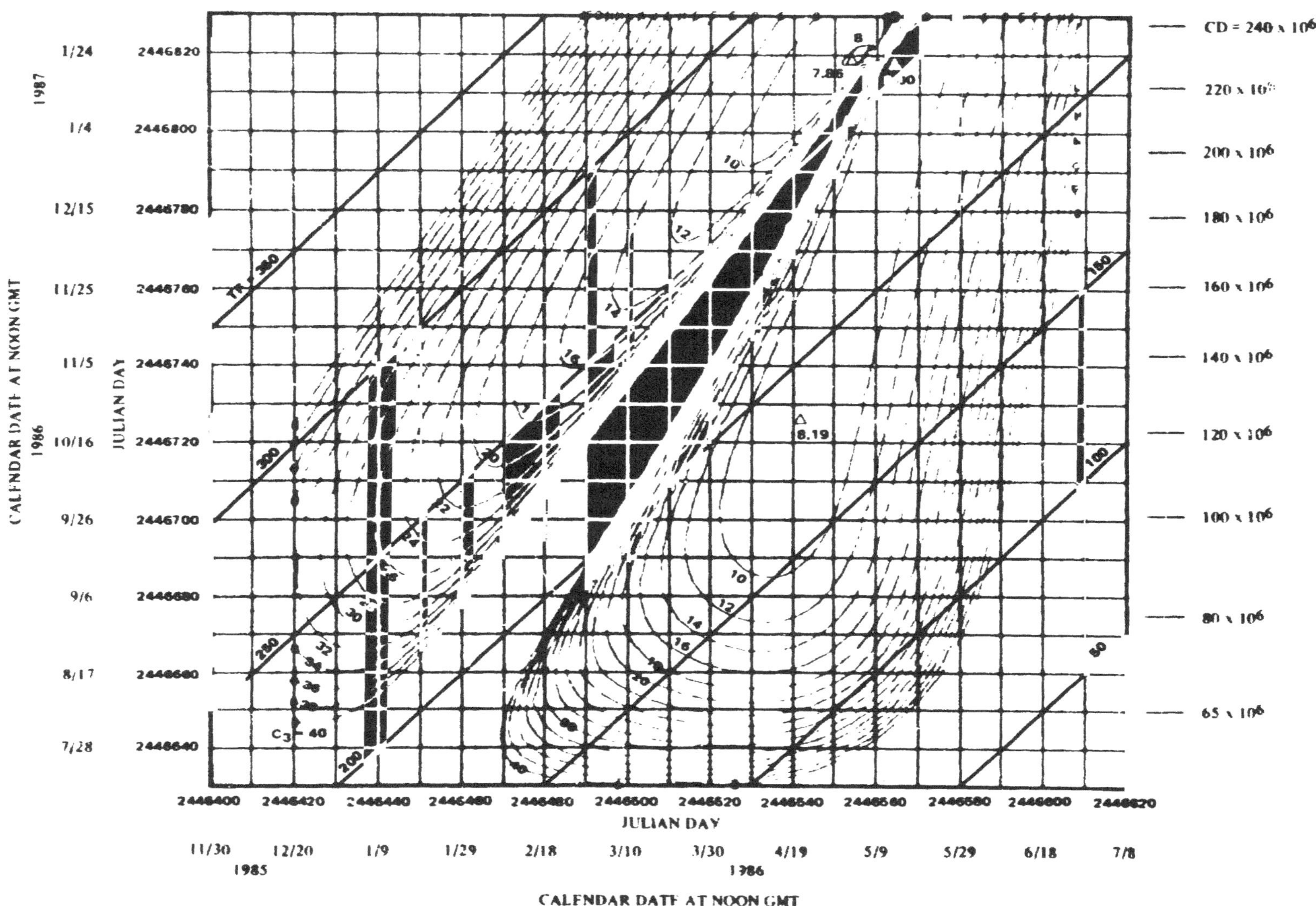

CONTOURS OF C_3 AND FLIGHT TIMES EARTH TO MARS 1985-86 DEPARTURE DATE

C_3 ♂ 1986

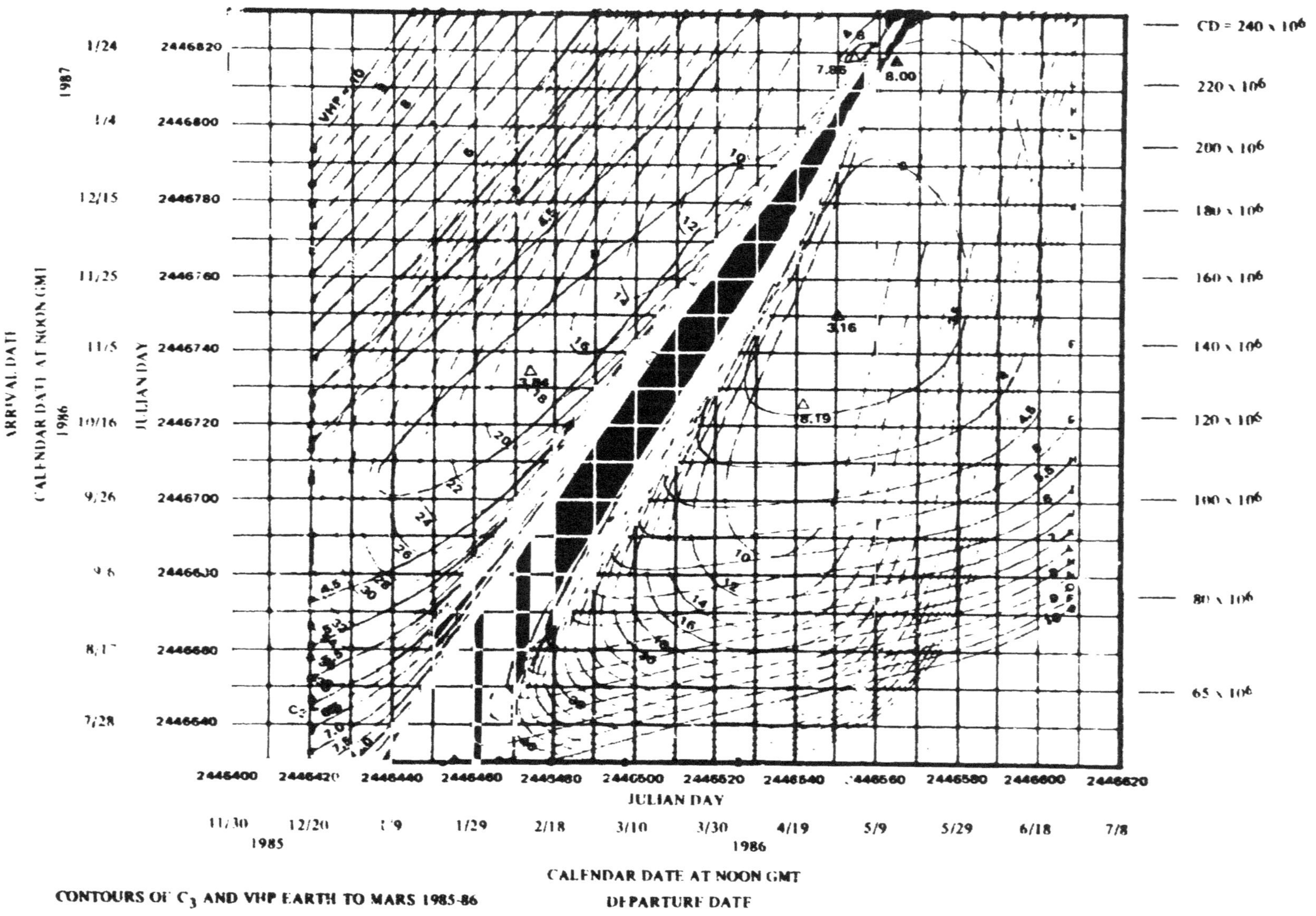

ARRIVAL DATE

CALENDAR DATE AT NOON GMT

JULIAN DAY

Calendar date	Julian day
1/24 (1987)	2446820
1/4 (1987)	2446800
12/15	2446780
11/25	2446760
11/5	2446740
10/16 (1986)	2446720
9/26	2446700
9/6	2446680
8/17	2446660
7/28	2446640

CD = 240 x 10^6
220 x 10^6
200 x 10^6
180 x 10^6
160 x 10^6
140 x 10^6
120 x 10^6
100 x 10^6
80 x 10^6
65 x 10^6

2446400 2446420 2446440 2446460 2446480 2446500 2446520 2446540 2446560 2446580 2446600 2446620

JULIAN DAY

11/30 12/20 1/9 1/29 2/18 3/10 3/30 4/19 5/9 5/29 6/18 7/8

1985 1986

CALENDAR DATE AT NOON GMT

DEPARTURE DATE

CONTOURS OF C_3 AND VHP EARTH TO MARS 1985-86

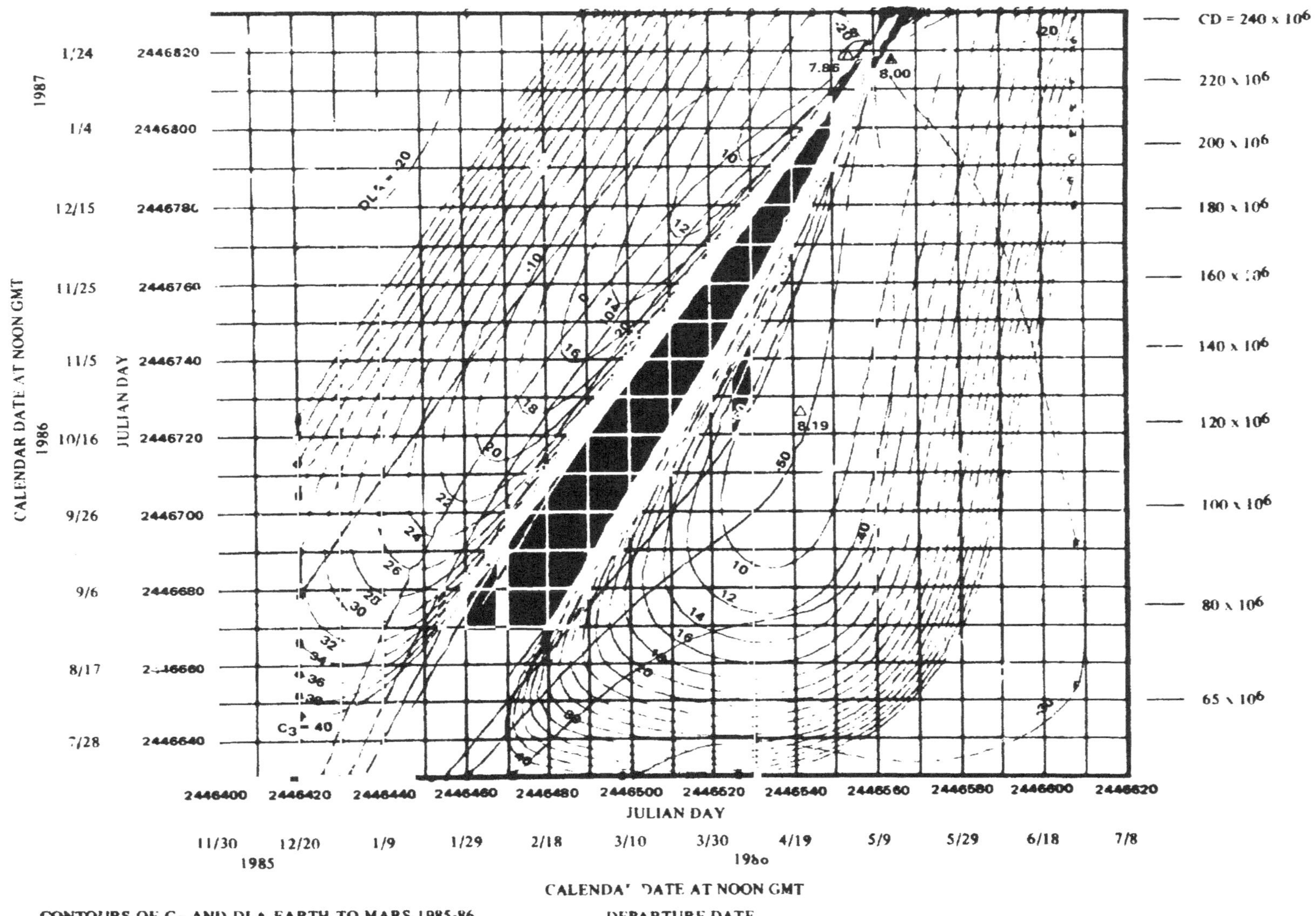

CONTOURS OF C_3 AND DLA EARTH TO MARS 1985-86

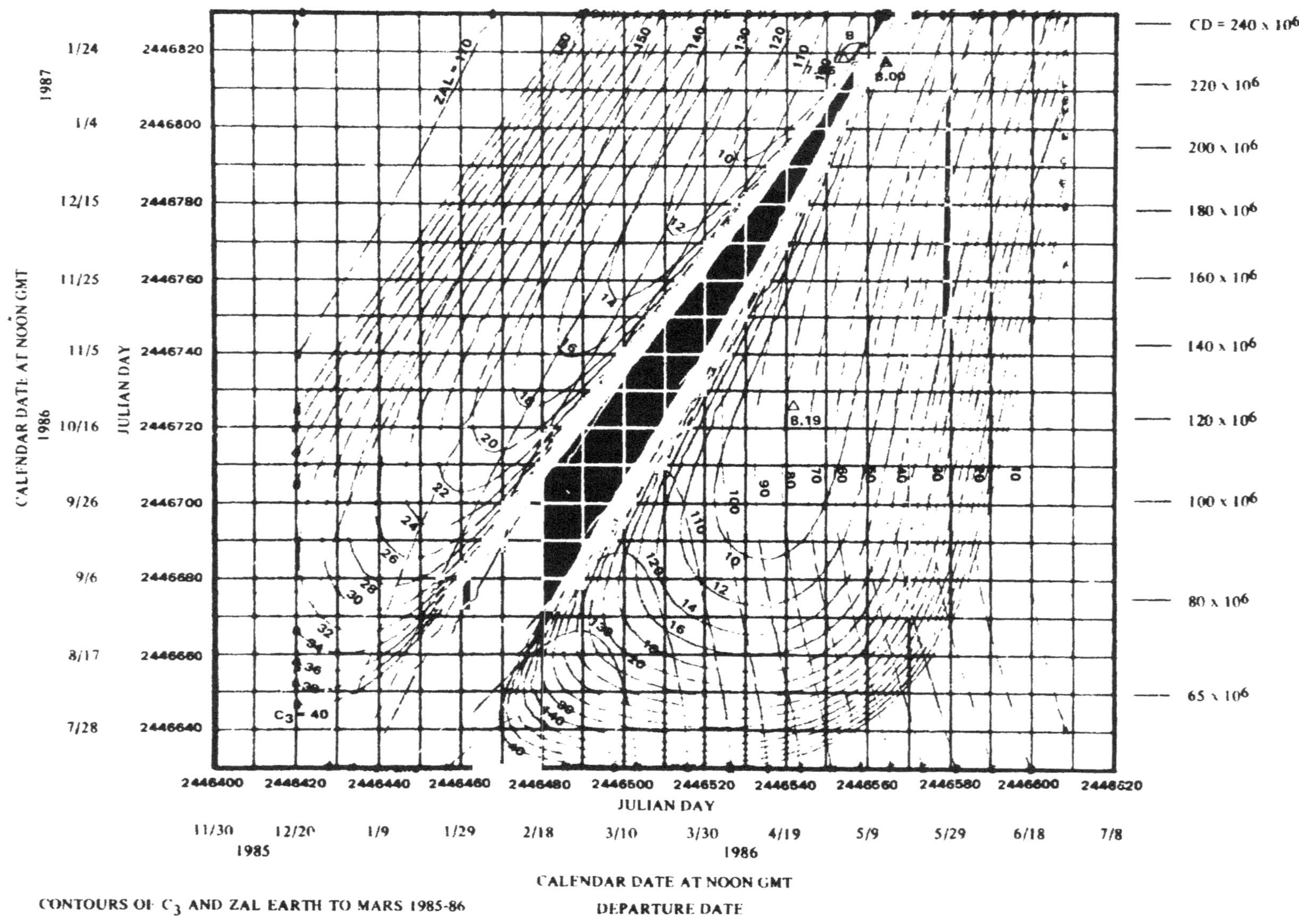

CONTOURS OF C_3 AND ZAL EARTH TO MARS 1985-86

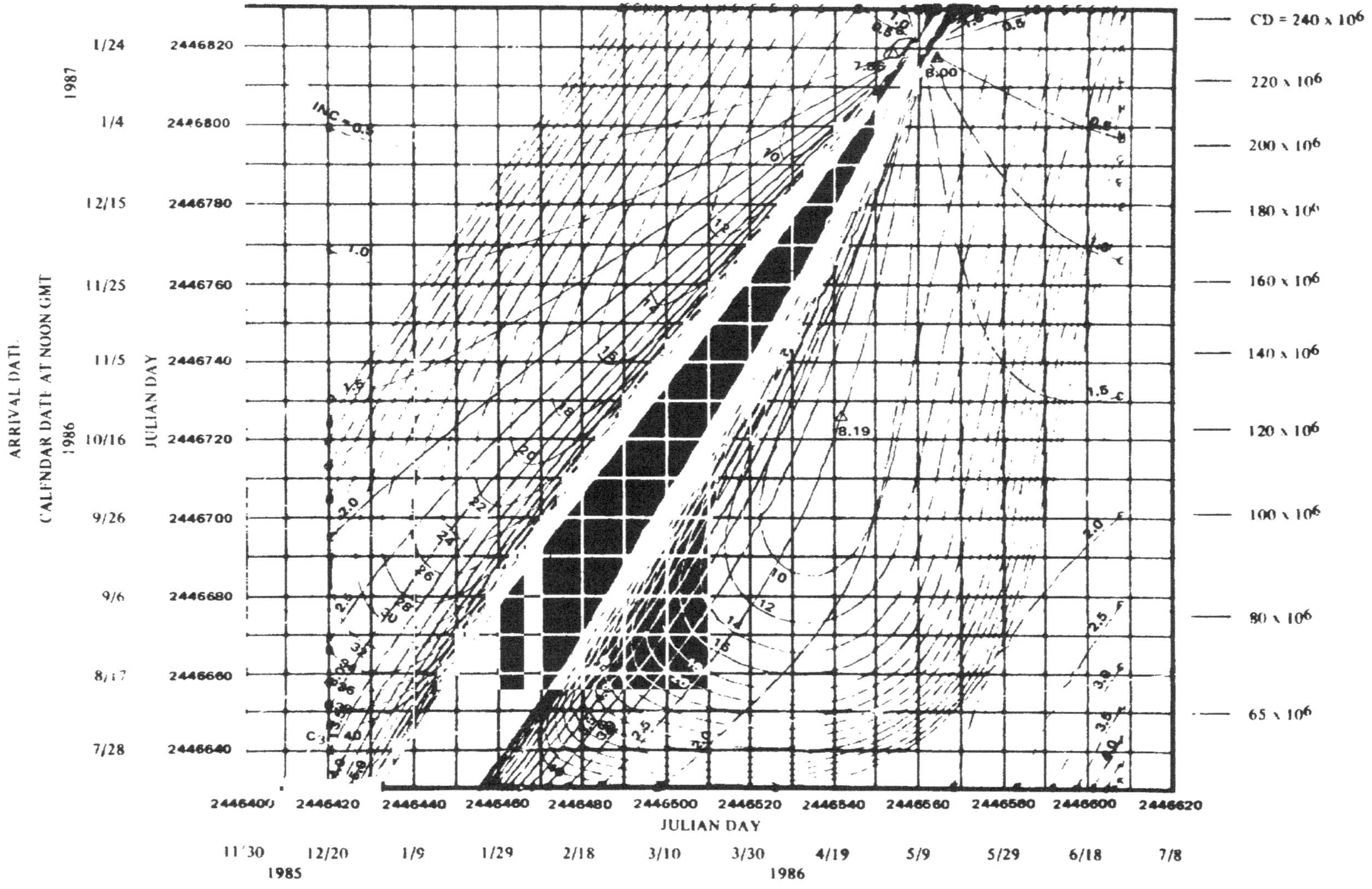

ARRIVAL DATE
CALENDAR DATE AT NOON GMT

JULIAN DAY

Calendar date	Julian day
1/24 (1987)	2446820
1/4	2446800
12/15 (1986)	2446780
11/25	2446760
11/5	2446740
10/16	2446720
9/26	2446700
9/6	2446680
8/17	2446660
7/28	2446640

CD = 240 x 10^6
220 x 10^6
200 x 10^6
180 x 10^6
160 x 10^6
140 x 10^6
120 x 10^6
100 x 10^6
80 x 10^6
65 x 10^6

INC = 0.5
1.0
1.5
2.0
2.5
C_3
7.85
8.00
8.19

2446400 2446420 2446440 2446460 2446480 2446500 2446520 2446540 2446560 2446580 2446600 2446620

JULIAN DAY

11/30 12/20 1/9 1/29 2/18 3/10 3/30 4/19 5/9 5/29 6/18 7/8
1985 1986

CALENDAR DATE AT NOON GMT
DEPARTURE DATE

CONTOURS OF C_3 AND INC EARTH TO MARS 1985-86

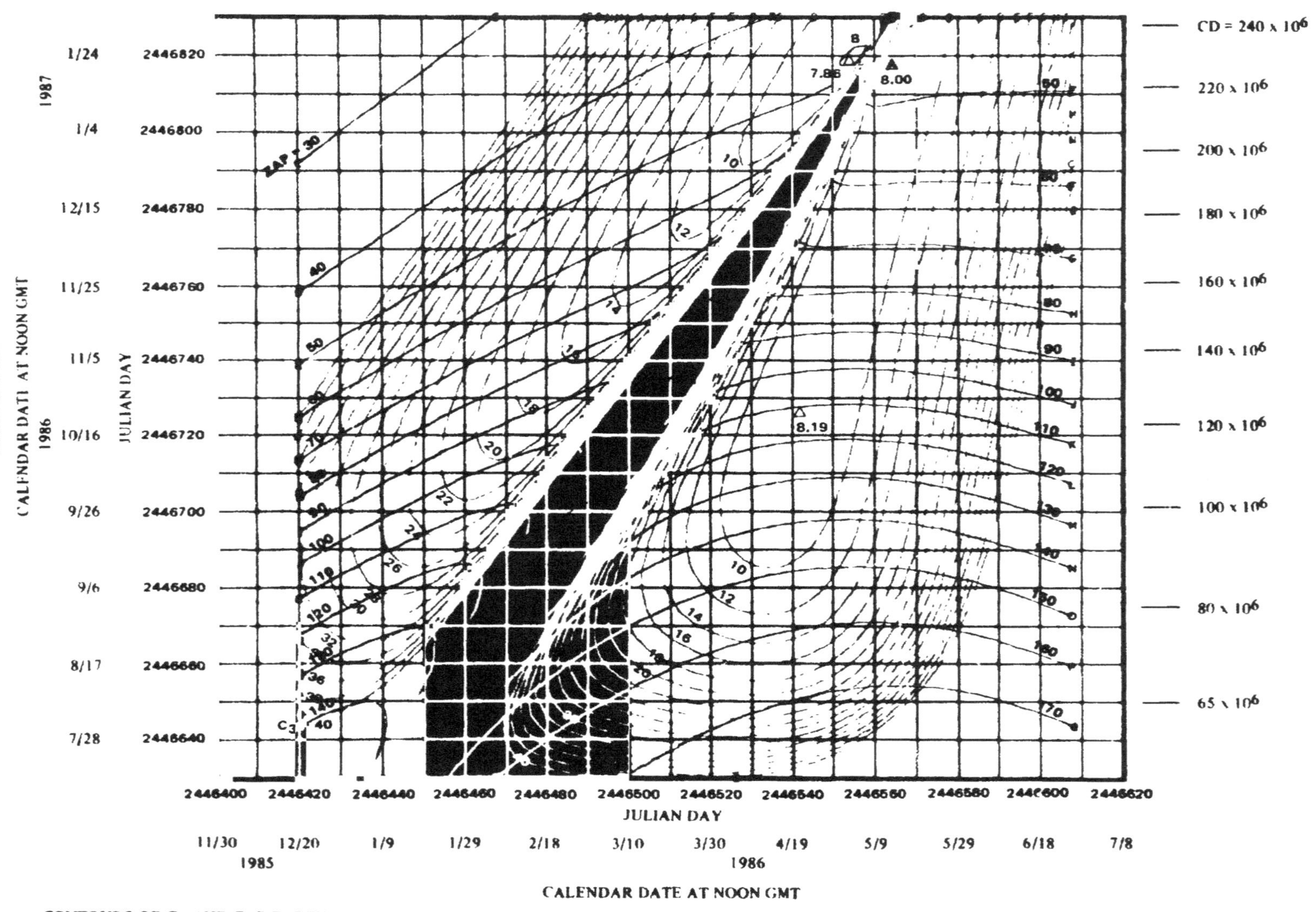

CONTOURS OF C_3 AND ZAP EARTH TO MARS 1985-86

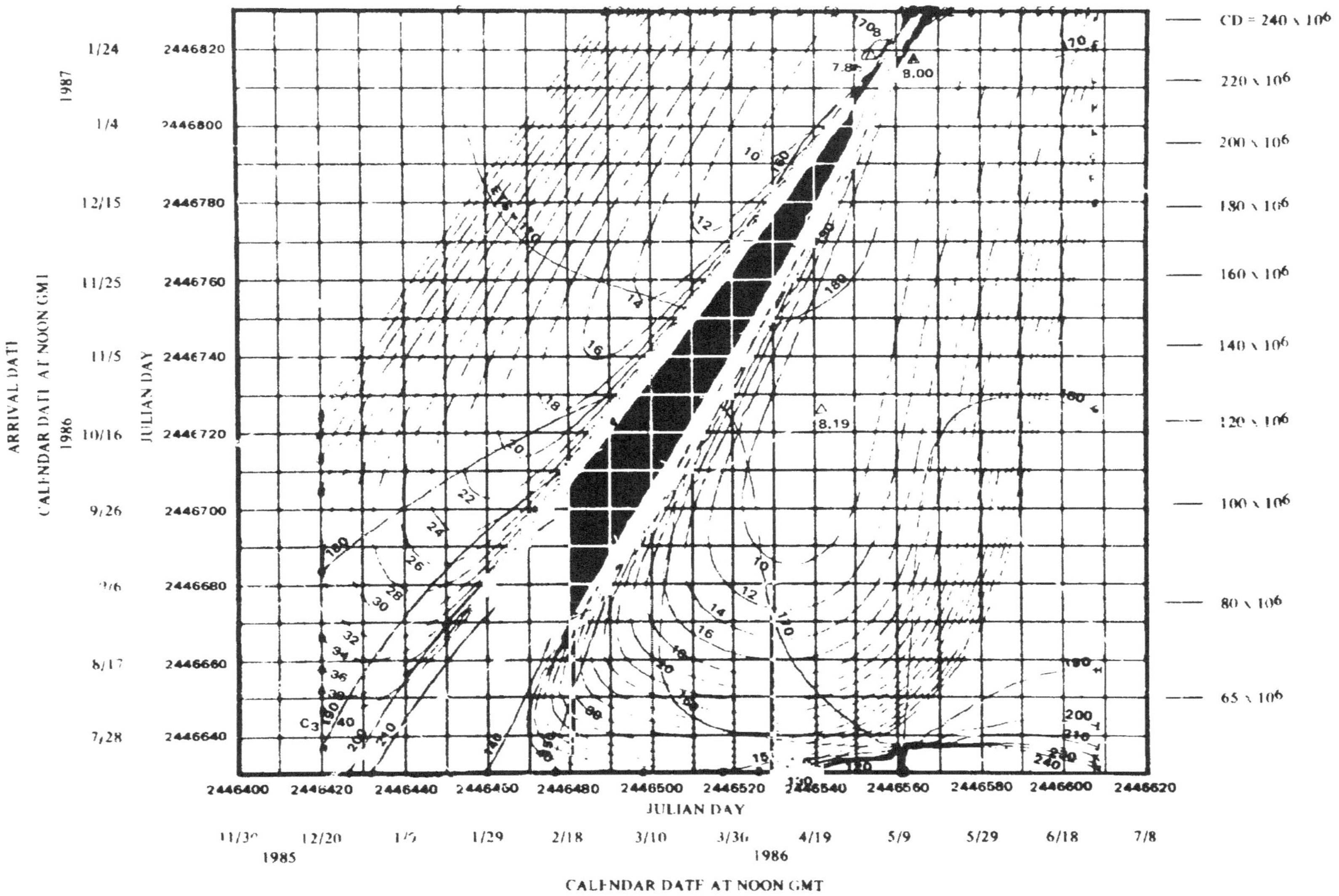

CD = 240 x 10^6
220 x 10^6
200 x 10^6
180 x 10^6
160 x 10^6
140 x 10^6
120 x 10^6
100 x 10^6
80 x 10^6
65 x 10^6

ARRIVAL DATE
CALENDAR DATE AT NOON GMT

		JULIAN DAY
1987	1/24	2446820
	1/4	2446800
	12/15	2446780
	11/25	2446760
	11/5	2446740
1986	10/16	2446720
	9/26	2446700
	9/6	2446680
	8/17	2446660
	7/28	2446640

JULIAN DAY	2446400	2446420	2446440	2446460	2446480	2446500	2446520	2446540	2446560	2446580	2446600	2446620
CALENDAR DATE AT NOON GMT	11/30	12/20	1/9	1/29	2/18	3/10	3/30	4/19	5/9	5/29	6/18	7/8
		1985					1986					

DEPARTURE DATE

CONTOURS OF C_3 AND ETS EARTH TO MARS 1985-86

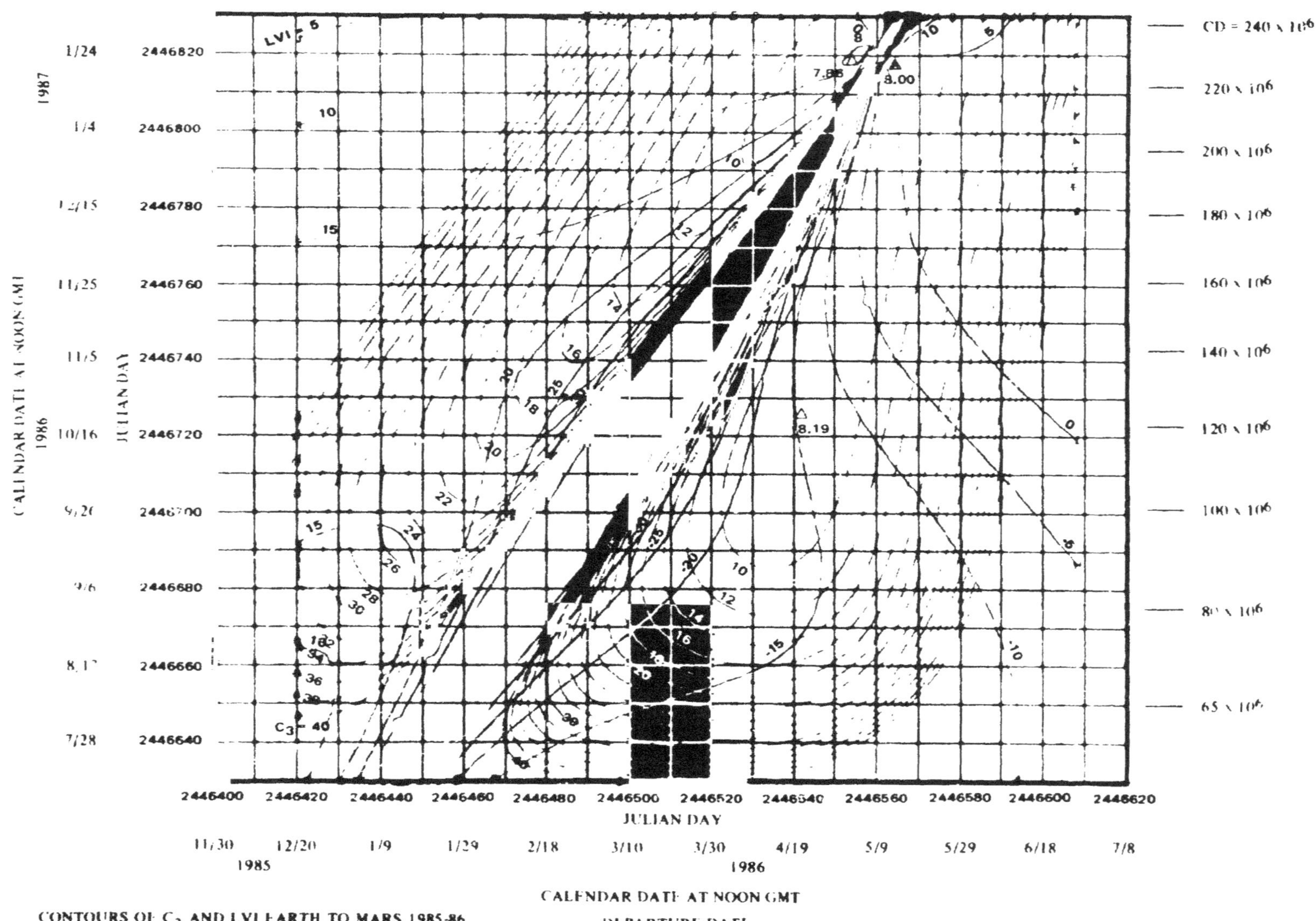

CONTOURS OF C_3 AND LVI EARTH TO MARS 1985-86

LVI
♂
1985

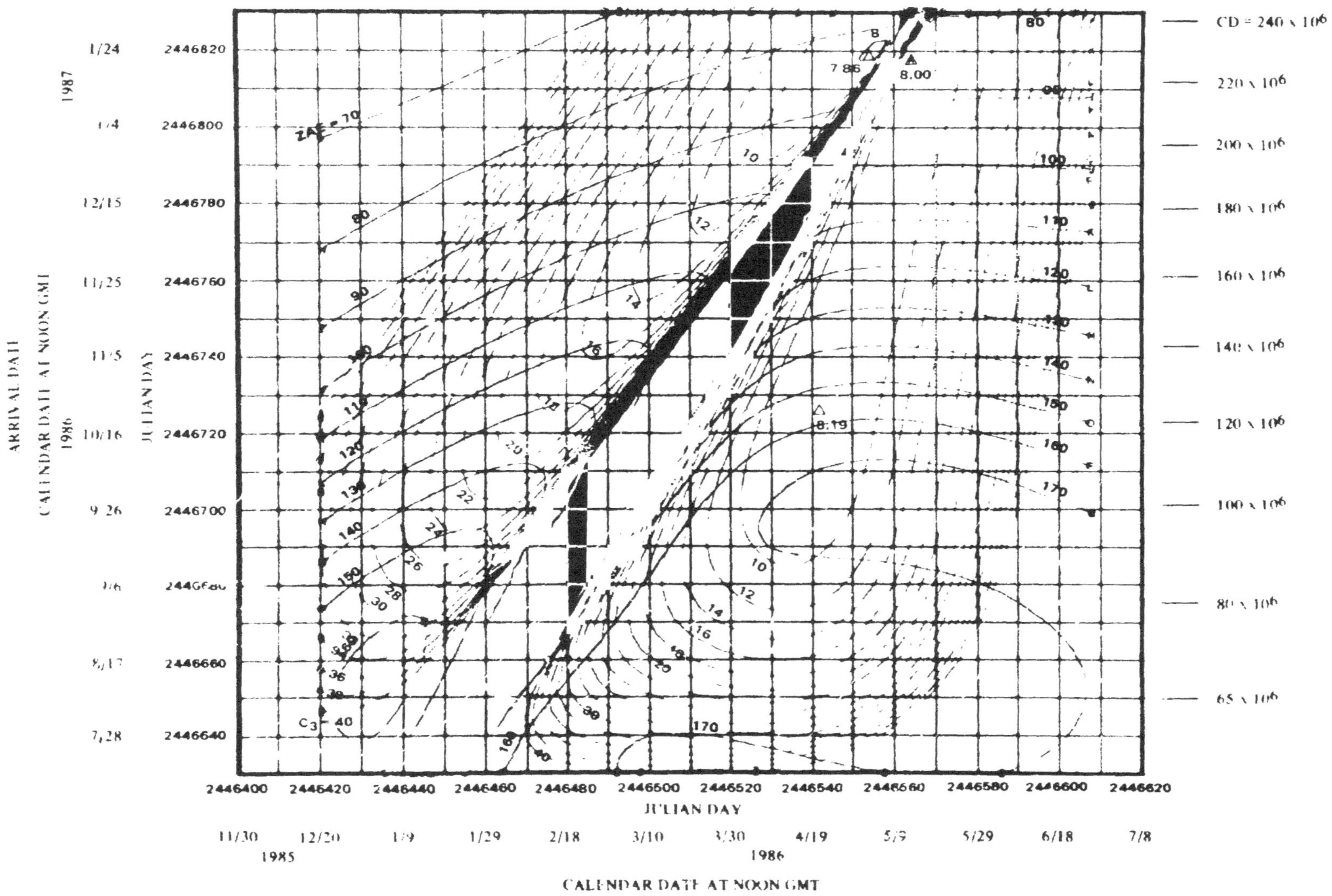

ARRIVAL DATE

CALENDAR DATE AT NOON GMT

JULIAN DAY

Arrival date	Julian day
1/24 (1987)	2446820
1/4	2446800
12/15 (1986)	2446780
11/25	2446760
11/5	2446740
10/16	2446720
9/26	2446700
9/6	2446680
8/17	2446660
7/28	2446640

$CD = 240 \times 10^6$

220×10^6

200×10^6

180×10^6

160×10^6

140×10^6

120×10^6

100×10^6

80×10^6

65×10^6

ZAE = 70

$C_3 = 40$

Julian day	2446400	2446420	2446440	2446460	2446480	2446500	2446520	2446540	2446560	2446580	2446600	2446620
Calendar date	11/30 (1985)	12/20	1/9 (1986)	1/29	2/18	3/10	3/30	4/19	5/9	5/29	6/18	7/8

JULIAN DAY

CALENDAR DATE AT NOON GMT

DEPARTURE DATE

CONTOURS OF C_3 AND ZAE EARTH TO MARS 1985-86

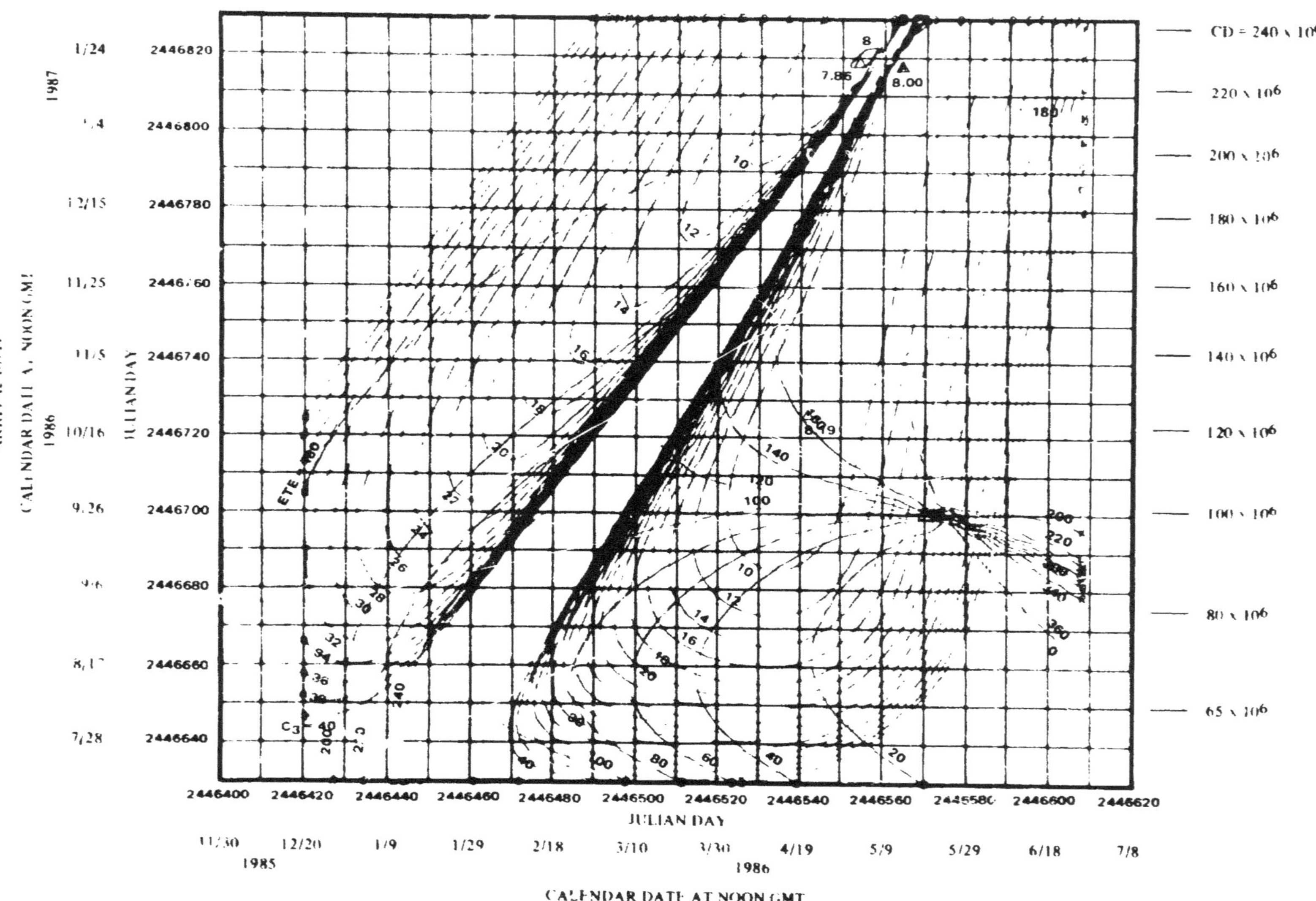

CONTOURS OF C_3 AND ETE EARTH TO MARS 1985 86

ETE
♂
1985

THA
♂
1985

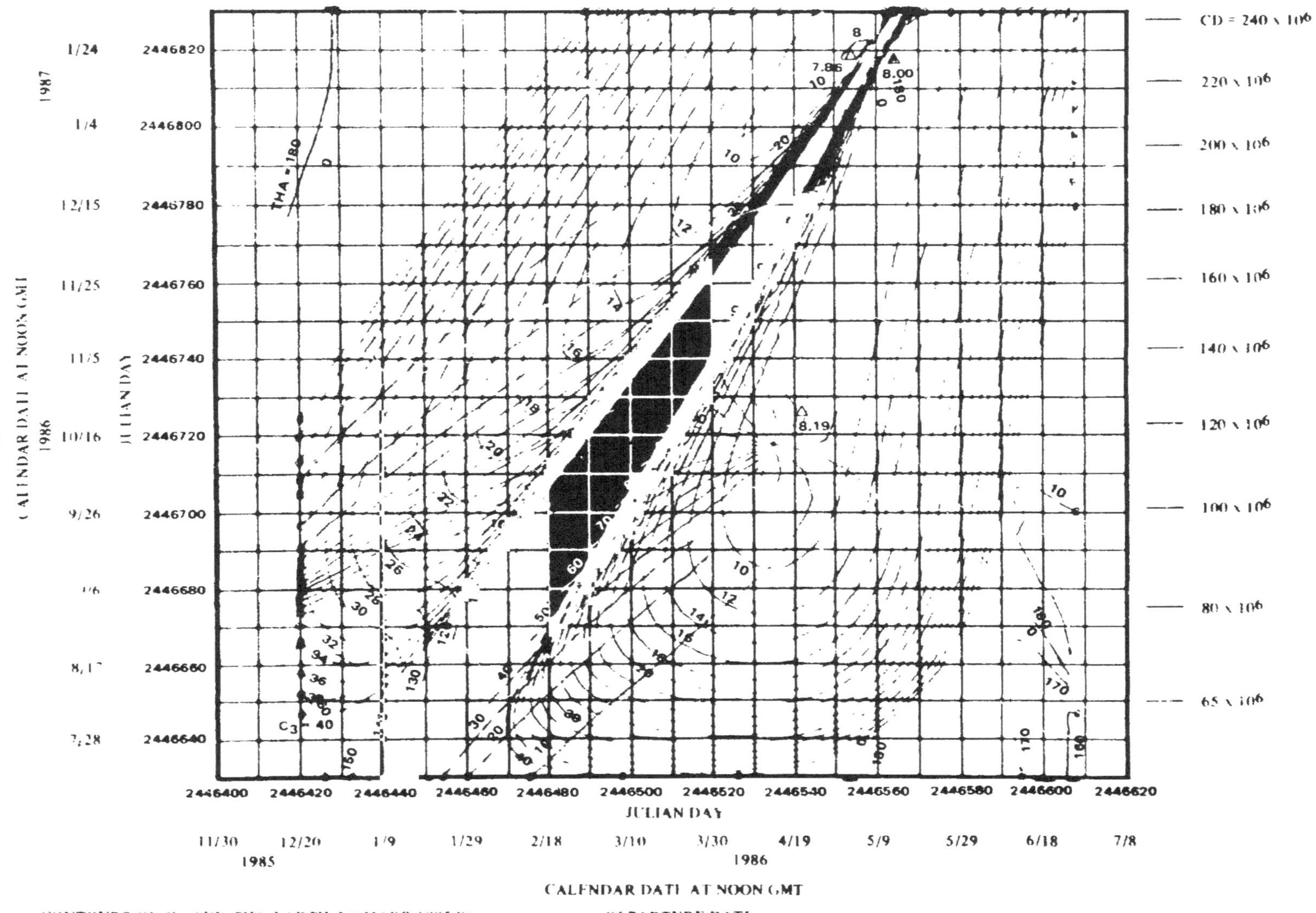

CONTOURS OF C_3 AND THA EARTH TO MARS 1985-86

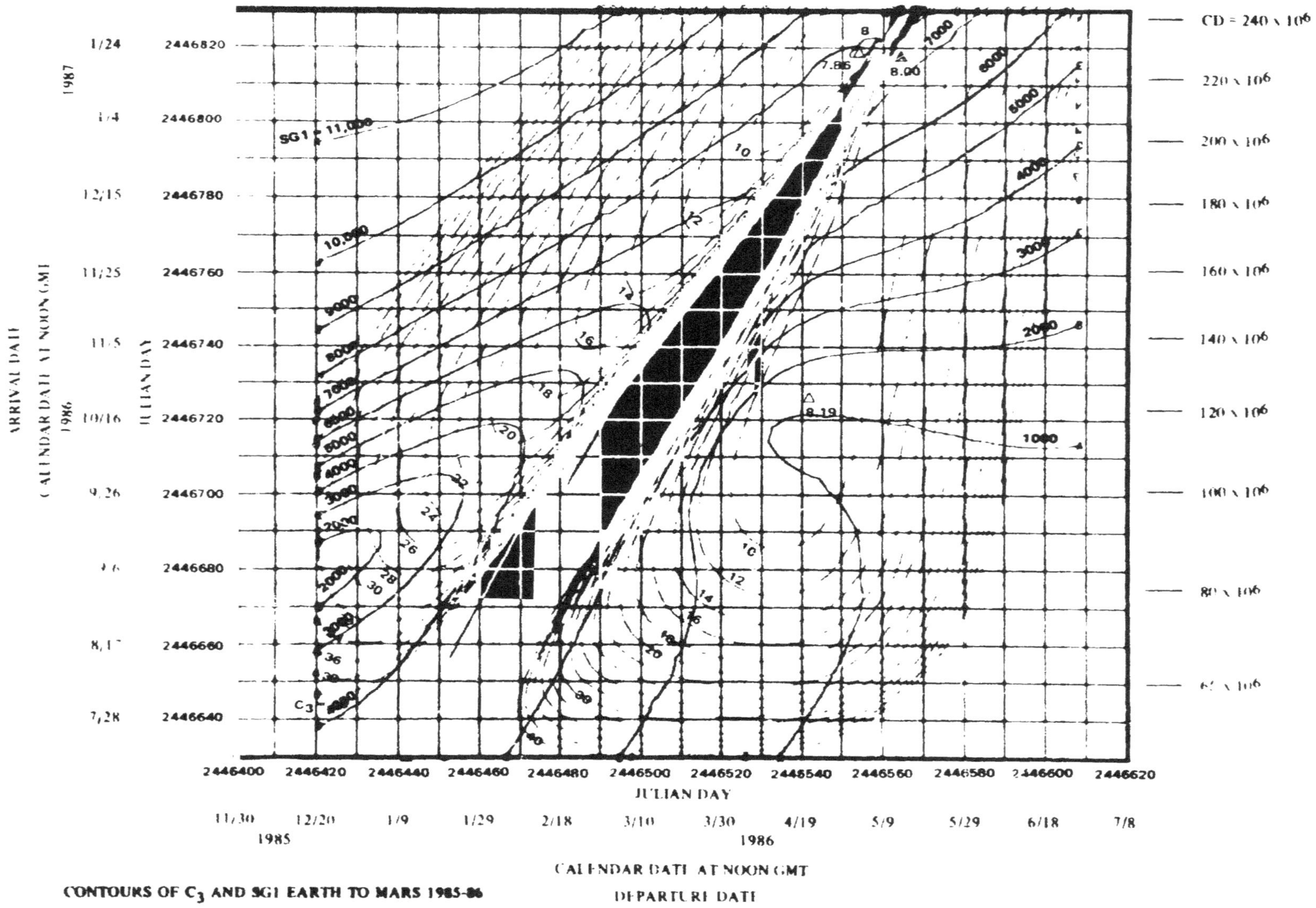

CONTOURS OF C_3 AND SG1 EARTH TO MARS 1985-86

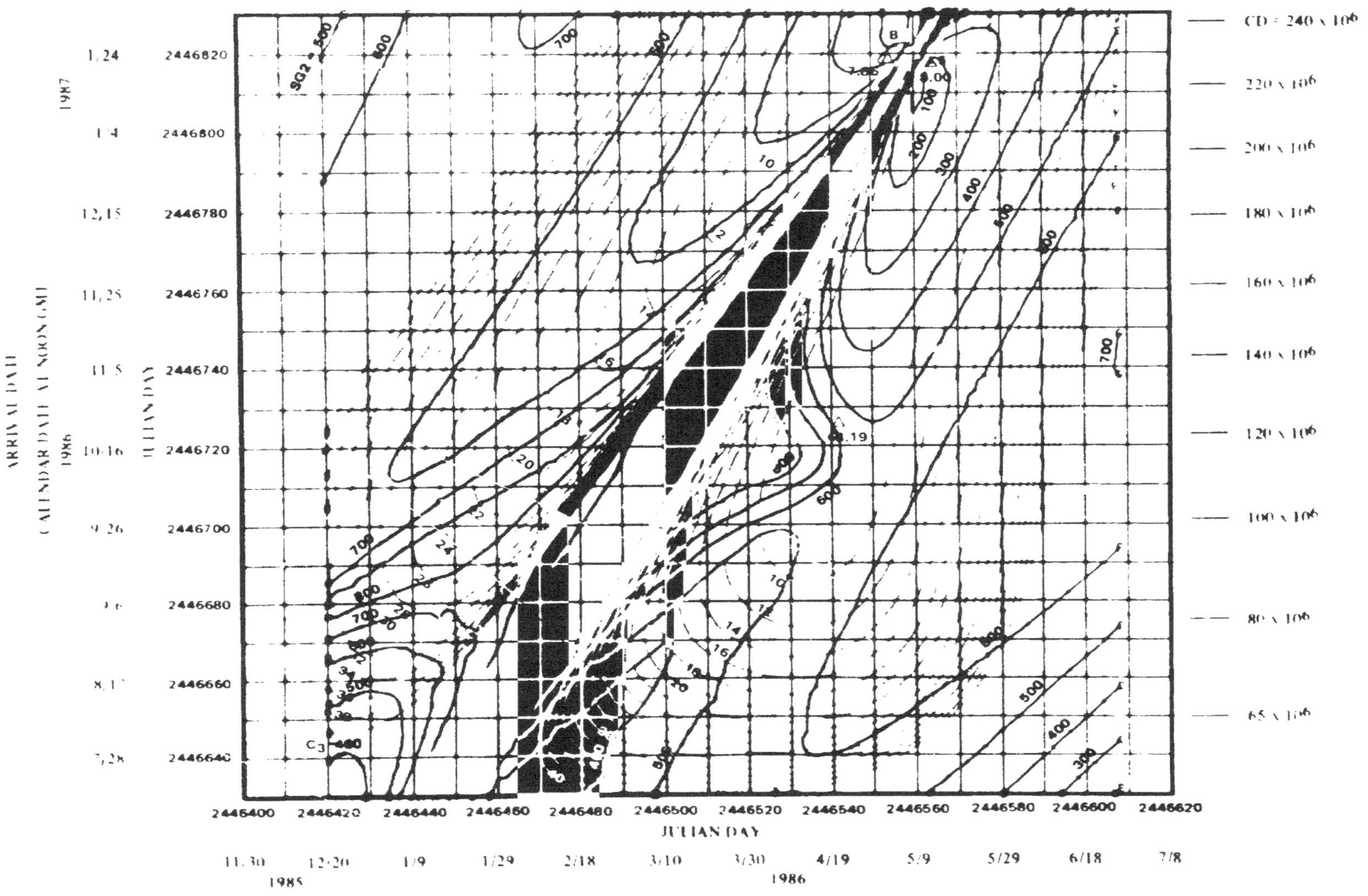

CD = 240 x 10^6
220 x 10^6
200 x 10^6
180 x 10^6
160 x 10^6
140 x 10^6
120 x 10^6
100 x 10^6
80 x 10^6
65 x 10^6

ARRIVAL DATE
CALENDAR DATE AT NOON GMT
JULIAN DAY

1987	1/24	2446820
	1/4	2446800
	12/15	2446780
	11/25	2446760
1986	11/5	2446740
	10/16	2446720
	9/26	2446700
	9/6	2446680
	8/17	2446660
	7/28	2446640

JULIAN DAY	2446400	2446420	2446440	2446460	2446480	2446500	2446520	2446540	2446560	2446580	2446600	2446620
CALENDAR DATE AT NOON GMT	11/30	12/20	1/9	1/29	2/18	3/10	3/30	4/19	5/9	5/29	6/18	7/8
	1985						1986					

DEPARTURE DATE

CONTOURS OF C_3 AND SG2 EARTH TO MARS 1985-86

ARRIVAL DATE
CALENDAR DATE AT NOON GMT

JULIAN DAY

JULIAN DAY

CALENDAR DATE AT NOON GMT
DEPARTURE DATE

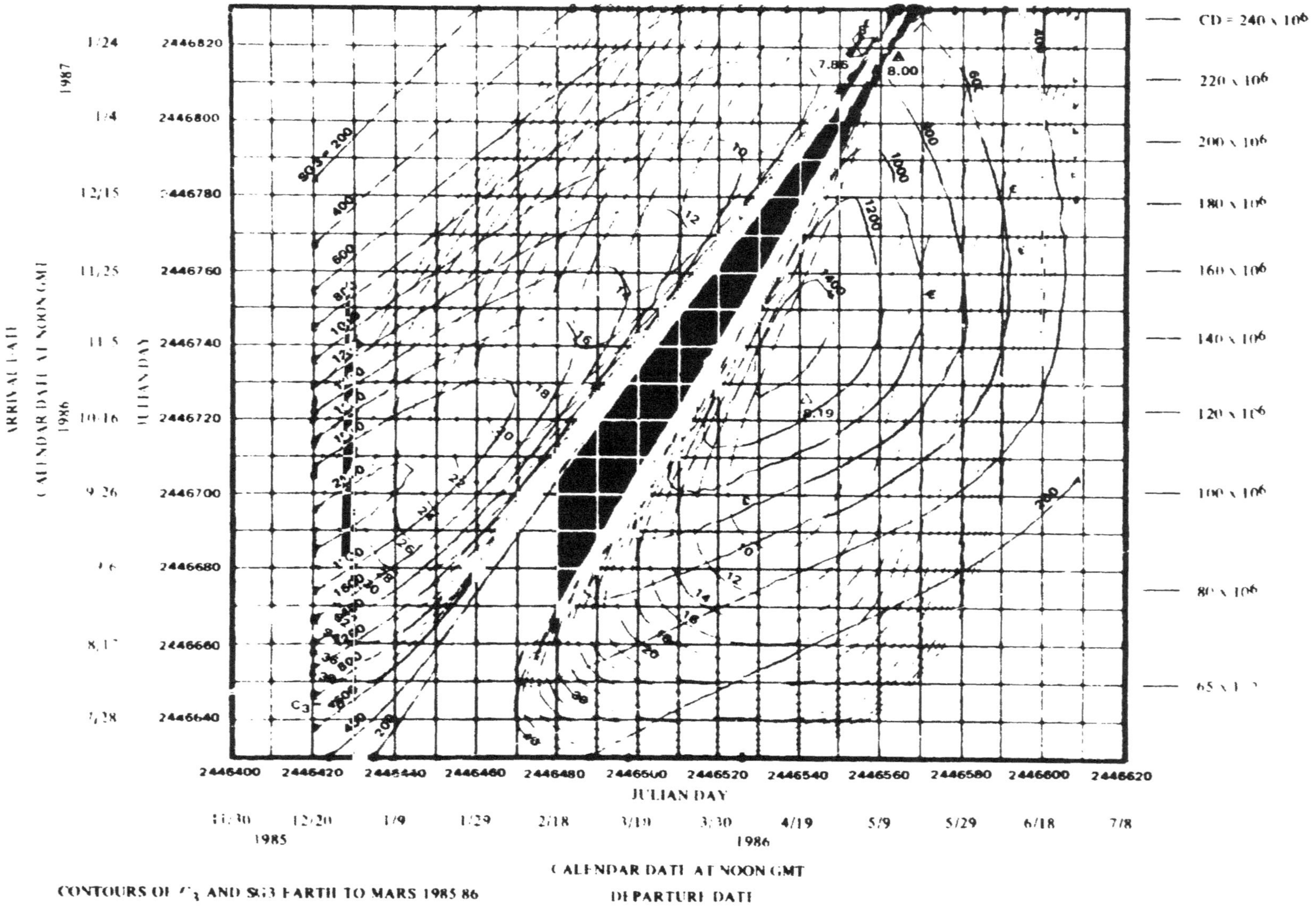

CONTOURS OF C_3 AND SG3 EARTH TO MARS 1985-86

SG3

1985

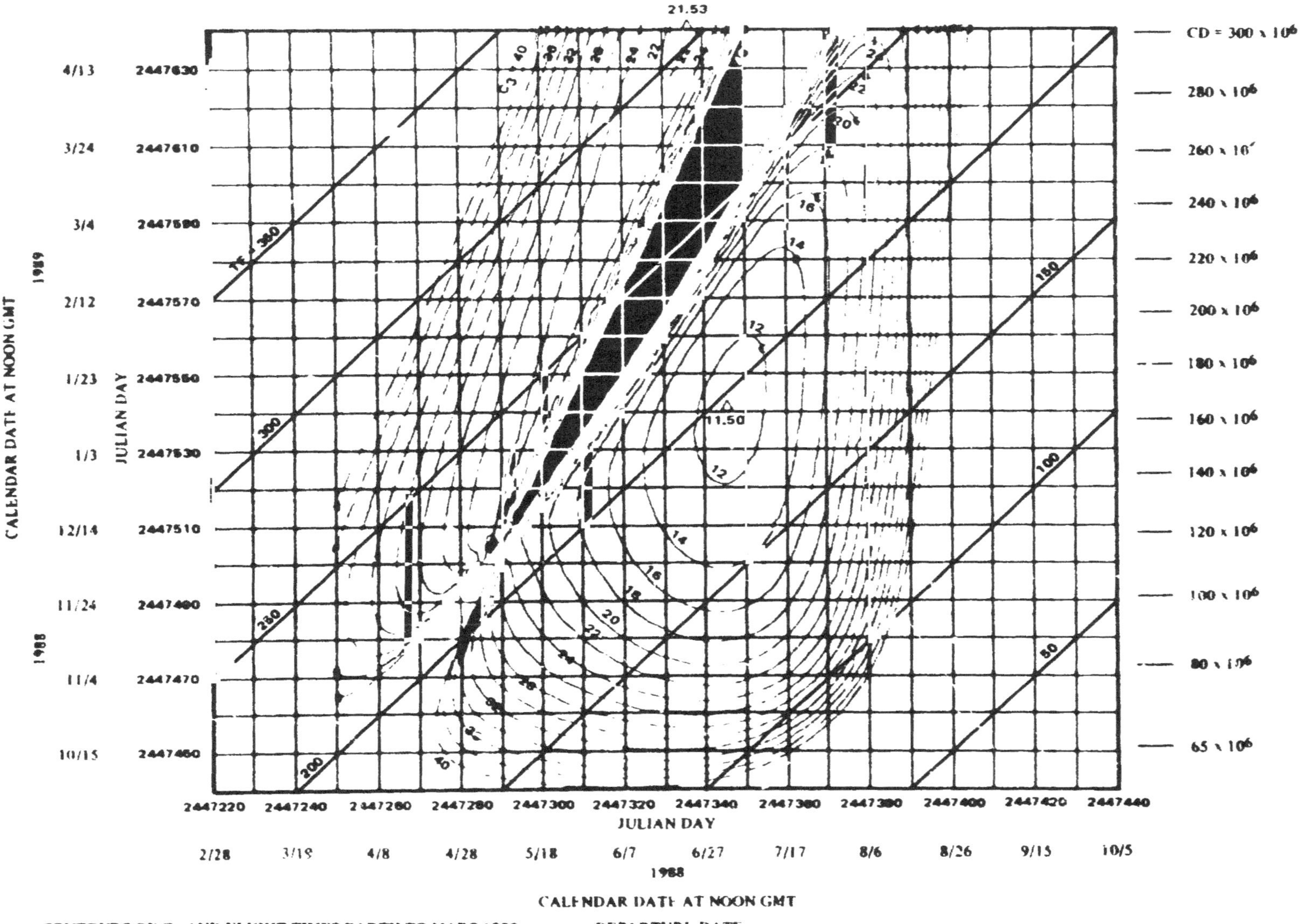

CONTOURS OF C_3 AND FLIGHT TIMES EARTH TO MARS 1988

C_3 ♂ 1988

VHP ♂ 1988

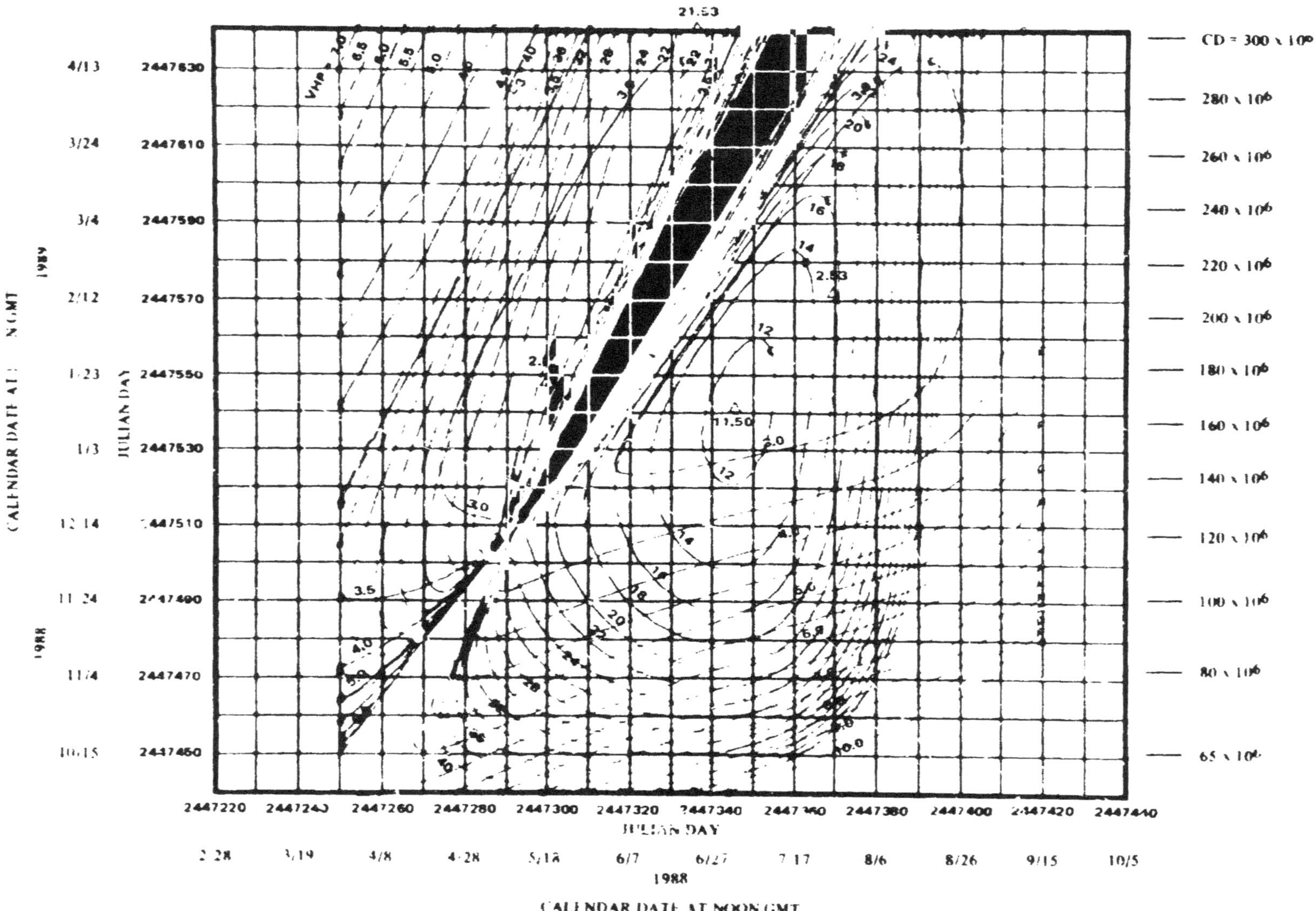

CONTOURS OF C_3 AND VHP EARTH TO MARS 1988

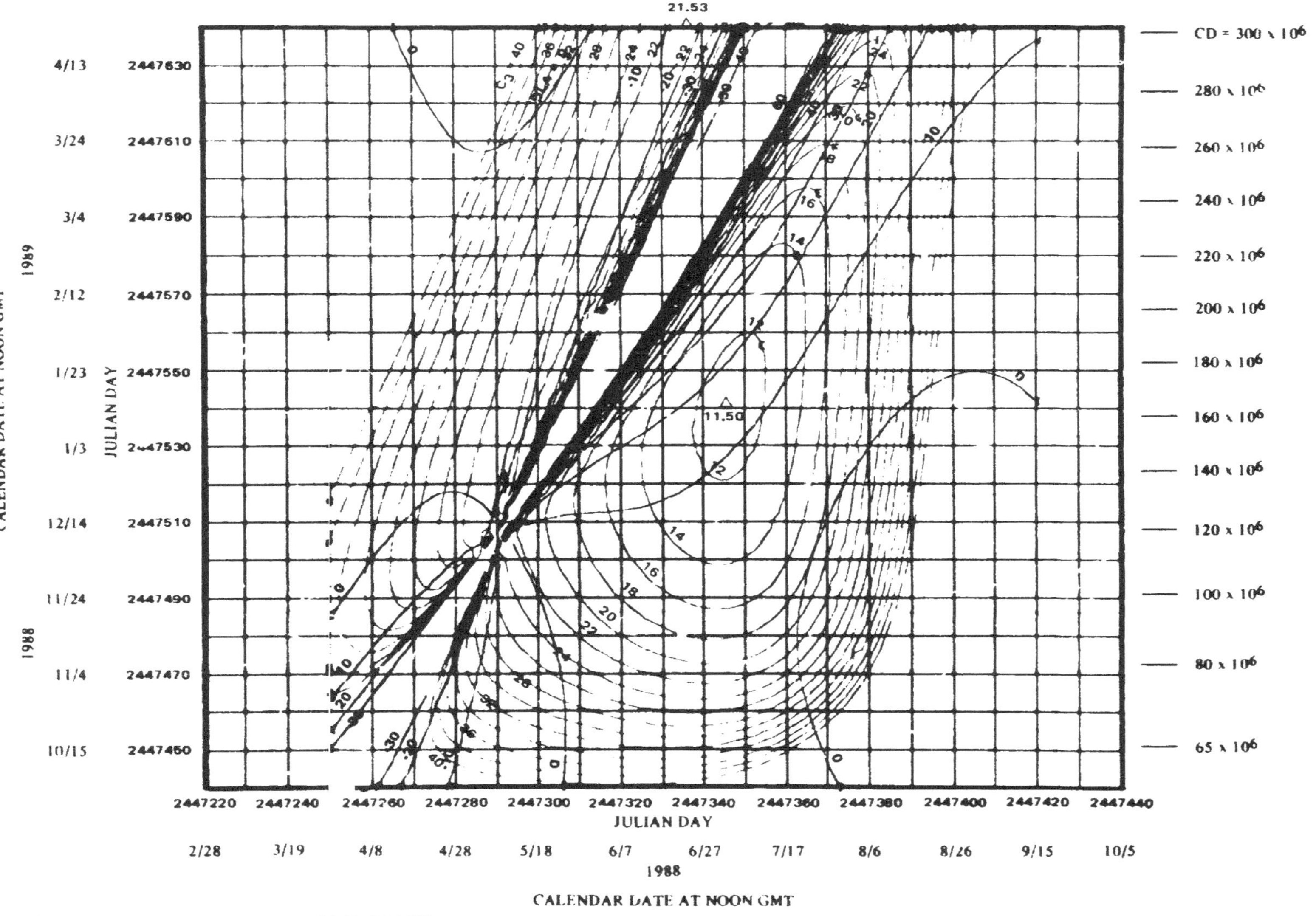

CONTOURS OF C_3 AND DLA EARTH TO MARS 1988

DLA ♂ 1988

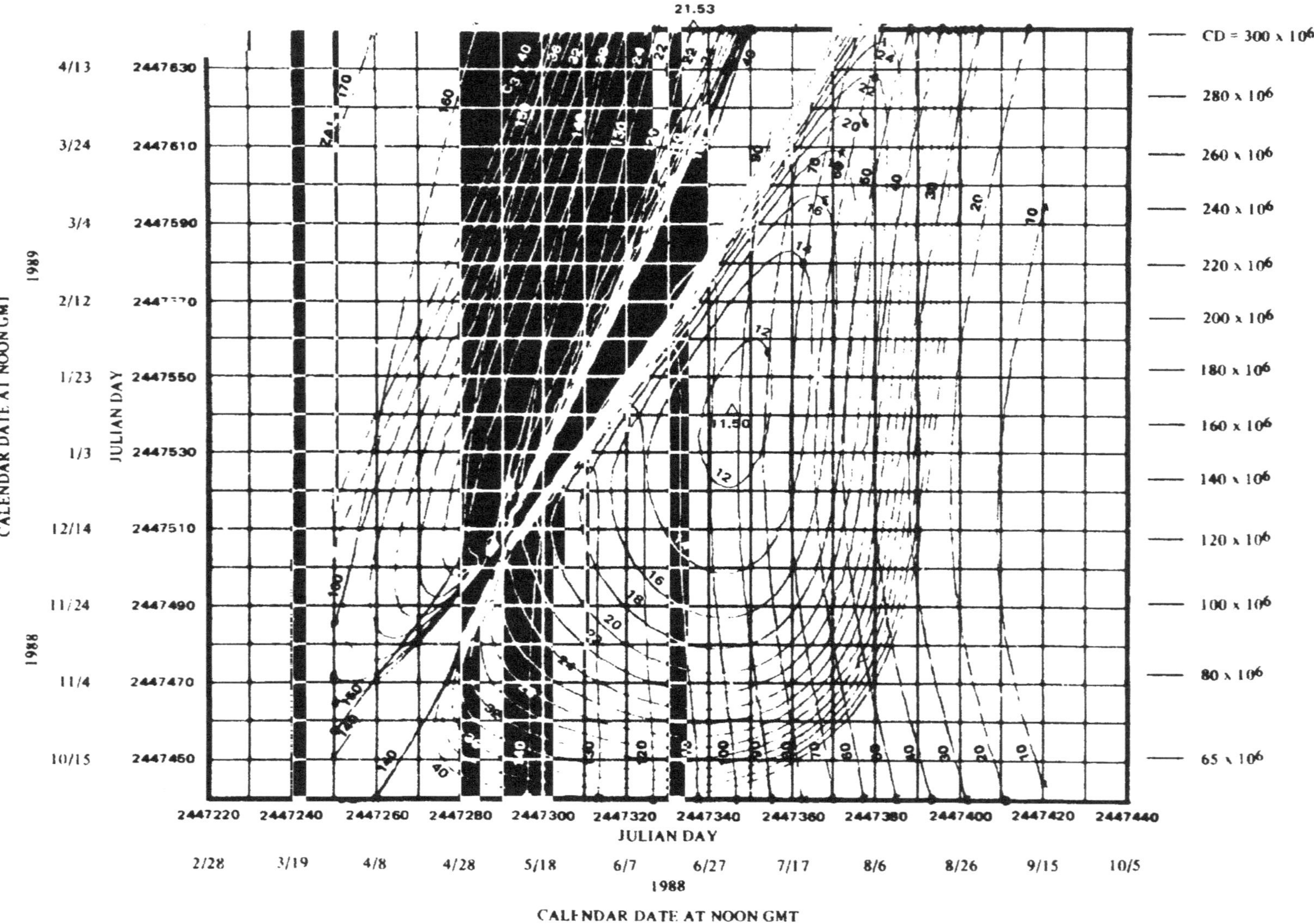

CONTOURS OF C_3 AND ZAL EARTH TO MARS 1988

ZAL ♂ 1988

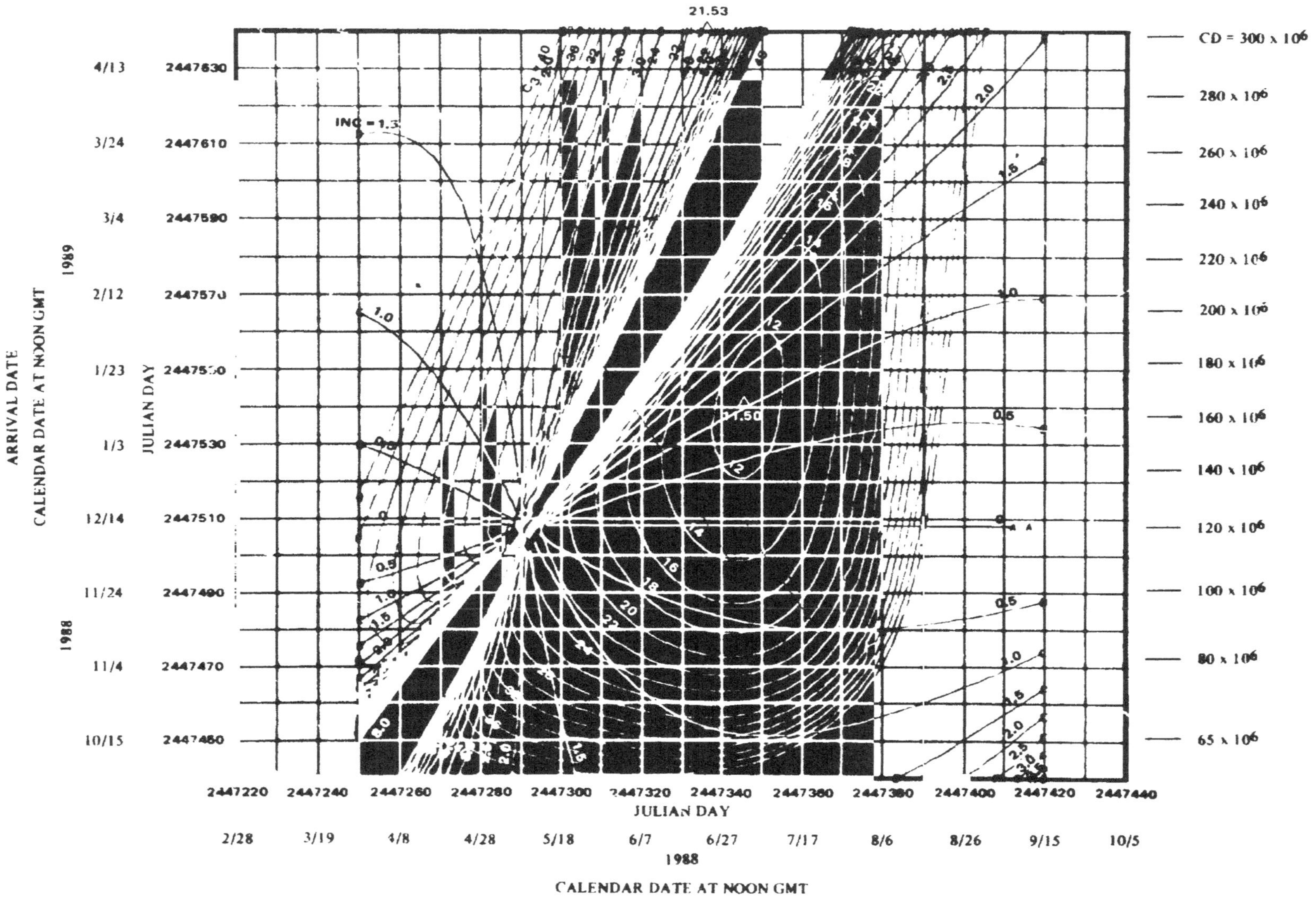

21.53

ARRIVAL DATE

CALENDAR DATE AT NOON GMT

1989

1988

Calendar date	Julian day	
4/13	2447630	
		CD = 300 x 10^6
		280 x 10^6
3/24	2447610	260 x 10^6
3/4	2447590	240 x 10^6
		220 x 10^6
2/12	2447570	200 x 10^6
1/23	2447550	180 x 10^6
		160 x 10^6
1/3	2447530	140 x 10^6
12/14	2447510	120 x 10^6
11/24	2447490	100 x 10^6
11/4	2447470	80 x 10^6
10/15	2447450	65 x 10^6

JULIAN DAY

INC = 1.3

C_3 = 40

21.50

2447220 2447240 2447260 2447280 2447300 2447320 2447340 2447360 2447380 2447400 2447420 2447440

JULIAN DAY

2/28 3/19 4/8 4/28 5/18 6/7 6/27 7/17 8/6 8/26 9/15 10/5

1988

CALENDAR DATE AT NOON GMT

DEPARTURE DATE

CONTOURS OF C_3 AND INC EARTH TO MARS 1988

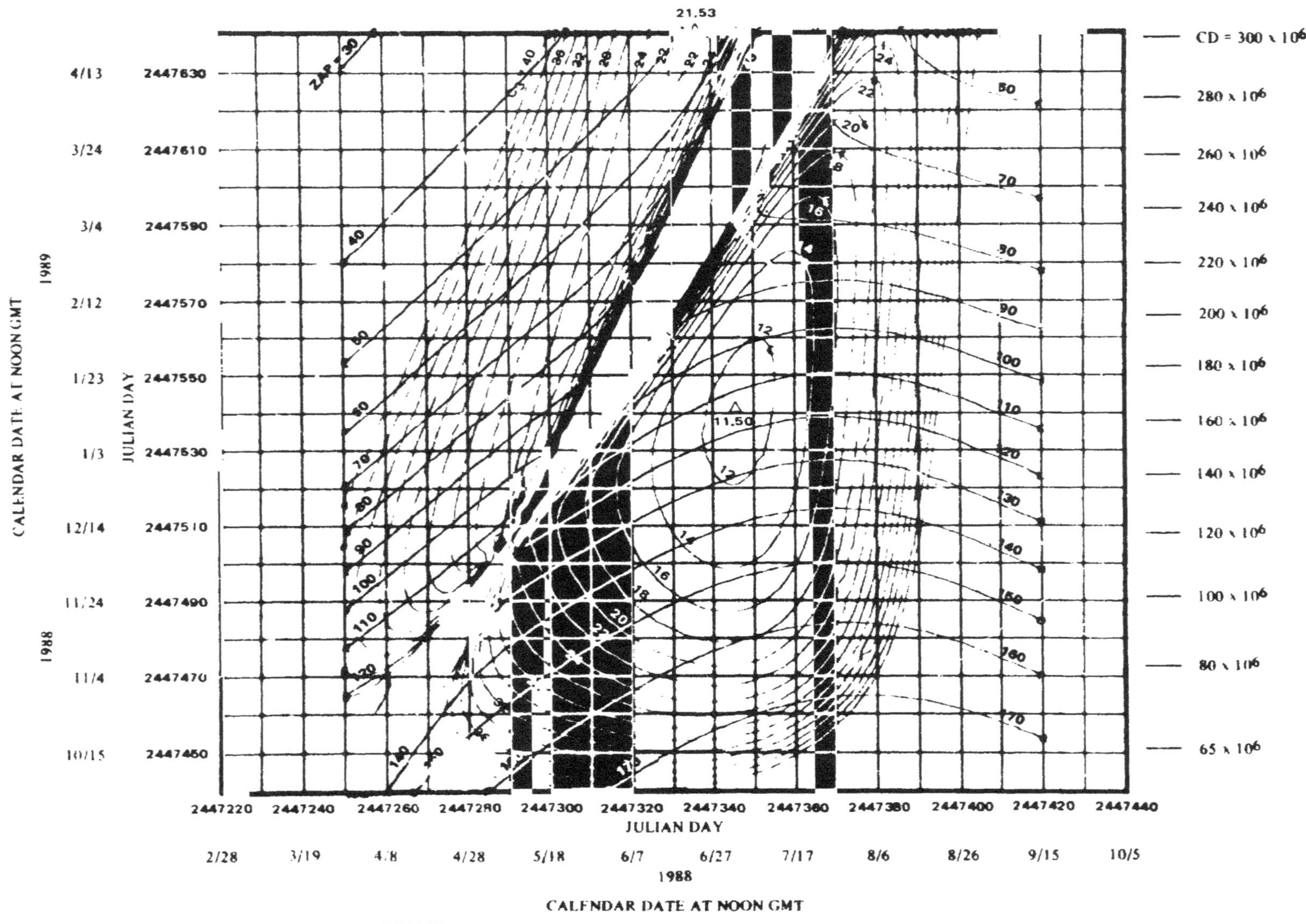

CONTOURS OF C_3 AND ZAP EARTH TO MARS 1988

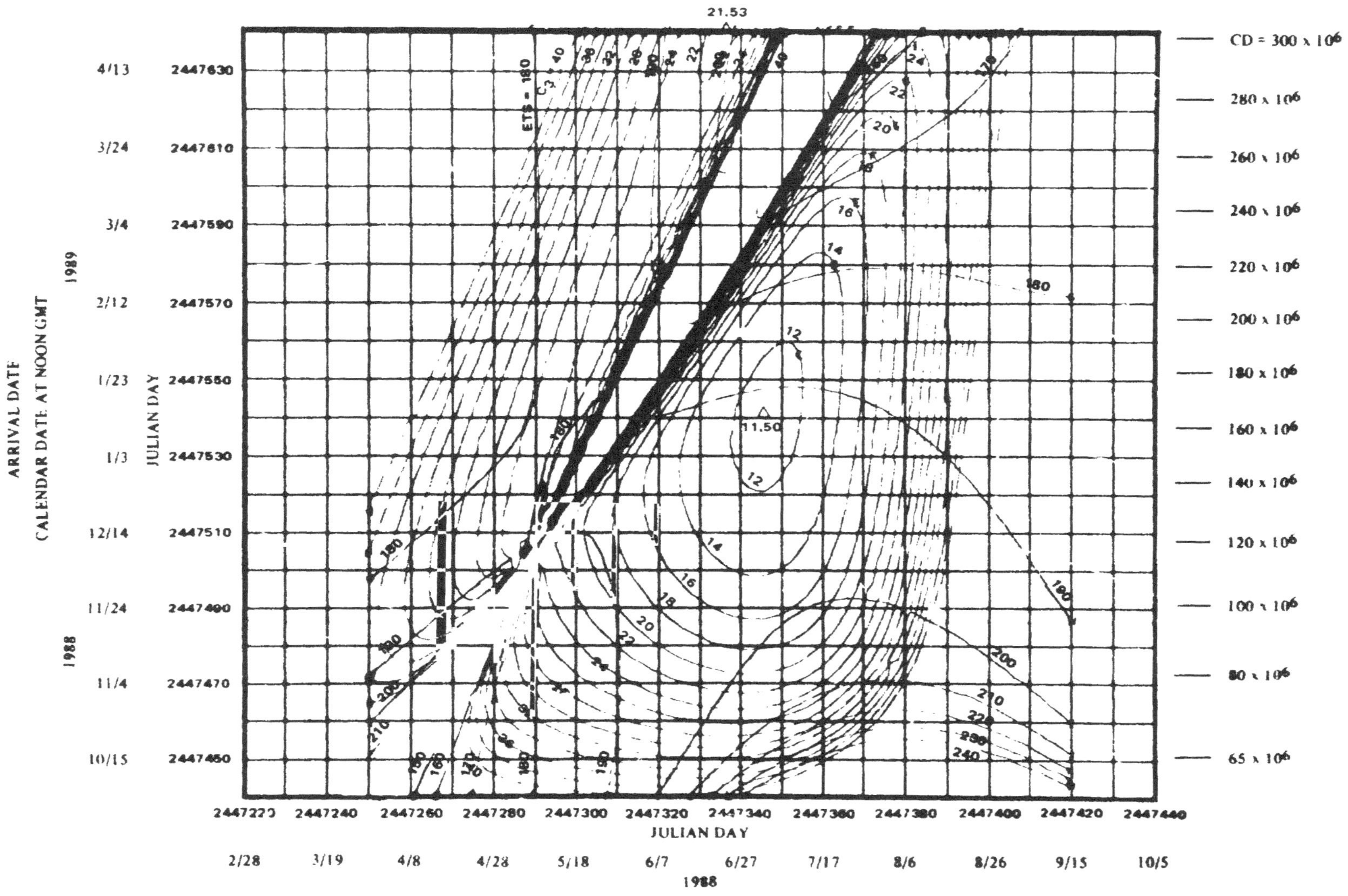

21.53

CD = 300×10^6
280×10^6
260×10^6
240×10^6
220×10^6
200×10^6
180×10^6
160×10^6
140×10^6
120×10^6
100×10^6
80×10^6
65×10^6

ARRIVAL DATE
CALENDAR DATE AT NOON GMT

JULIAN DAY

1989: 4/13 (2447630), 3/24 (2447610), 3/4 (2447590), 2/12 (2447570), 1/23 (2447550), 1/3 (2447530)
1988: 12/14 (2447510), 11/24 (2447490), 11/4 (2447470), 10/15 (2447450)

2447220 2447240 2447260 2447280 2447300 2447320 2447340 2447360 2447380 2447400 2447420 2447440

JULIAN DAY

2/28 3/19 4/8 4/28 5/18 6/7 6/27 7/17 8/6 8/26 9/15 10/5

1988

CALENDAR DATE AT NOON GMT

DEPARTURE DATE

CONTOURS OF C_3 AND ETS EARTH TO MARS 1988

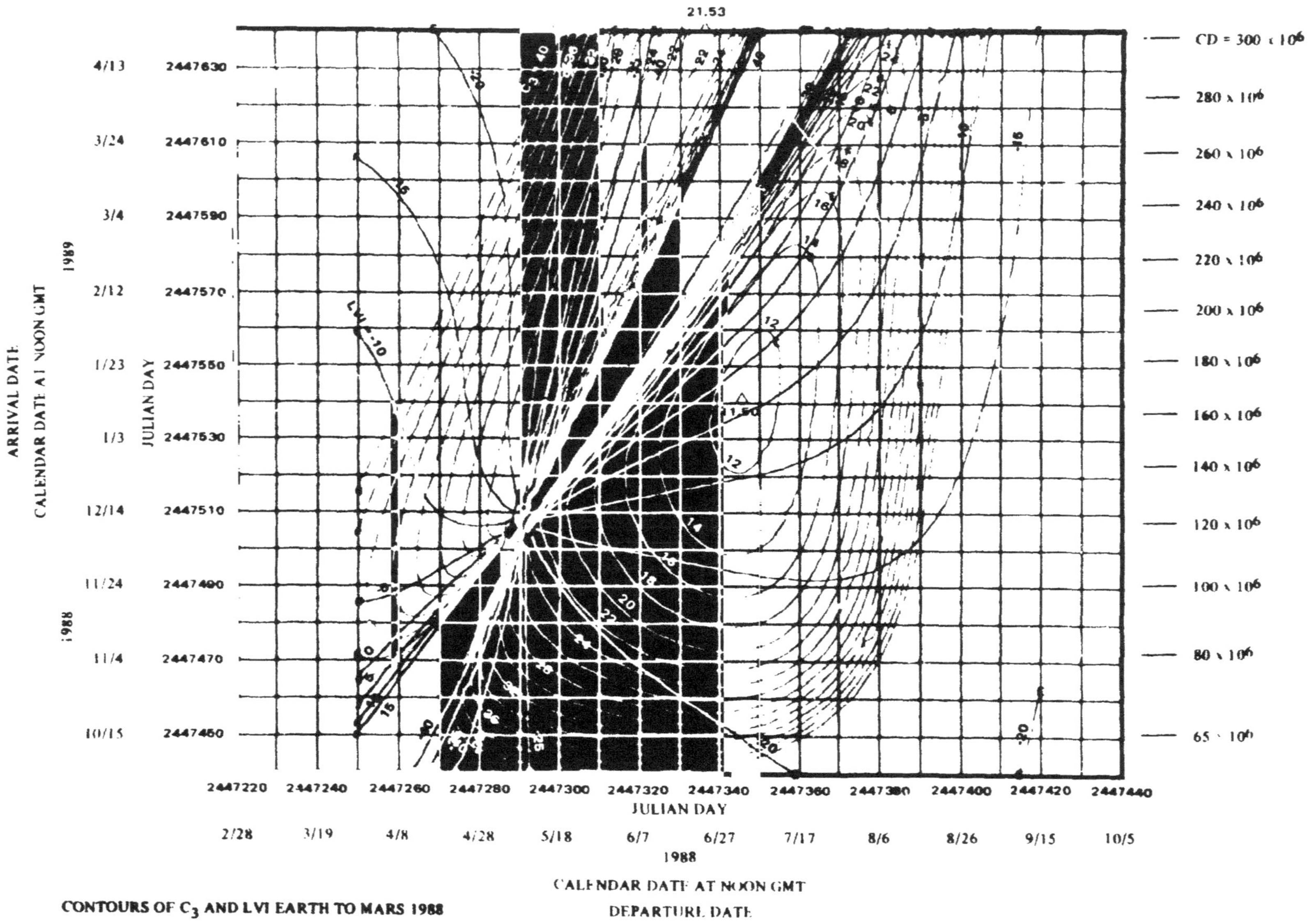

21.53

ARRIVAL DATE
CALENDAR DATE AT NOON GMT

1989		
	4/13	2447630
	3/24	2447610
	3/4	2447590
	2/12	2447570
	1/23	2447550
	1/3	2447530
1988	12/14	2447510
	11/24	2447490
	11/4	2447470
	10/15	2447450

JULIAN DAY

CD = 300 x 10^6
280 x 10^6
260 x 10^6
240 x 10^6
220 x 10^6
200 x 10^6
180 x 10^6
160 x 10^6
140 x 10^6
120 x 10^6
100 x 10^6
80 x 10^6
65 x 10^6

LVI = 10

2447220 2447240 2447260 2447280 2447300 2447320 2447340 2447360 2447380 2447400 2447420 2447440

JULIAN DAY

2/28 3/19 4/8 4/28 5/18 6/7 6/27 7/17 8/6 8/26 9/15 10/5

1988

CALENDAR DATE AT NOON GMT

DEPARTURE DATE

CONTOURS OF C_3 AND LVI EARTH TO MARS 1988

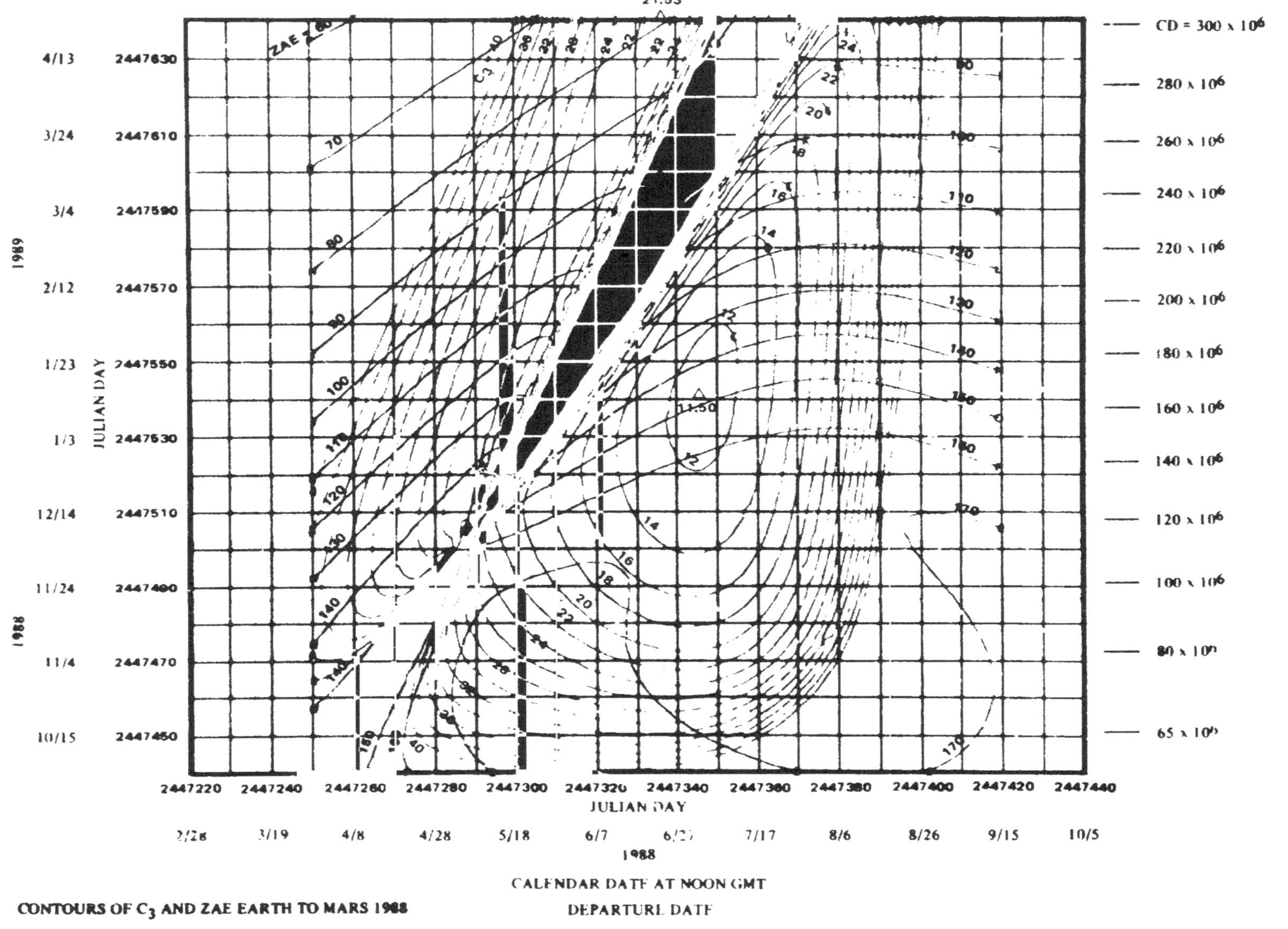

CONTOURS OF C_3 AND ZAE EARTH TO MARS 1988

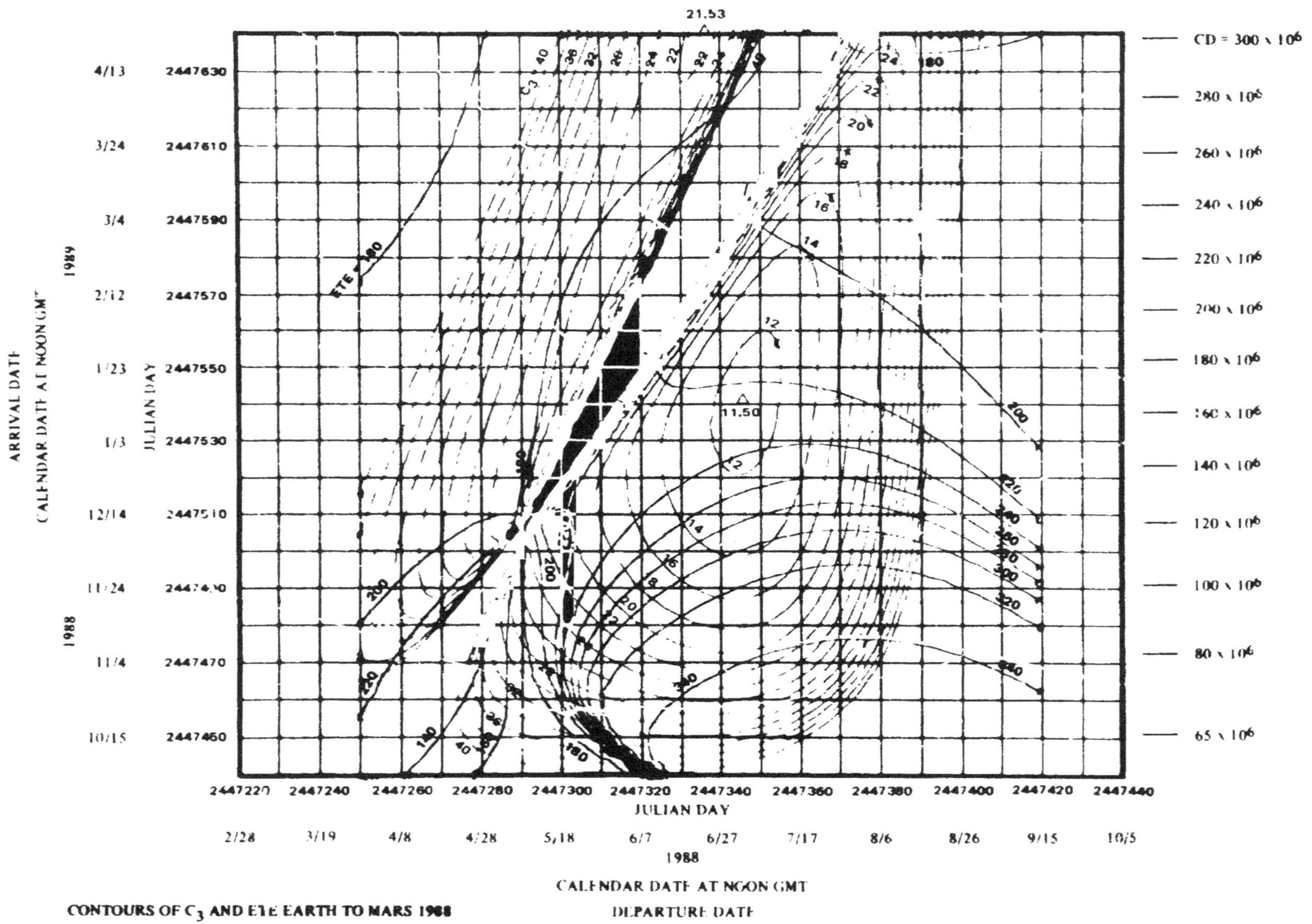

21.53

ARRIVAL DATE CALENDAR DATE AT NOON GMT	JULIAN DAY
4/13 (1989)	2447630
3/24	2447610
3/4	2447590
2/12	2447570
1/23	2447550
1/3	2447530
12/14 (1988)	2447510
11/24	2447490
11/4	2447470
10/15	2447450

CD = 300 x 10^6
280 x 10^6
260 x 10^6
240 x 10^6
220 x 10^6
200 x 10^6
180 x 10^6
160 x 10^6
140 x 10^6
120 x 10^6
100 x 10^6
80 x 10^6
65 x 10^6

C_3 40 36 32 28 24 22 20 18 16 14 12 11.50

ETE = 180 200 220 240 260 280 300 320 340

JULIAN DAY	2447220	2447240	2447260	2447280	2447300	2447320	2447340	2447360	2447380	2447400	2447420	2447440
CALENDAR DATE AT NOON GMT (1988)	2/28	3/19	4/8	4/28	5/18	6/7	6/27	7/17	8/6	8/26	9/15	10/5

DEPARTURE DATE

CONTOURS OF C_3 AND ETE EARTH TO MARS 1988

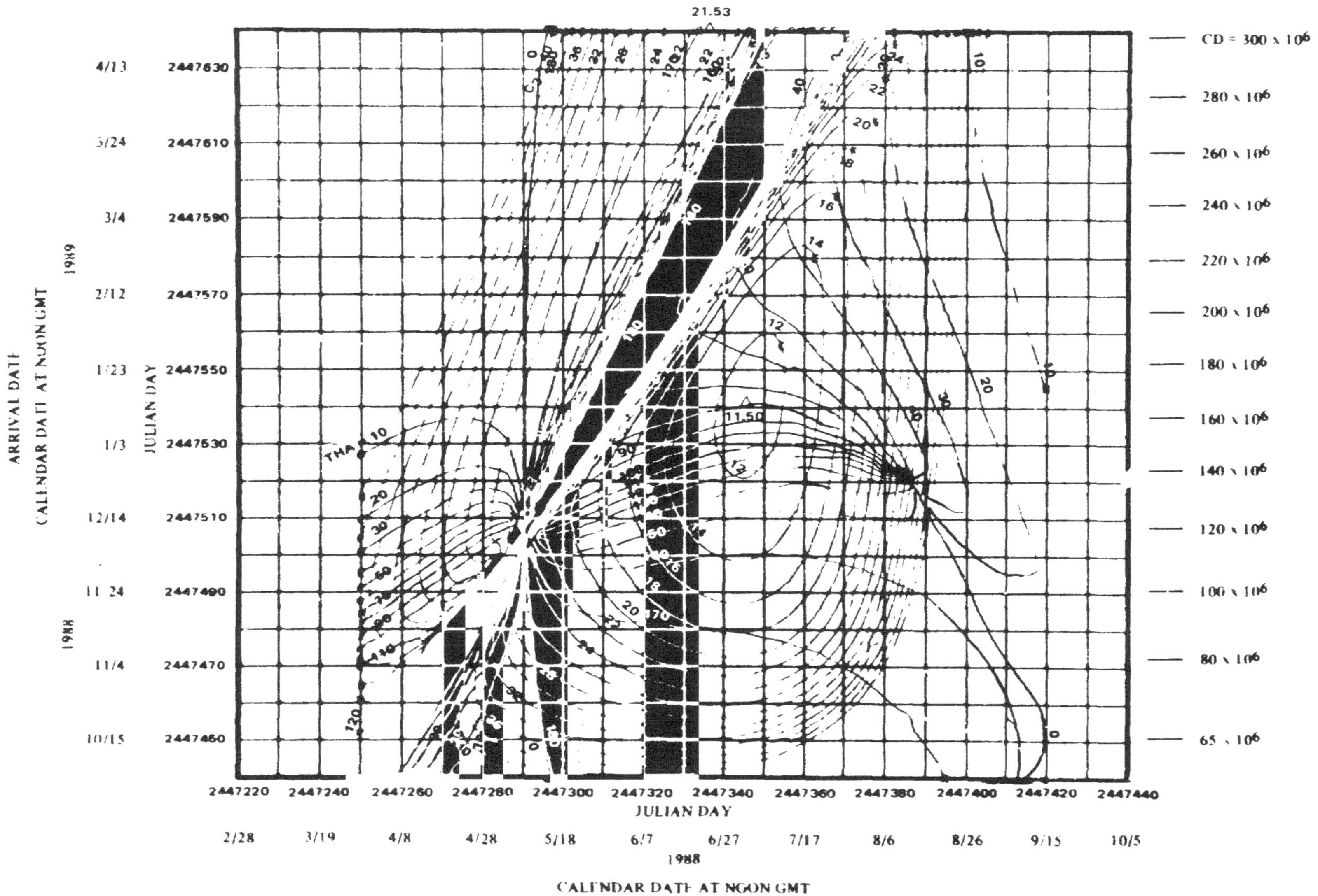

21.53

CD = 300 x 10^6
280 x 10^6
260 x 10^6
240 x 10^6
220 x 10^6
200 x 10^6
180 x 10^6
160 x 10^6
140 x 10^6
120 x 10^6
100 x 10^6
80 x 10^6
65 x 10^6

ARRIVAL DATE
CALENDAR DATE AT NOON GMT

1989	JULIAN DAY
4/13	2447630
3/24	2447610
3/4	2447590
2/12	2447570
1/23	2447550
1/3	2447530
12/14	2447510
11/24	2447490
11/4	2447470
10/15	2447450

1988

2447220 2447240 2447260 2447280 2447300 2447320 2447340 2447360 2447380 2447400 2447420 2447440

JULIAN DAY

2/28 3/19 4/8 4/28 5/18 6/7 6/27 7/17 8/6 8/26 9/15 10/5

1988

CALENDAR DATE AT NOON GMT

DEPARTURE DATE

CONTOURS OF C_3 AND THA EARTH TO MARS 1988

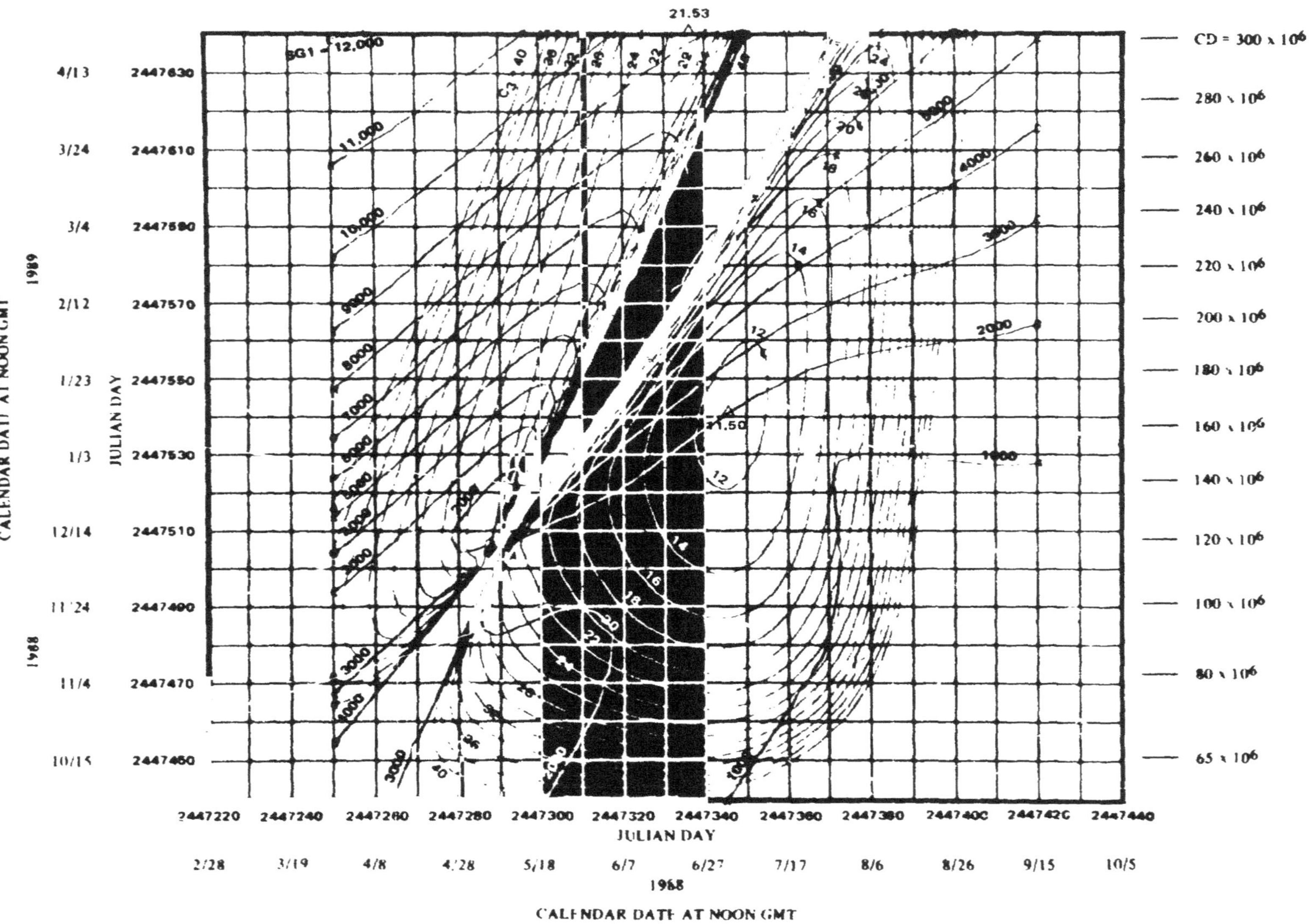

CONTOURS OF C_3 AND SG1 EARTH TO MARS 1988

SG1 ♂ 1988

SG2 ♂ 1988

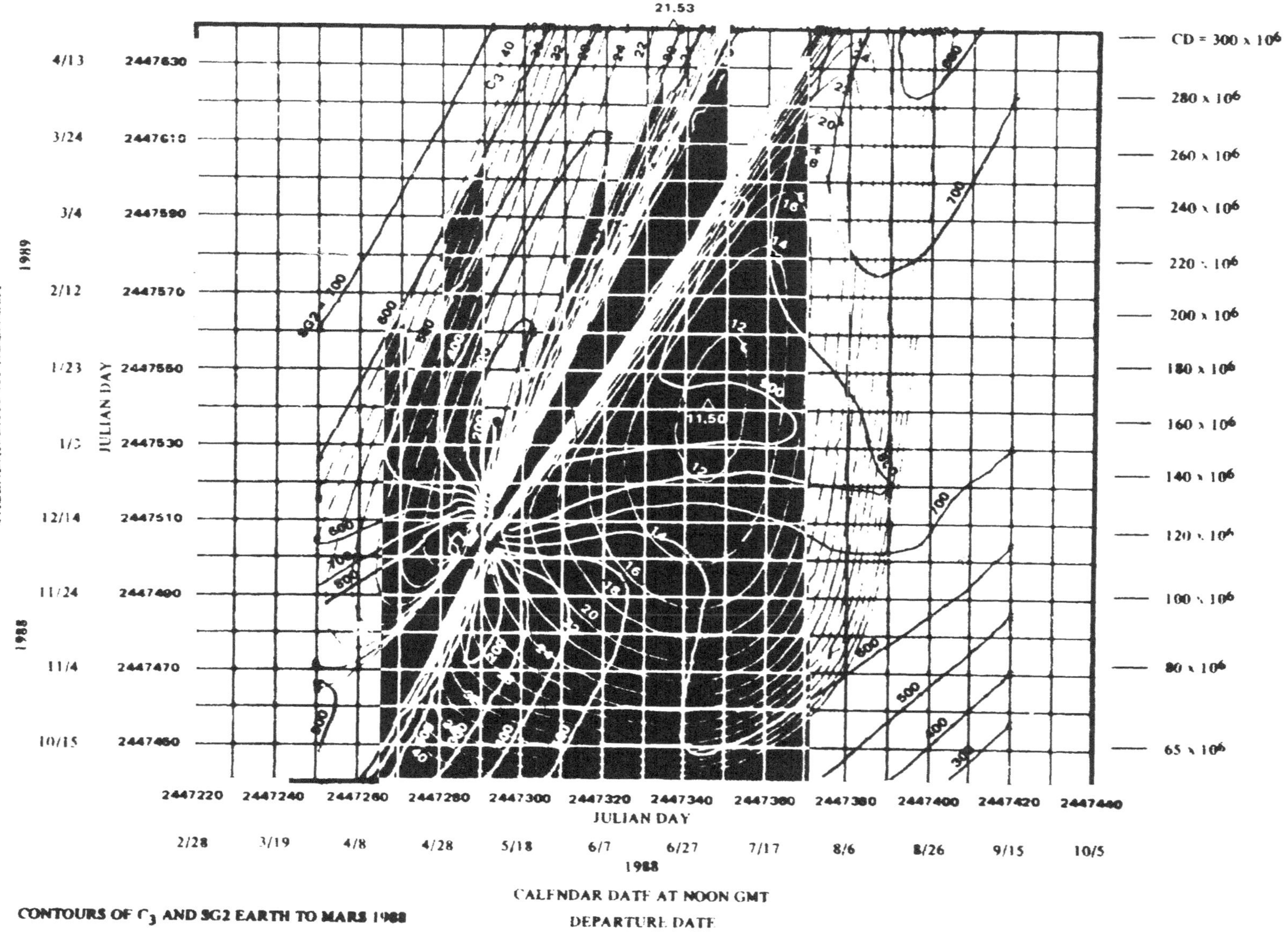

CONTOURS OF C_3 AND SG2 EARTH TO MARS 1988

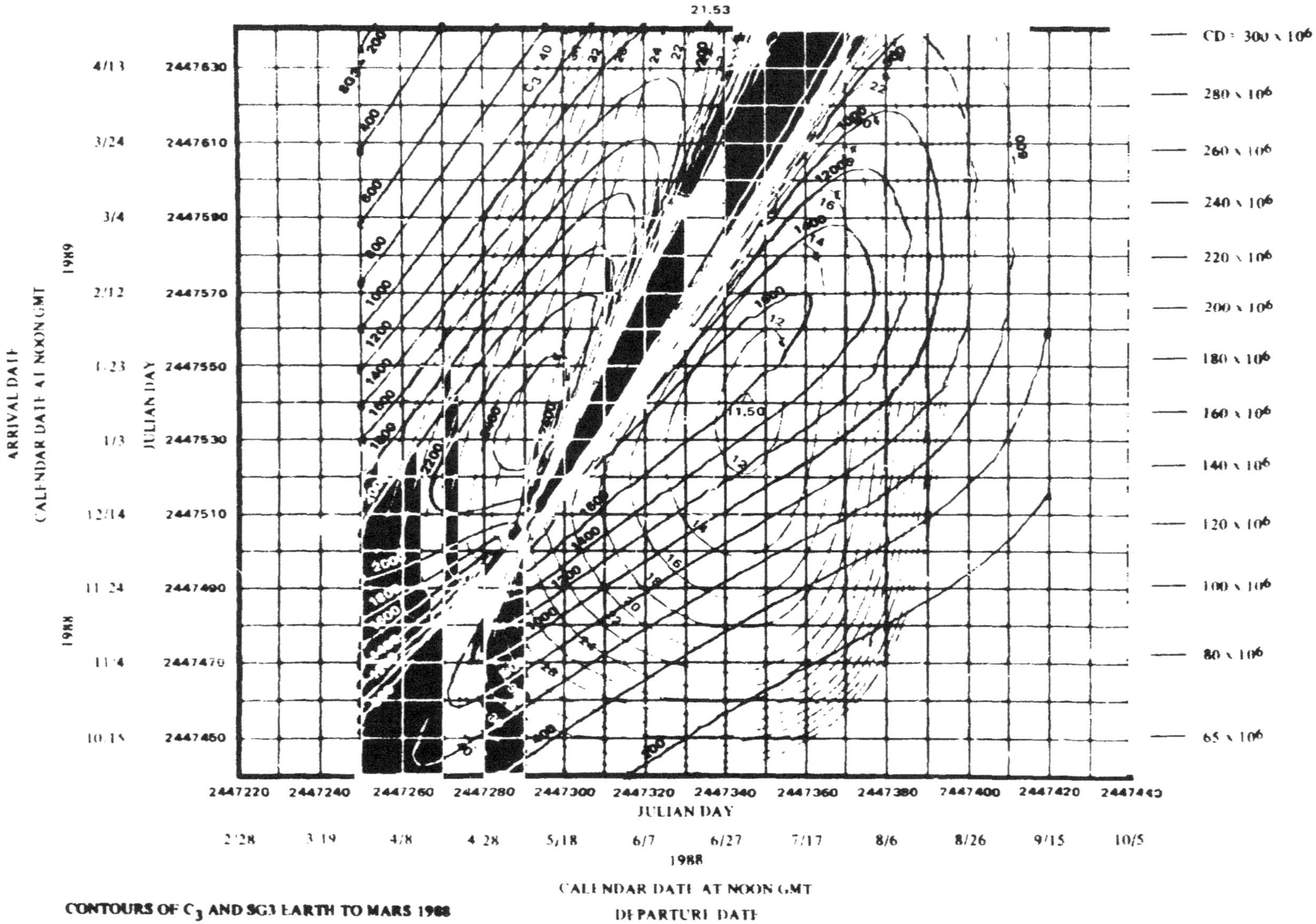

21.53

ARRIVAL DATE
CALENDAR DATE AT NOON GMT

	JULIAN DAY
4/13	2447630
3/24	2447610
3/4	2447590
2/12	2447570
1/23	2447550
1/3	2447530
12/14	2447510
11/24	2447490
11/4	2447470
10/15	2447450

1989

1988

CD = 300 x 10^6
280 x 10^6
260 x 10^6
240 x 10^6
220 x 10^6
200 x 10^6
180 x 10^6
160 x 10^6
140 x 10^6
120 x 10^6
100 x 10^6
80 x 10^6
65 x 10^6

JULIAN DAY	2447220	2447240	2447260	2447280	2447300	2447320	2447340	2447360	2447380	2447400	2447420	2447440
1988	2/28	3/19	4/8	4/28	5/18	6/7	6/27	7/17	8/6	8/26	9/15	10/5

CALENDAR DATE AT NOON GMT
DEPARTURE DATE

CONTOURS OF C_3 AND SG3 EARTH TO MARS 1988

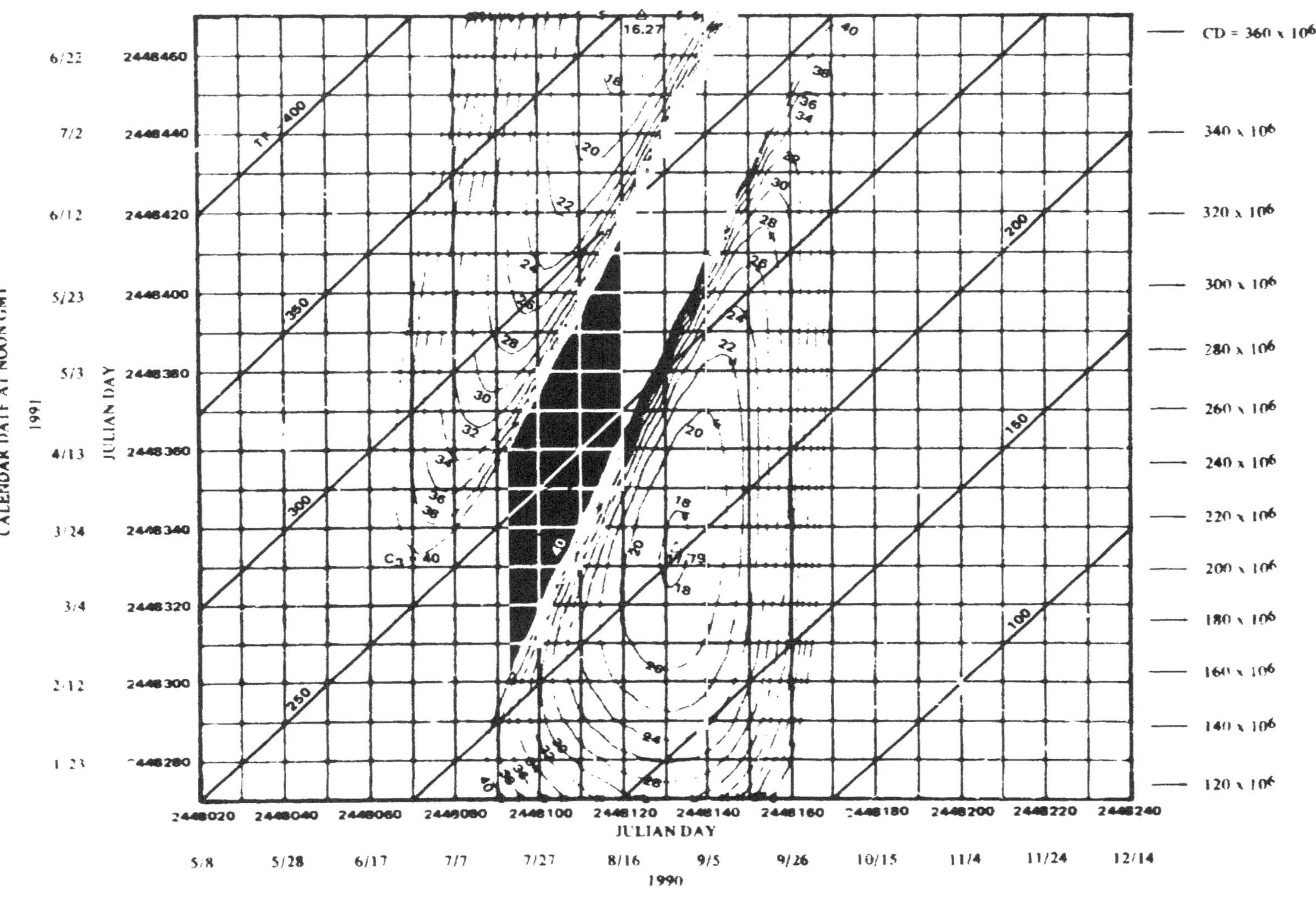

CONTOURS OF C_3 AND FLIGHT TIMES EARTH TO MARS 1990

C_3 ♂ 1990

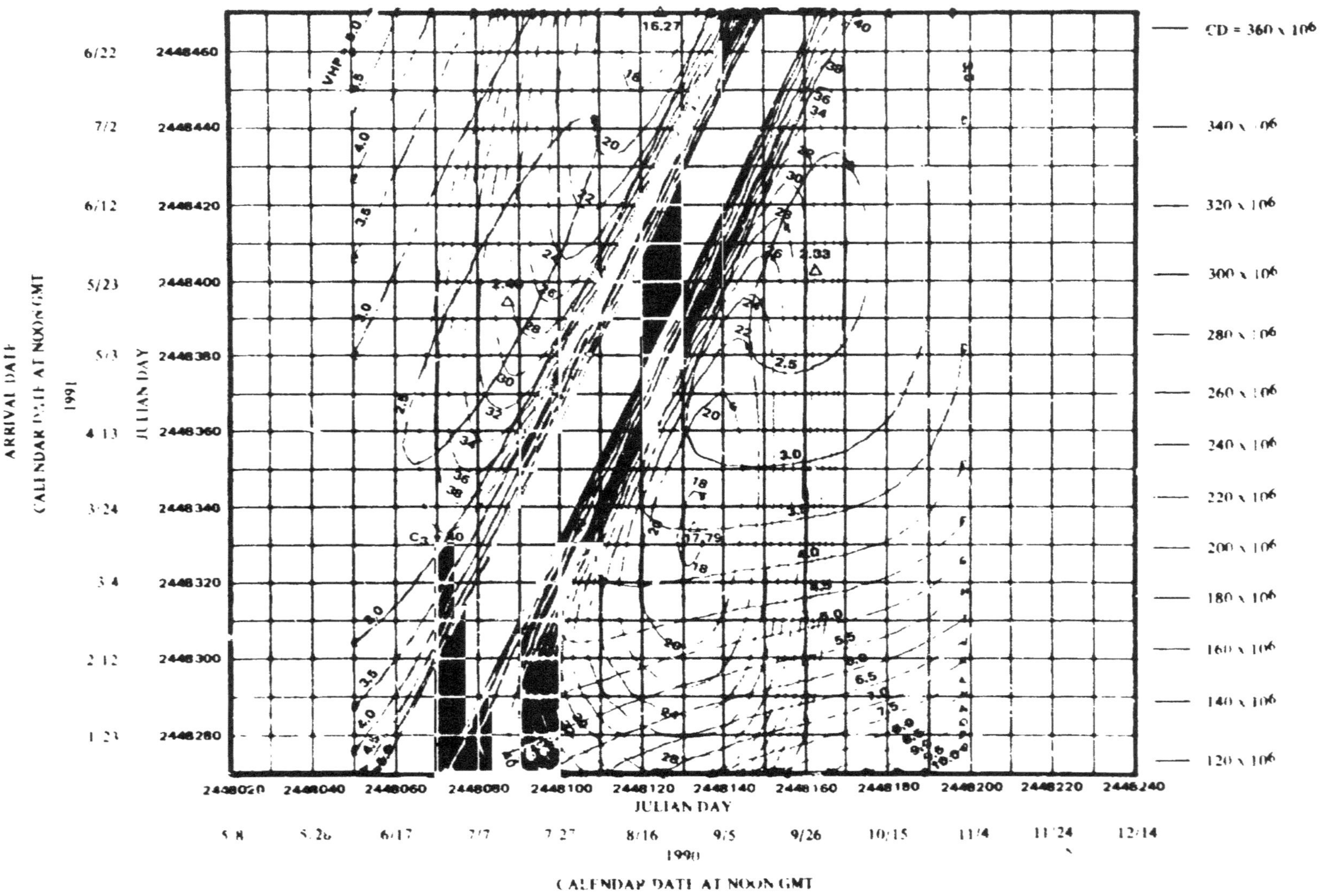

VHP

1990

ARRIVAL DATE
CALENDAR DATE AT NOON GMT
1991

Arrival date	Julian day	CD
6/22	2448460	CD = 360 x 10^6
7/2	2448440	340 x 10^6
6/12	2448420	320 x 10^6
5/23	2448400	300 x 10^6
5/3	2448380	280 x 10^6
4/13	2448360	260 x 10^6
3/24	2448340	240 x 10^6
3/4	2448320	220 x 10^6
2/12	2448300	200 x 10^6
1/23	2448280	180 x 10^6
		160 x 10^6
		140 x 10^6
		120 x 10^6

JULIAN DAY

2448020 2448040 2448060 2448080 2448100 2448120 2448140 2448160 2448180 2448200 2448220 2448240

JULIAN DAY

5/8 5/26 6/17 7/7 7/27 8/16 9/5 9/26 10/15 11/4 11/24 12/14

1990

CALENDAR DATE AT NOON GMT

DEPARTURE DATE

CONTOURS OF C_3 AND VHP EARTH TO MARS 1990

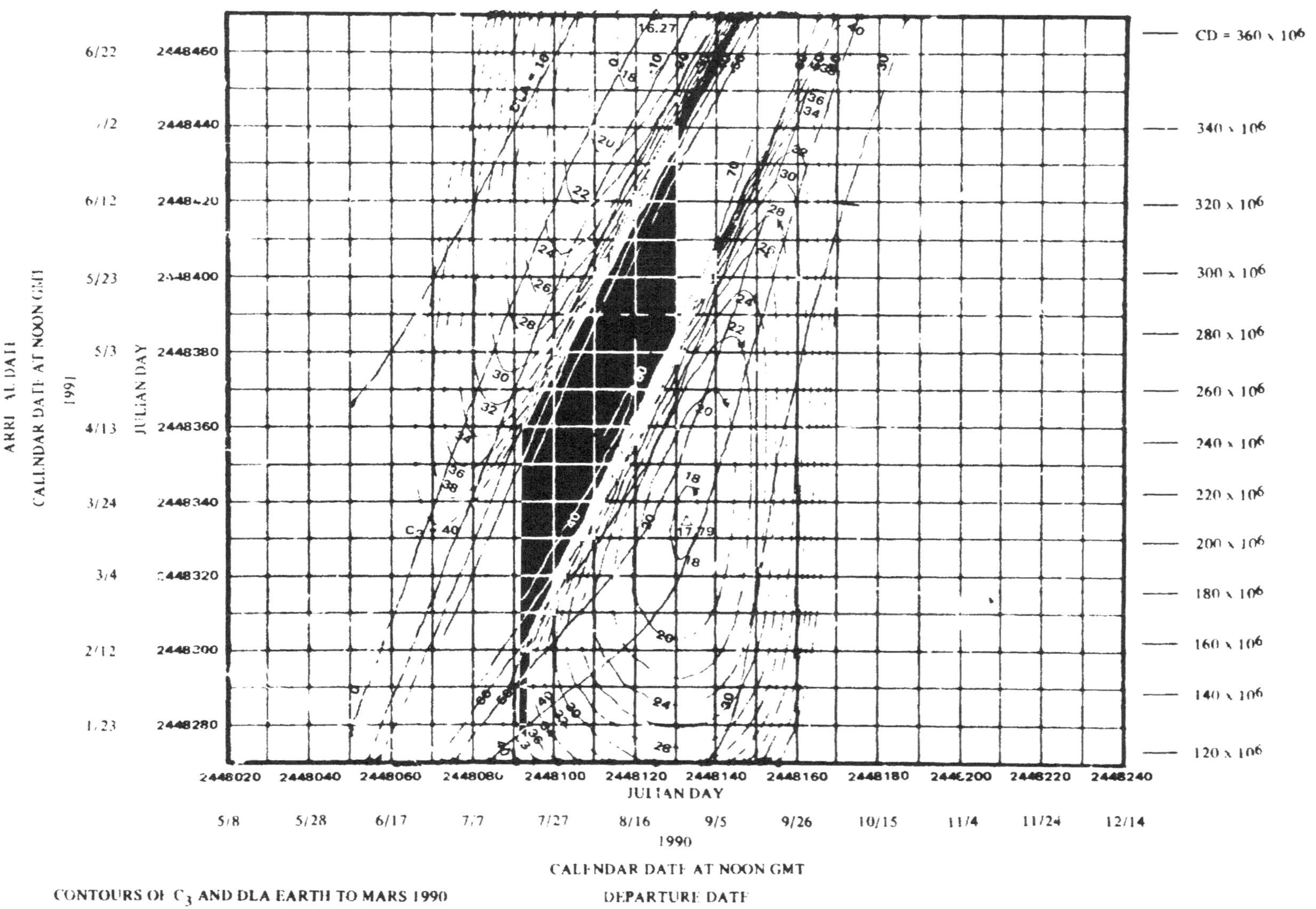

ARRIVAL DATE
CALENDAR DATE AT NOON GMT
1991

	JULIAN DAY	
6/22	2448460	CD = 360 x 10^6
7/2	2448440	340 x 10^6
6/12	2448420	320 x 10^6
5/23	2448400	300 x 10^6
5/3	2448380	280 x 10^6
4/13	2448360	260 x 10^6
3/24	2448340	240 x 10^6
3/4	2448320	220 x 10^6
2/12	2448300	200 x 10^6
1/23	2448280	180 x 10^6
		160 x 10^6
		140 x 10^6
		120 x 10^6

JULIAN DAY	2448020	2448040	2448060	2448080	2448100	2448120	2448140	2448160	2448180	2448200	2448220	2448240
1990	5/8	5/28	6/17	7/7	7/27	8/16	9/5	9/26	10/15	11/4	11/24	12/14

CALENDAR DATE AT NOON GMT
DEPARTURE DATE

CONTOURS OF C_3 AND DLA EARTH TO MARS 1990

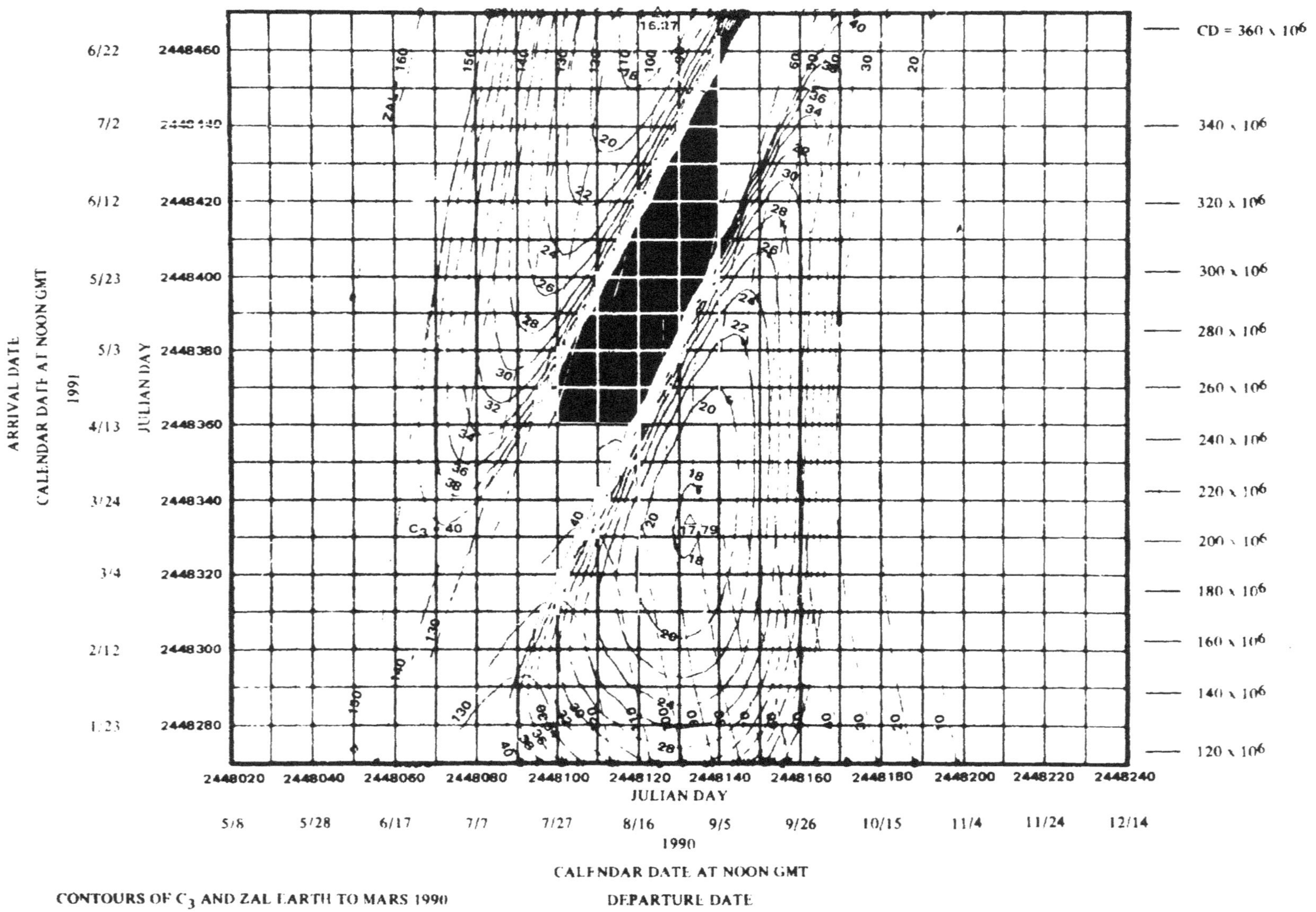

ZAL

1990

ARRIVAL DATE
CALENDAR DATE AT NOON GMT
1991

Calendar date	Julian day
6/22	2448460
7/2	2448440
6/12	2448420
5/23	2448400
5/3	2448380
4/13	2448360
3/24	2448340
3/4	2448320
2/12	2448300
1/23	2448280

CD = 360×10^6
340×10^6
320×10^6
300×10^6
280×10^6
260×10^6
240×10^6
220×10^6
200×10^6
180×10^6
160×10^6
140×10^6
120×10^6

Julian day	Calendar date
2448020	5/8
2448040	5/28
2448060	6/17
2448080	7/7
2448100	7/27
2448120	8/16
2448140	9/5
2448160	9/26
2448180	10/15
2448200	11/4
2448220	11/24
2448240	12/14

JULIAN DAY
1990
CALENDAR DATE AT NOON GMT
DEPARTURE DATE

CONTOURS OF C_3 AND ZAL EARTH TO MARS 1990

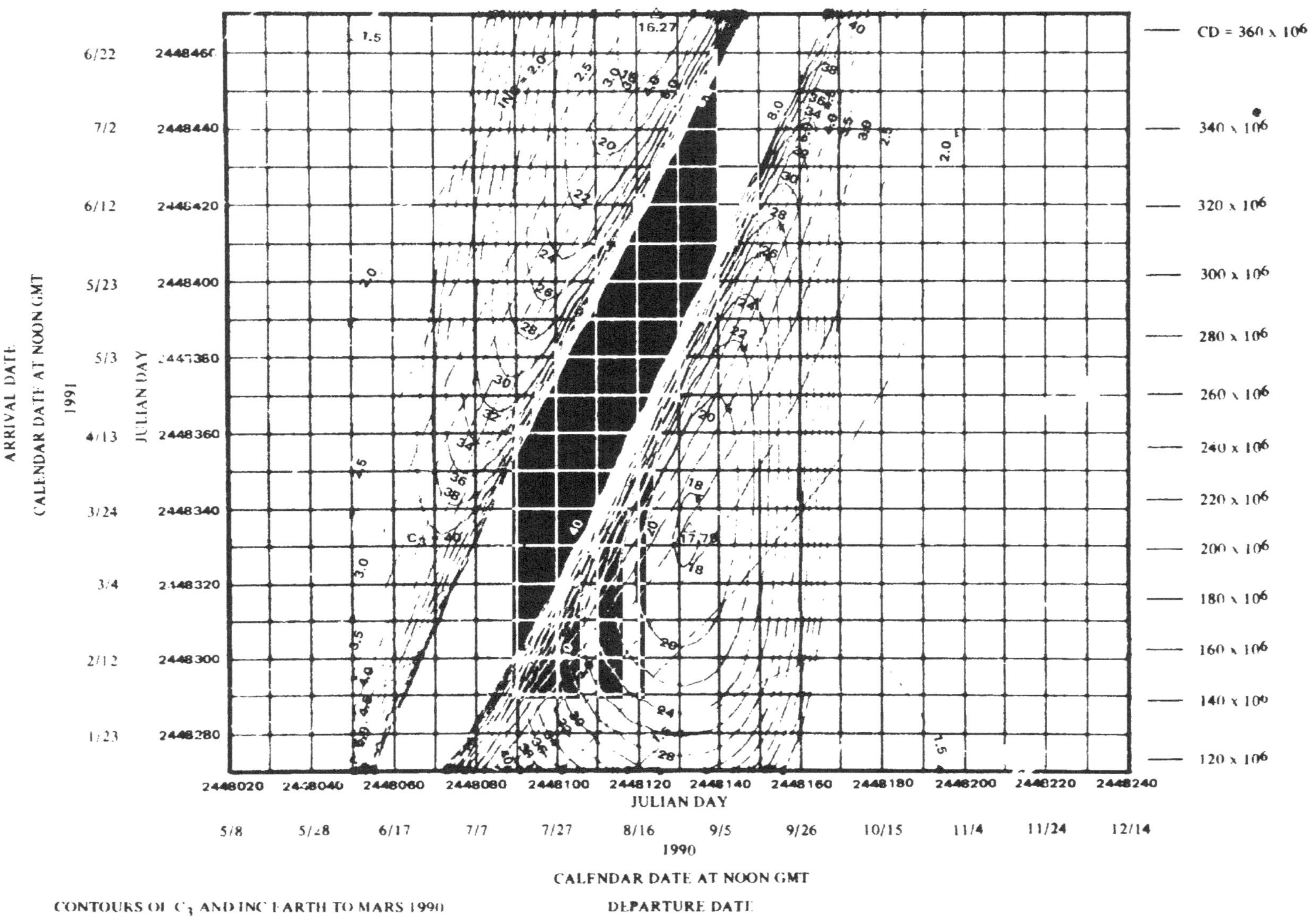

ARRIVAL DATE
CALENDAR DATE AT NOON GMT
1991

Calendar date	Julian day
6/22	2448460
7/2	2448440
6/12	2448420
5/23	2448400
5/3	2447380
4/13	2448360
3/24	2448340
3/4	2448320
2/12	2448300
1/23	2448280

CD = 360 x 10^6
340 x 10^6
320 x 10^6
300 x 10^6
280 x 10^6
260 x 10^6
240 x 10^6
220 x 10^6
200 x 10^6
180 x 10^6
160 x 10^6
140 x 10^6
120 x 10^6

2448020 2448040 2448060 2448080 2448100 2448120 2448140 2448160 2448180 2448200 2448220 2448240
JULIAN DAY

5/8 5/28 6/17 7/7 7/27 8/16 9/5 9/26 10/15 11/4 11/24 12/14
1990

CALENDAR DATE AT NOON GMT
DEPARTURE DATE

CONTOURS OF C_3 AND INC EARTH TO MARS 1990

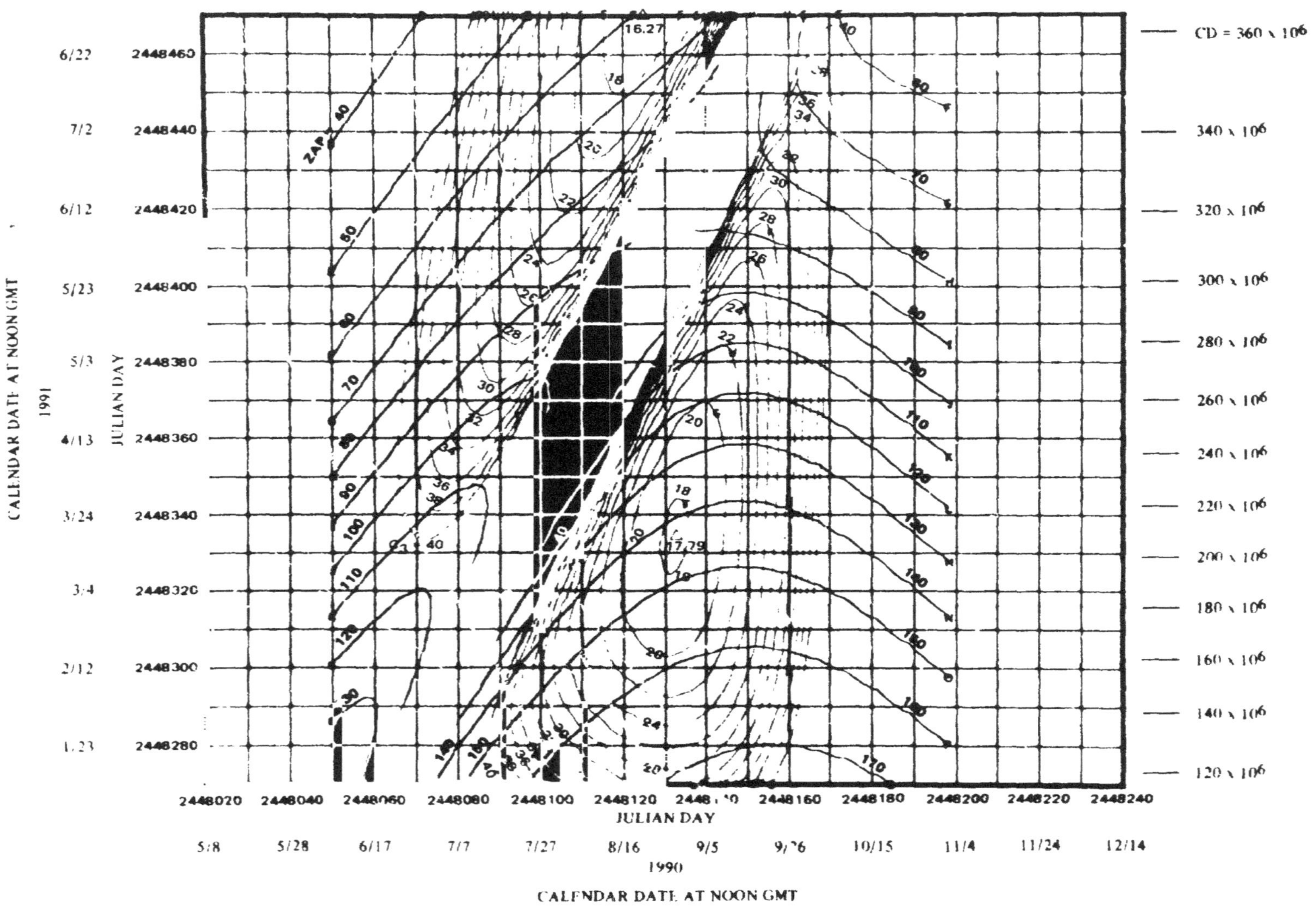

ONTOURS OF C_3 AND ZAP EARTH TO MARS 1990

ZAP ♂ 1990

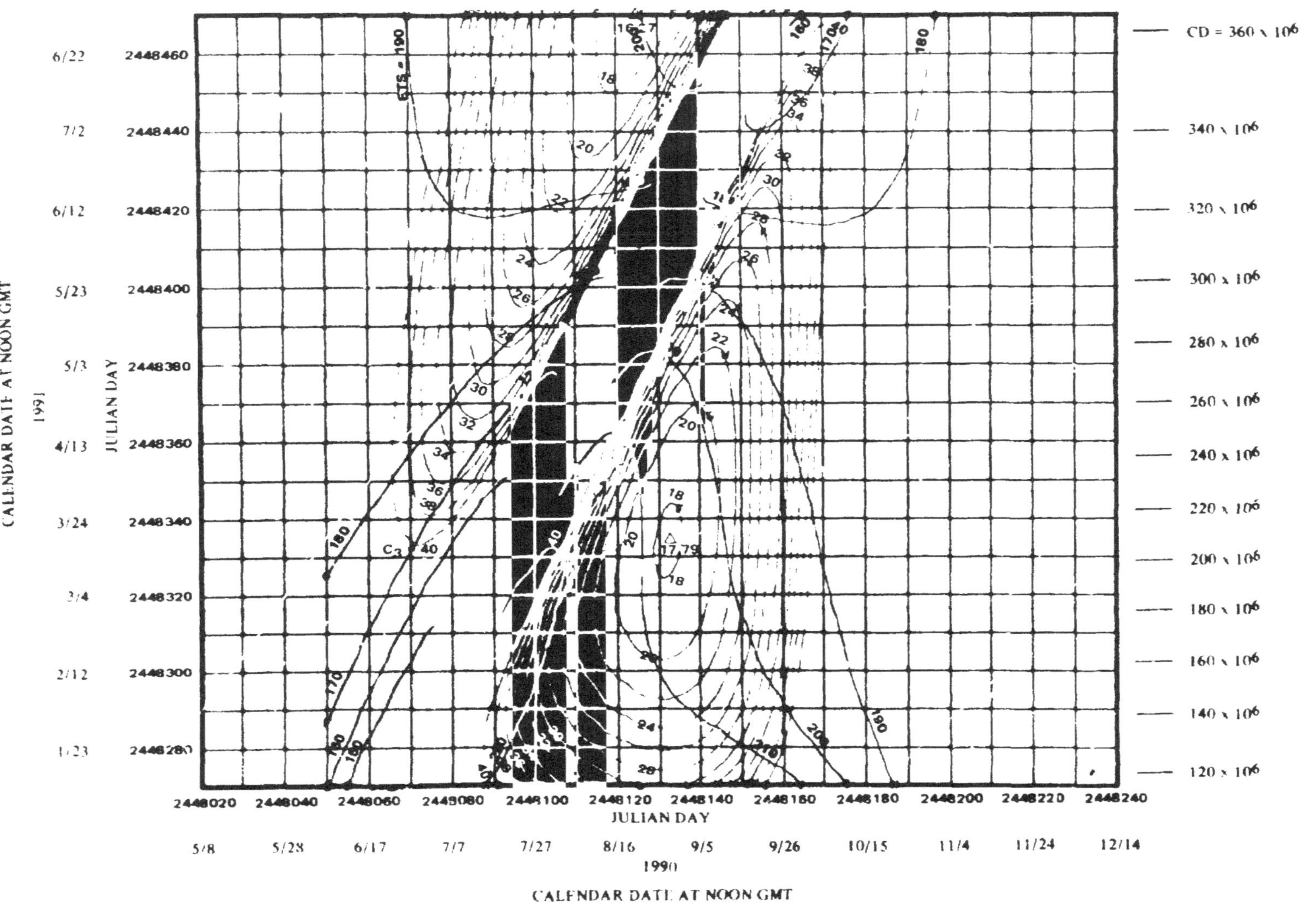

CONTOURS OF C_3 AND ETS EARTH TO MARS 1990

ETS ♂ 1990

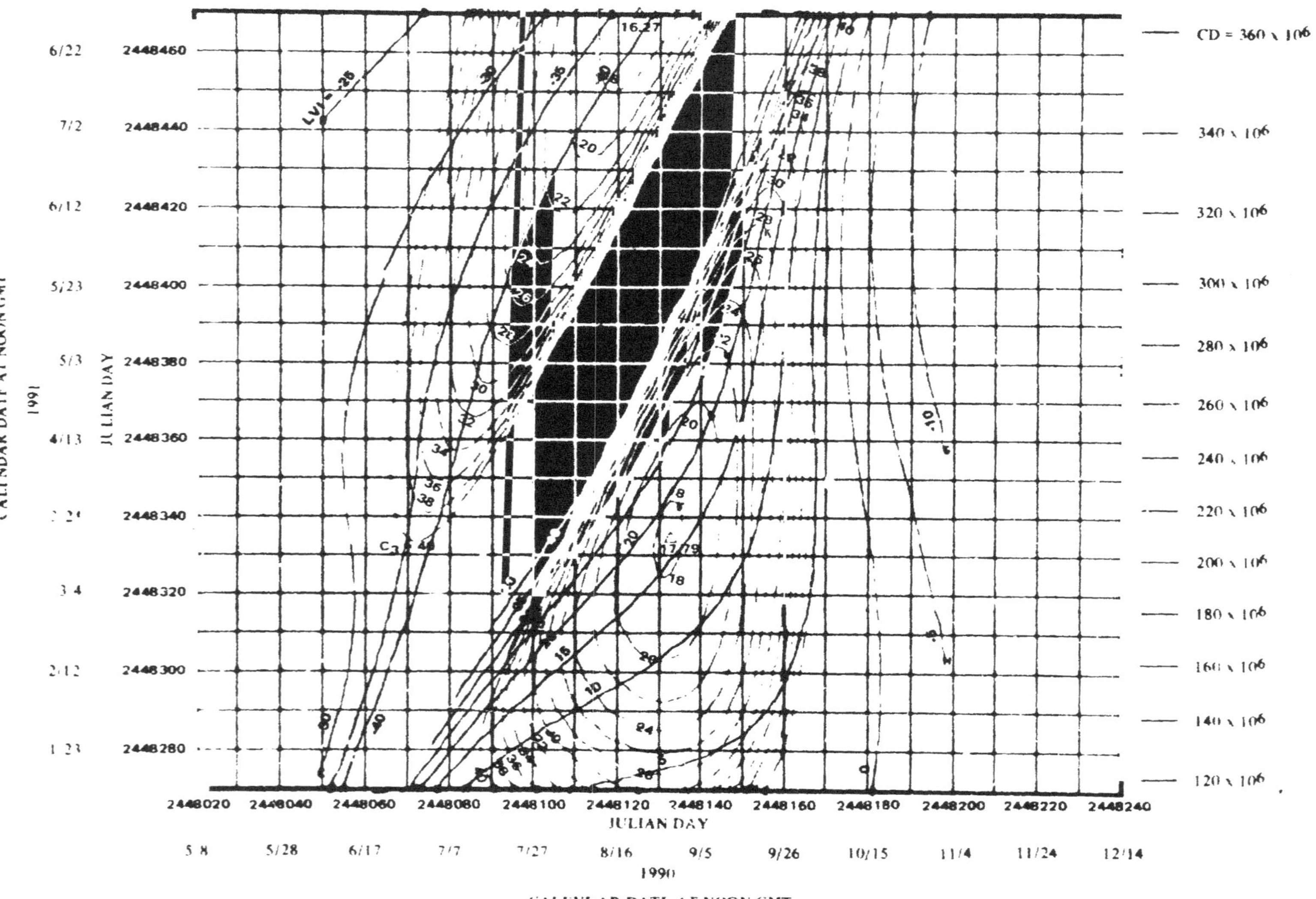

CONTOURS OF C_3 AND LVI EARTH TO MARS 1990

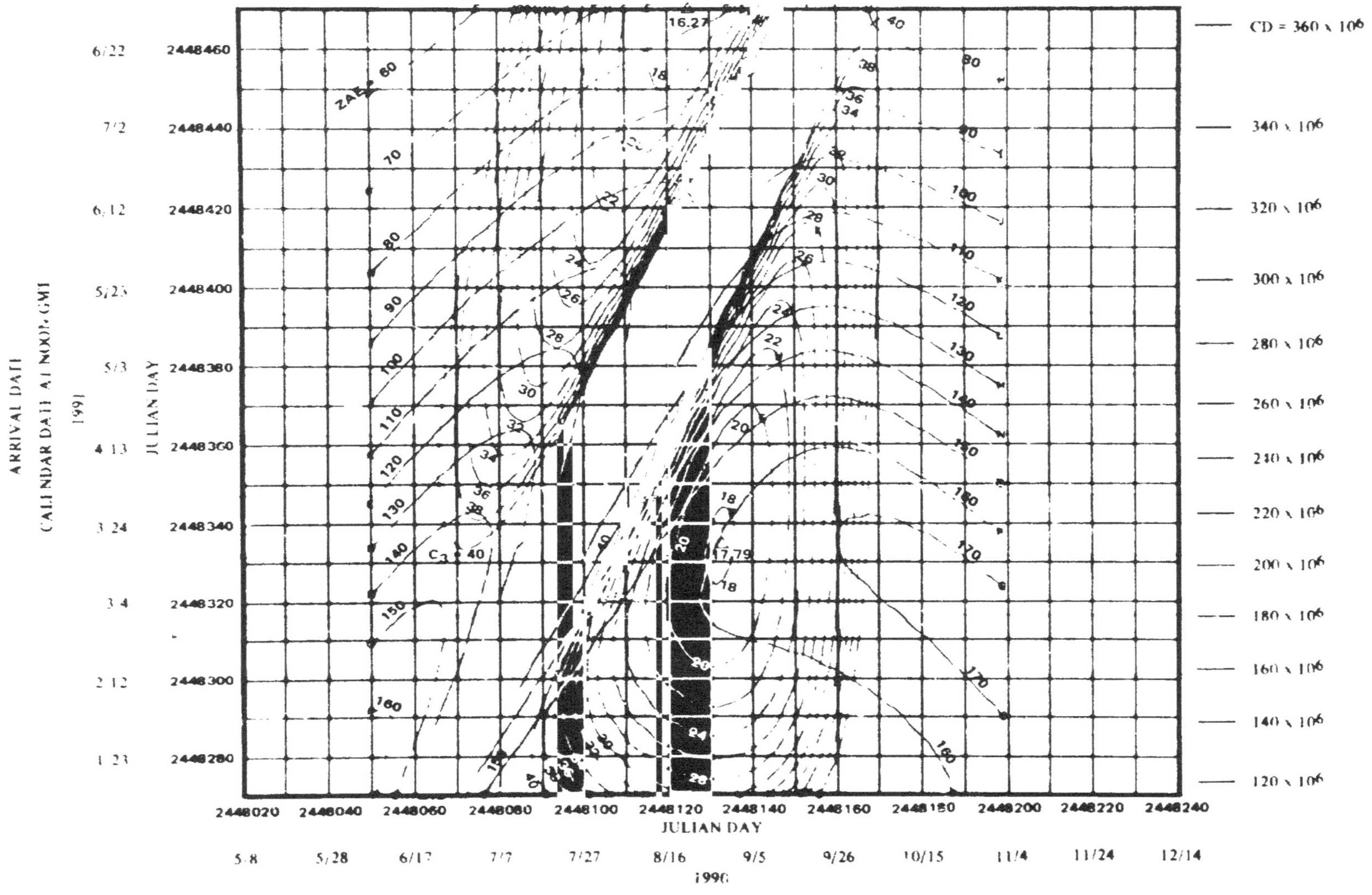

CD = 360 x 10^6
340 x 10^6
320 x 10^6
300 x 10^6
280 x 10^6
260 x 10^6
240 x 10^6
220 x 10^6
200 x 10^6
180 x 10^6
160 x 10^6
140 x 10^6
120 x 10^6

ARRIVAL DATE
CALENDAR DATE AT NOON GMT
1991

Calendar date	Julian day
6/22	2448460
7/2	2448440
6/12	2448420
5/23	2448400
5/3	2448380
4/13	2448360
3/24	2448340
3/4	2448320
2/12	2448300
1/23	2448280

JULIAN DAY

Julian day	2448020	2448040	2448060	2448080	2448100	2448120	2448140	2448160	2448180	2448200	2448220	2448240
Calendar date	5/8	5/28	6/17	7/7	7/27	8/16	9/5	9/26	10/15	11/4	11/24	12/14

1990

CALENDAR DATE AT NOON GMT

DEPARTURE DATE

CONTOURS OF C_3 AND ZAE EARTH TO MARS 1990

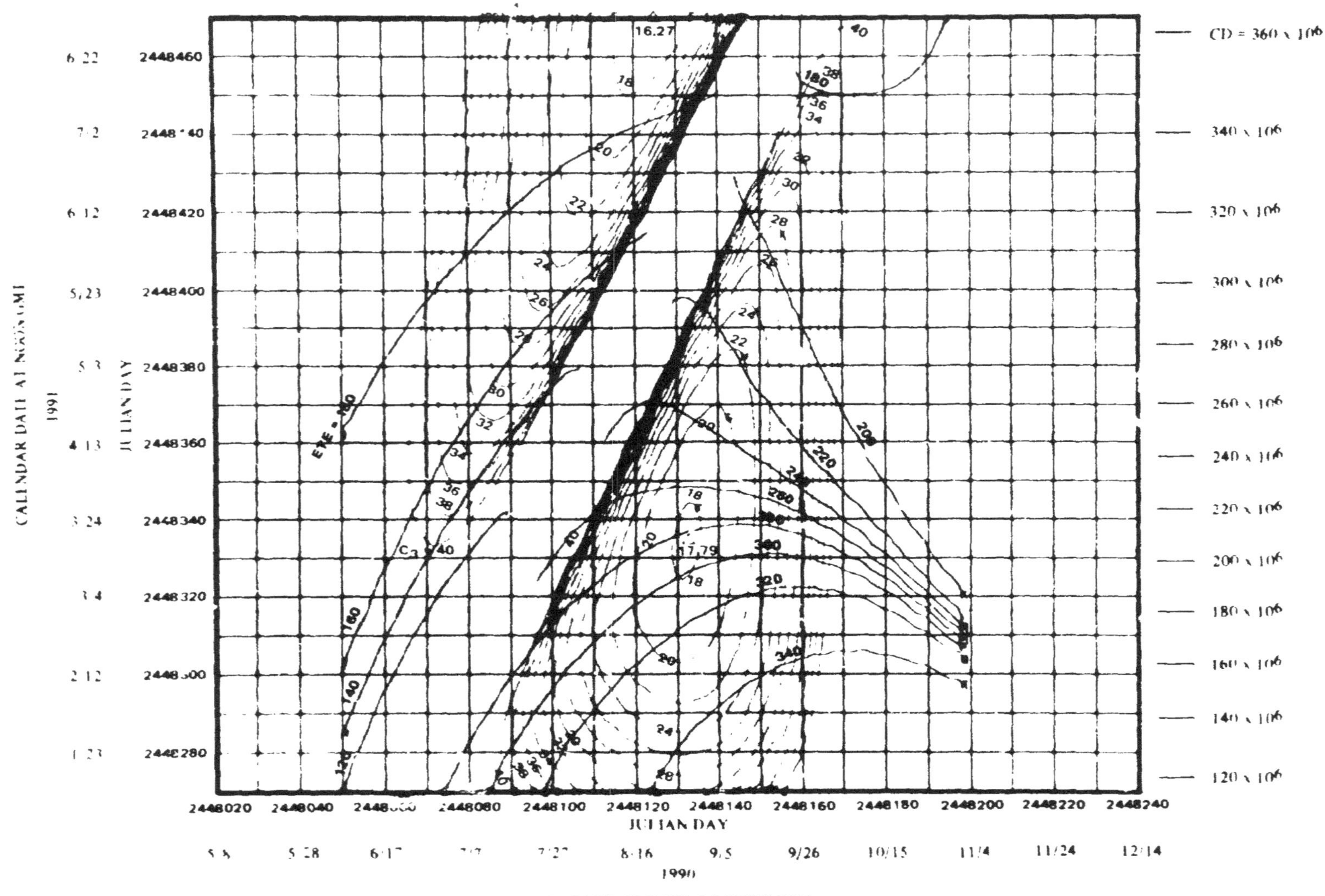

CONTOURS OF C_3 AND ETE EARTH TO MARS 1990

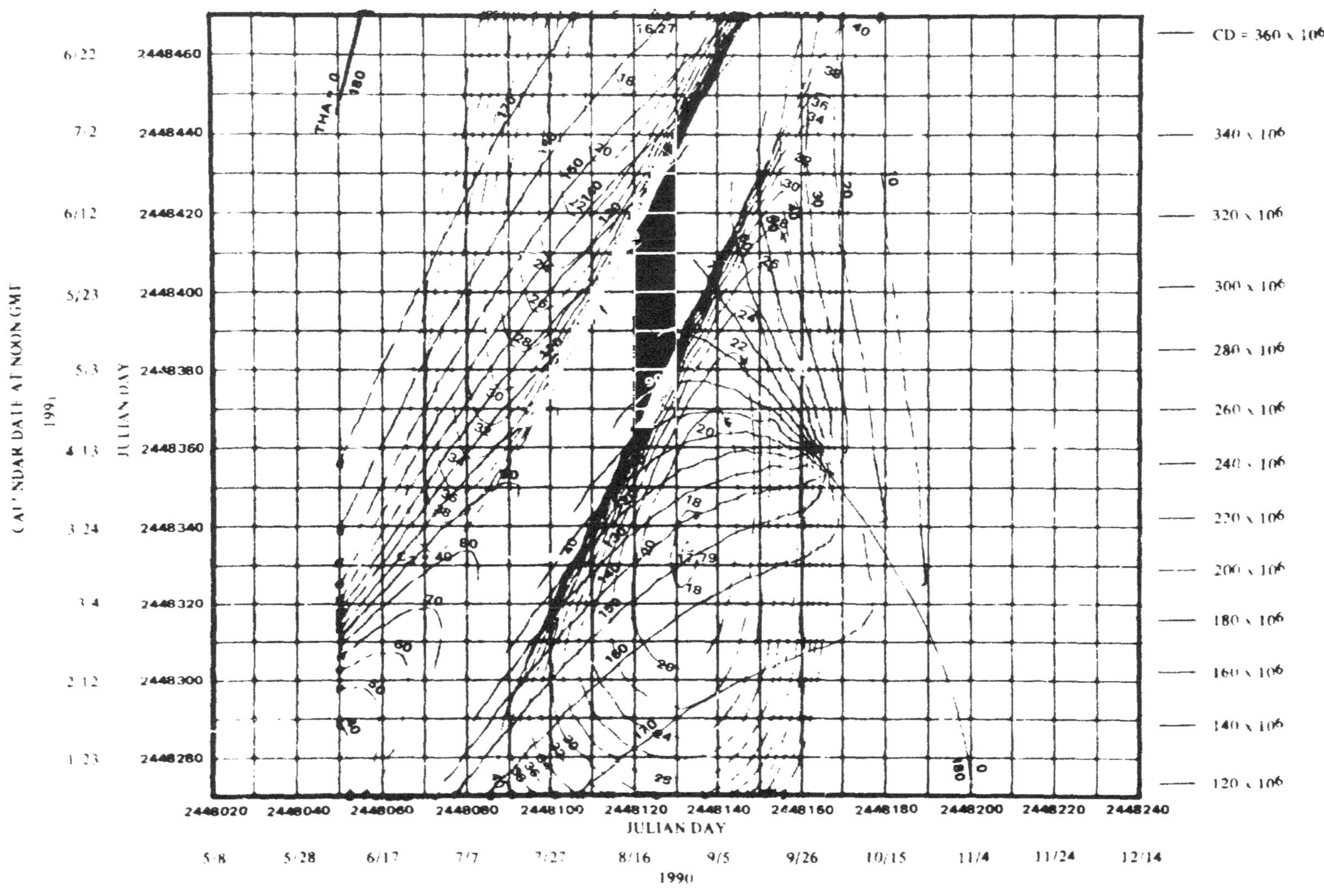

CONTOURS OF C_3 AND THA EARTH TO MARS 1990

THA
♂
1990

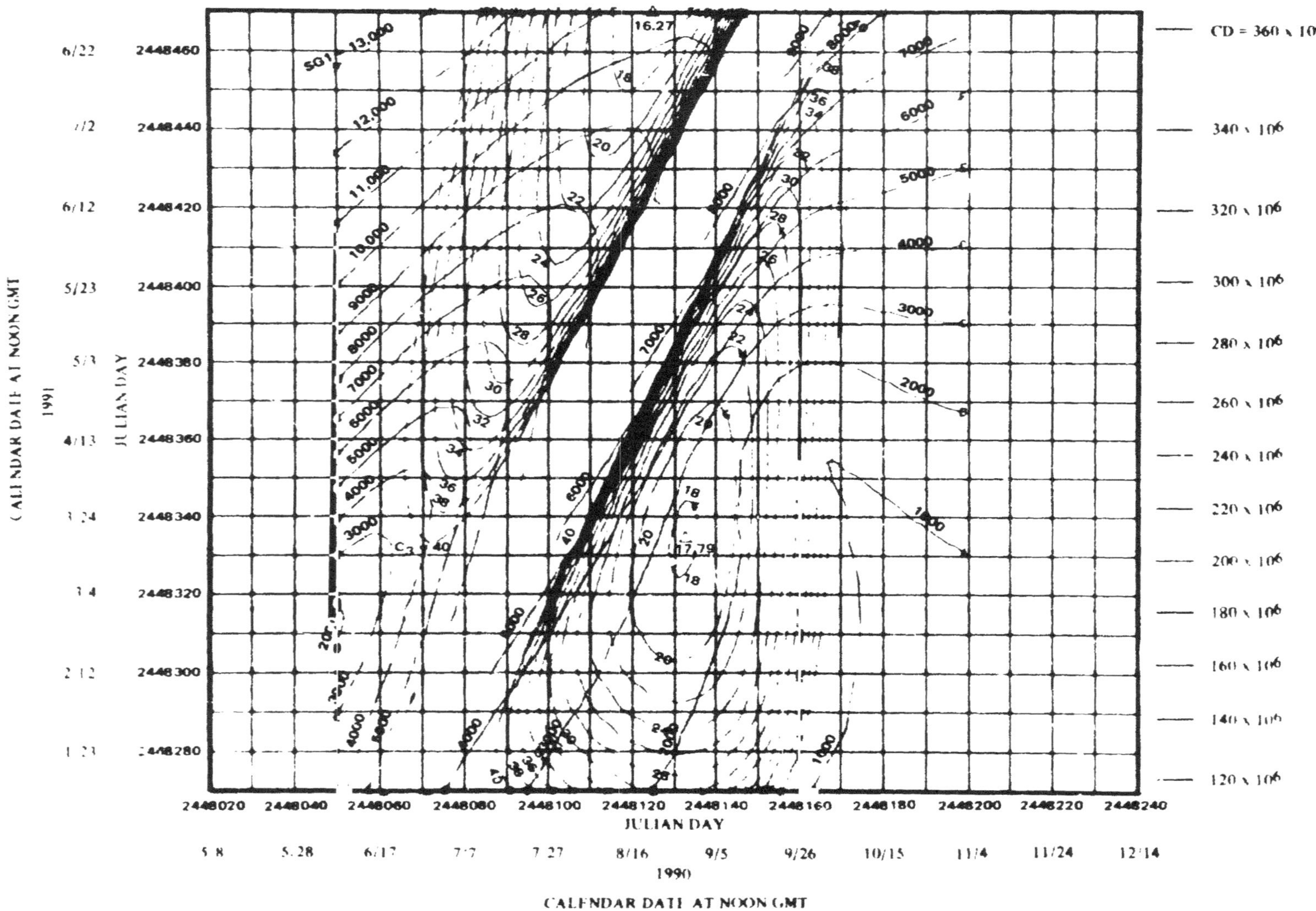

CONTOURS OF C_3 AND SG1 EARTH TO MARS 1990

SG1 ♂ 1990

ARRIVAL DATE
CALENDAR DATE AT NOON GMT
1991

CALENDAR DATE	JULIAN DAY
6/22	2448460
7/2	2448440
6/12	2448420
5/23	2448400
5/3	2448380
4/13	2448360
3/24	2448340
3/4	2448320
2/12	2448300
1/23	2448280

CD = 360 x 10^6
340 x 10^6
320 x 10^6
300 x 10^6
280 x 10^6
260 x 10^6
240 x 10^6
220 x 10^6
200 x 10^6
180 x 10^6
160 x 10^6
140 x 10^6
120 x 10^6

JULIAN DAY	2448020	2448040	2448060	2448080	2448100	2448120	2448140	2448160	2448180	2448200	2448220	2448240
CALENDAR DATE (1990)	5/8	5/28	6/17	7/7	7/27	8/16	9/5	9/26	10/15	11/4	11/24	12/14

CALENDAR DATE AT NOON GMT
DEPARTURE DATE

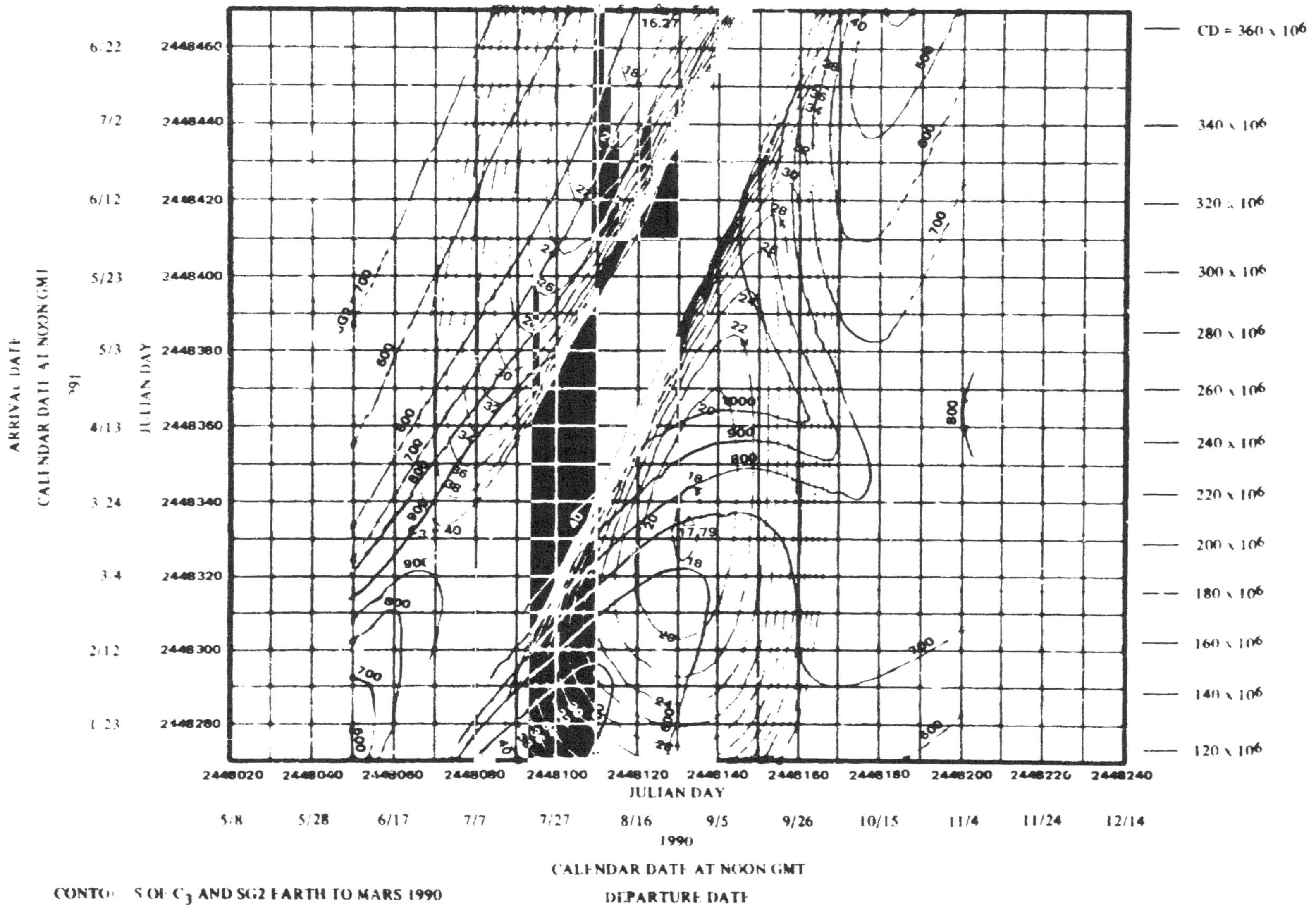

CONTO S OF C_3 AND SG2 EARTH TO MARS 1990

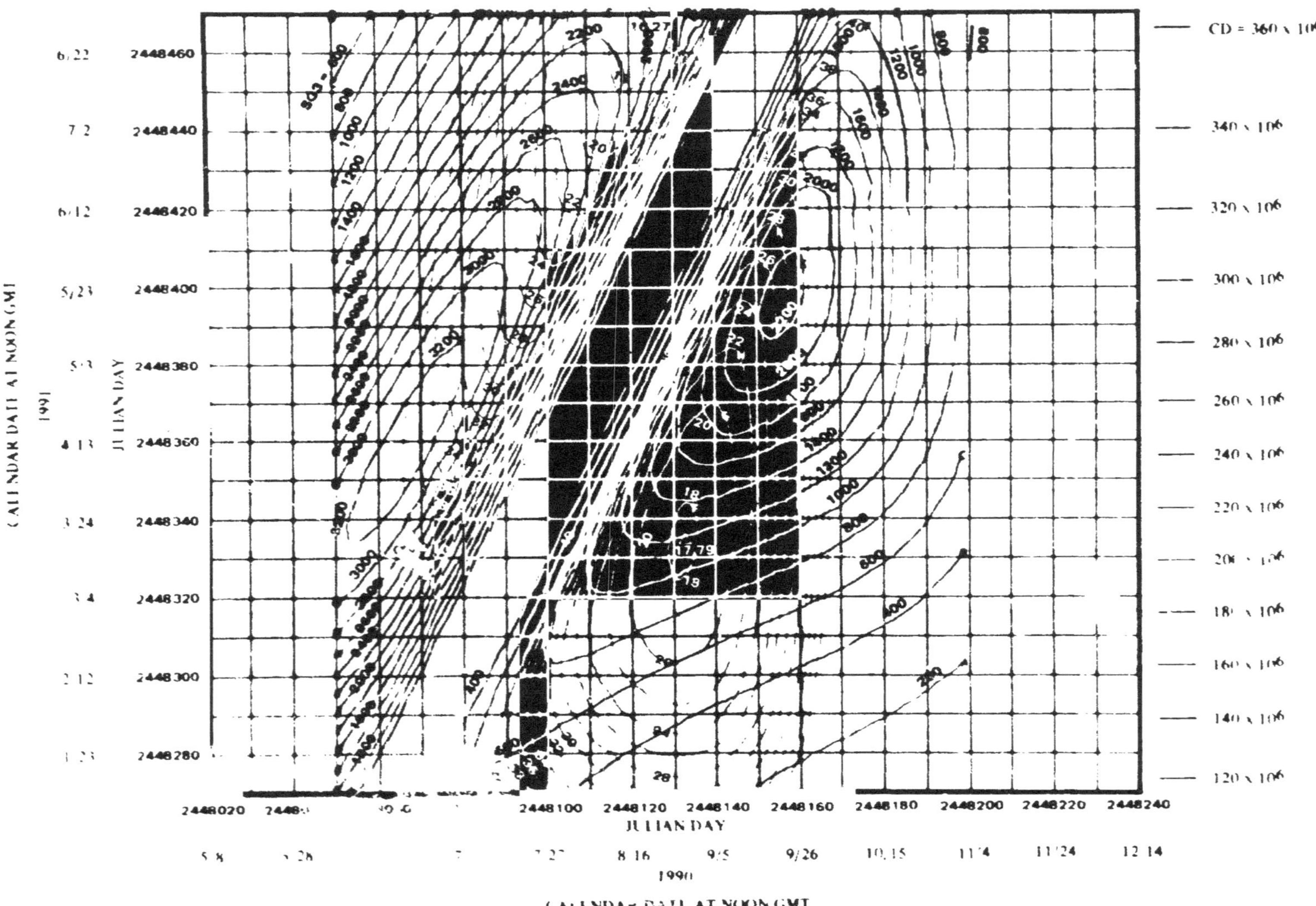

CONTOURS OF C_3 AND SG3 EARTH TO MARS 1990

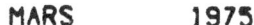
MARS 1975
40.
30.
20.
10.
0.
-10.
-20.
-30.
-40.
DECLIN
75/01/01
75/01/21
75/02/10
75/03/02
75/03/22
75/04/11
75/05/01
75/05/21
75/06/10
75/06/30
75/07/20
75/08/09
75/08/29
75/09/18
75/10/08
75/10/28
75/11/17
75/12/07
75/12/27
76/01/16
YR/M/D

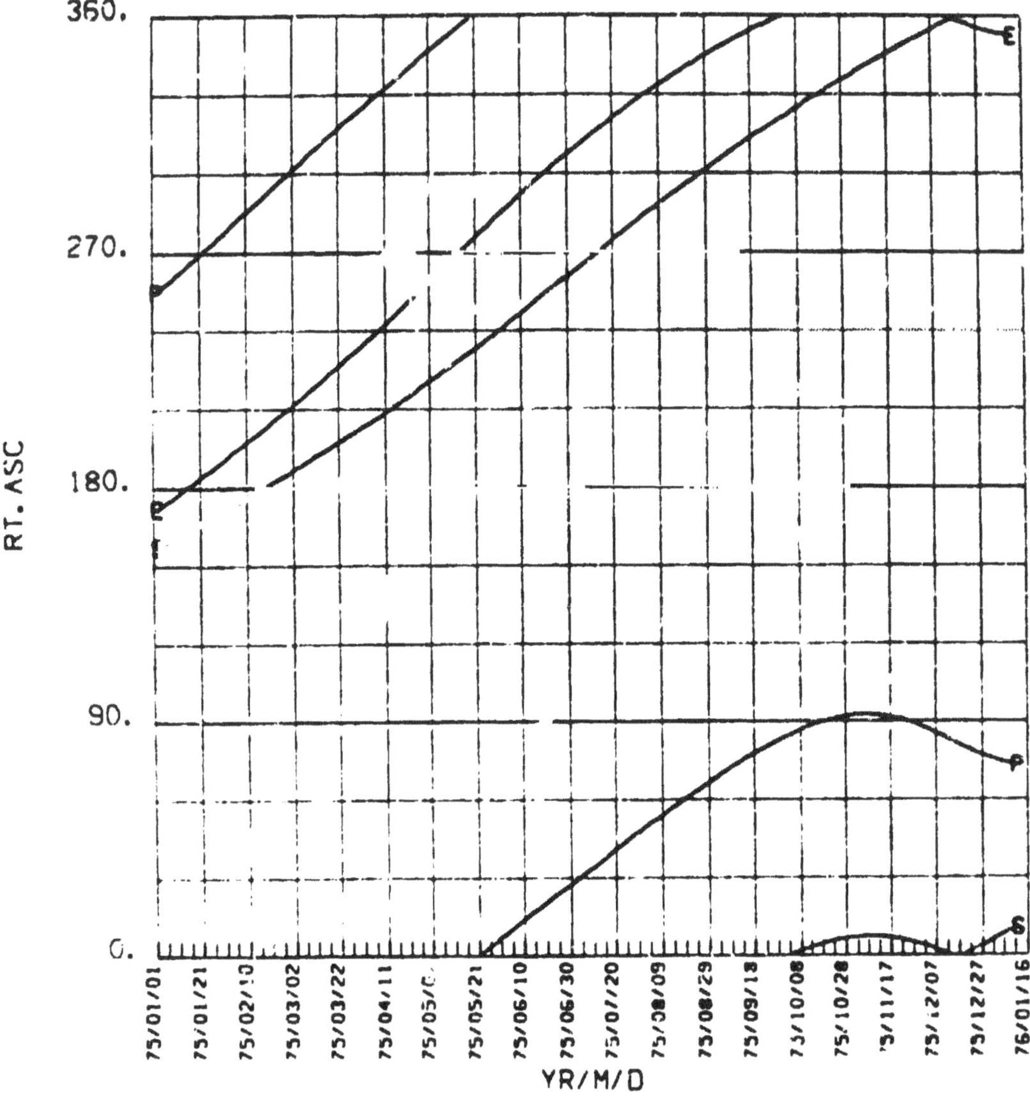

MARS 1975
RT. ASC
360.
270.
180.
90.
0.
75/01/01
75/01/21
75/02/10
75/03/02
75/03/22
75/04/11
75/05/01
75/05/21
75/06/10
75/06/30
75/07/20
75/08/09
75/08/29
75/09/18
75/10/08
75/10/28
75/11/17
75/12/07
75/12/27
76/01/16
YR/M/D

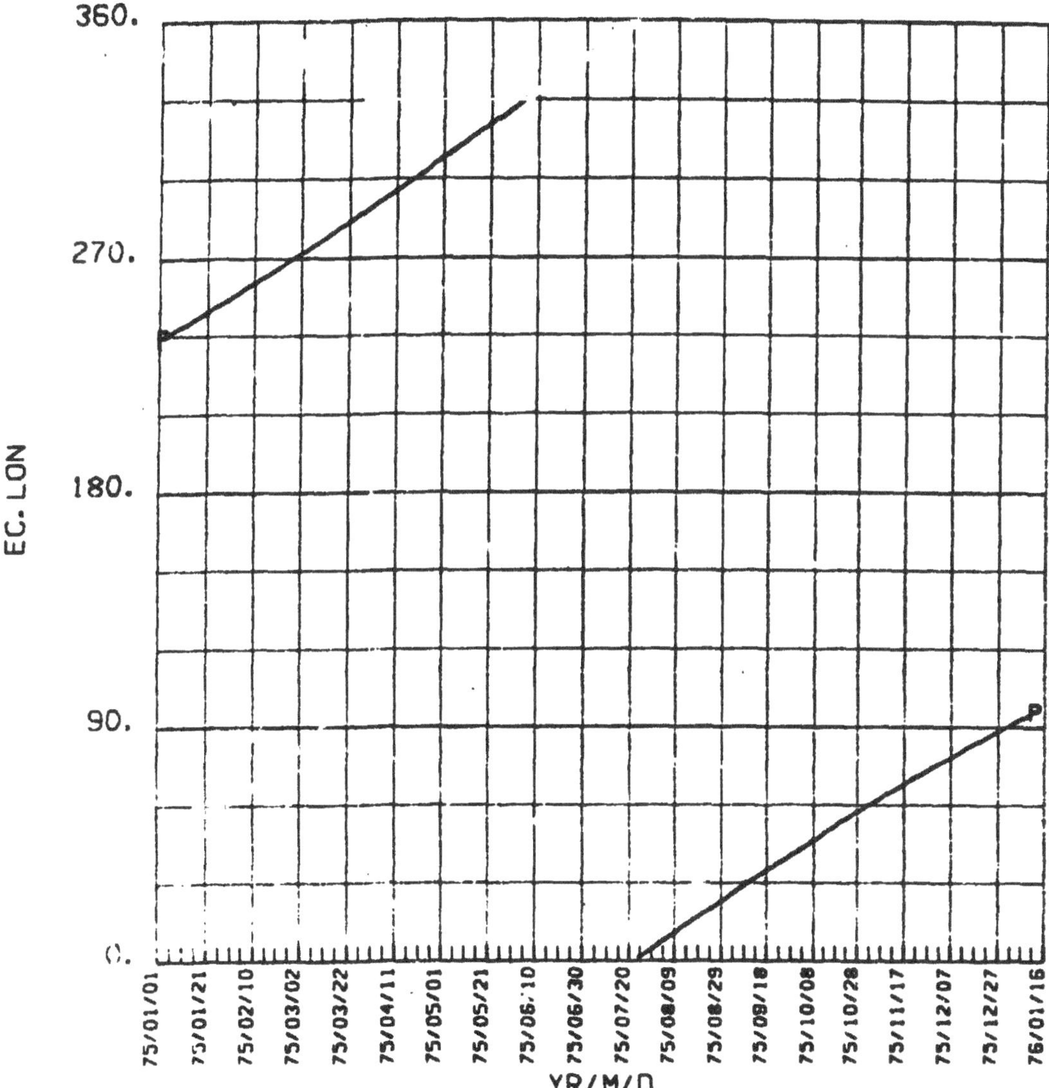
MARS 1975
360.
270.
180.
90.
0.
EC. LON
75/01/01
75/01/21
75/02/10
75/03/02
75/03/22
75/04/11
75/05/01
75/05/21
75/06/10
75/06/30
75/07/20
75/08/09
75/08/29
75/09/18
75/10/08
75/10/28
75/11/17
75/12/07
75/12/27
76/01/16
YR/M/D

MARS 1975

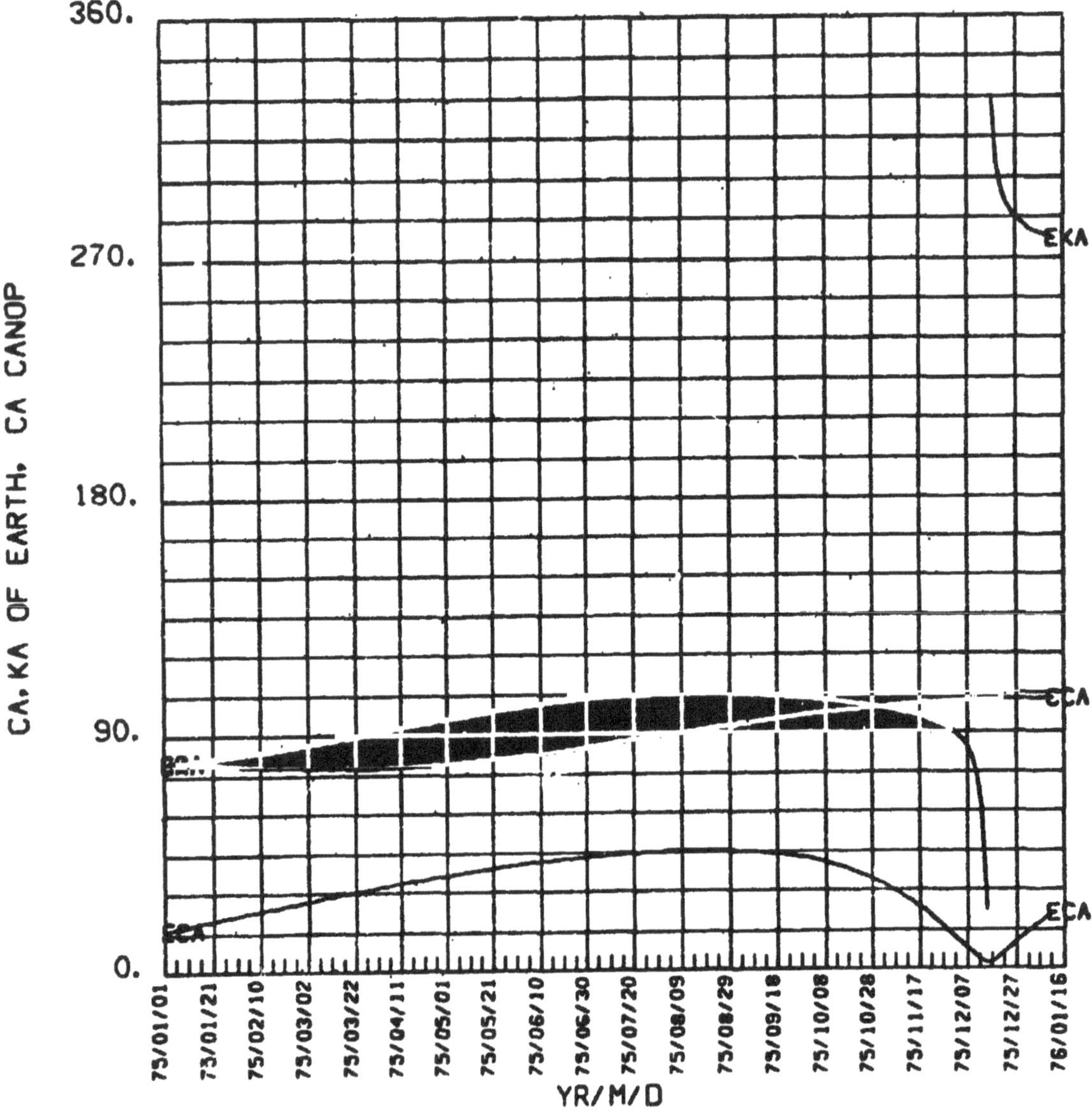

360.
270.
180.
90.
0.
CA.KA OF EARTH, CA CANOP
EKA
ECA
ECA
75/01/01
75/01/21
75/02/10
75/03/02
75/03/22
75/04/11
75/05/01
75/05/21
75/06/10
75/06/30
75/07/20
75/08/09
75/08/29
75/09/18
75/10/08
75/10/28
75/11/17
75/12/07
75/12/27
76/01/16
YR/M/D

MARS 1975

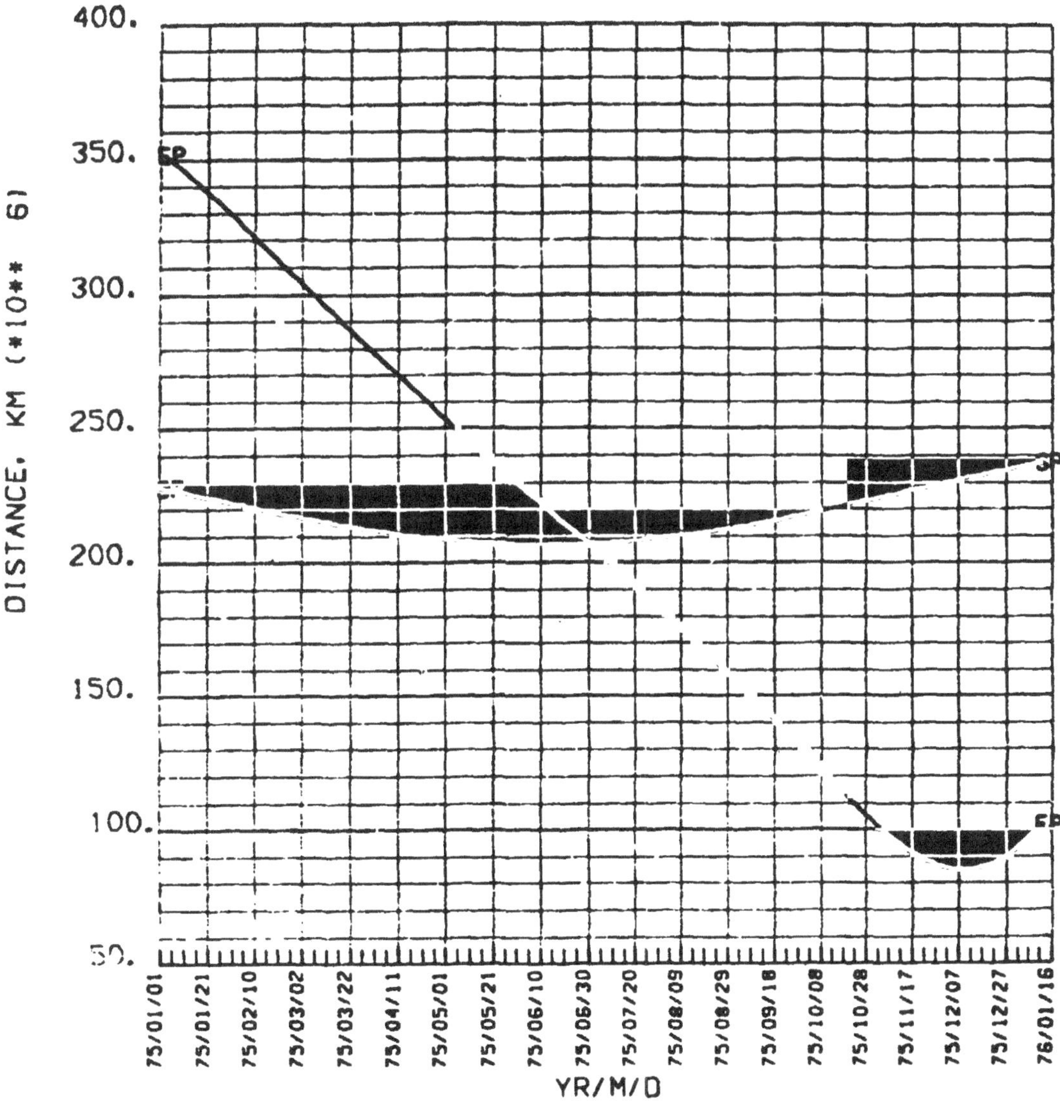
400.
350.
300.
250.
200.
150.
100.
50.
DISTANCE, KM (*10** 6)
SP
SP
SP
75/01/01
75/01/21
75/02/10
75/03/02
75/03/22
75/04/11
75/05/01
75/05/21
75/06/10
75/06/30
75/07/20
75/08/09
75/08/29
75/09/18
75/10/08
75/10/28
75/11/17
75/12/07
75/12/27
76/01/16
YR/M/D

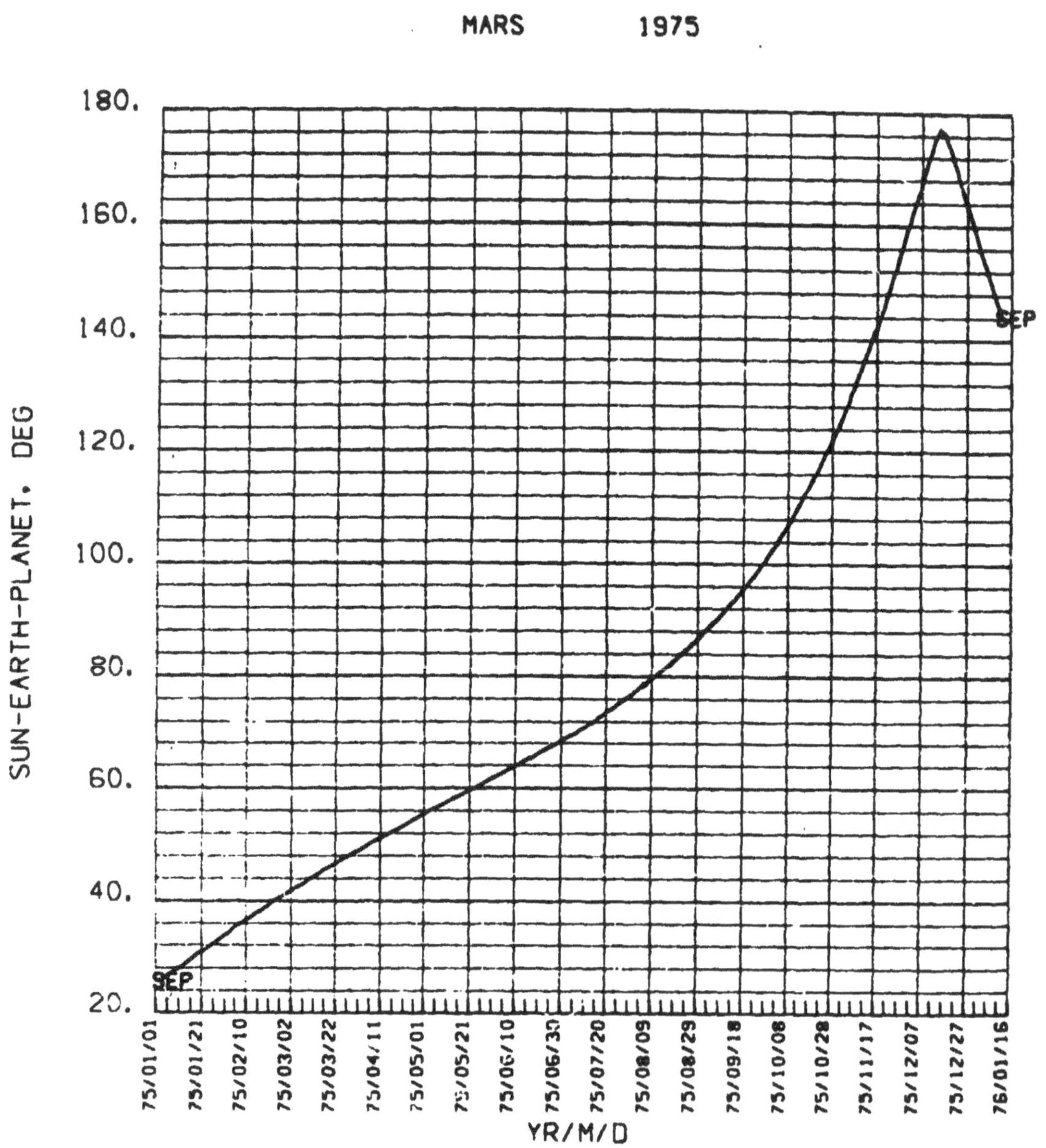

MARS

1975
180.
160.
140.
120.
100.
80.
60.
40.
20.
SUN-EARTH-PLANET. DEG
SEP
SEP
75/01/01
75/01/21
75/02/10
75/03/02
75/03/22
75/04/11
75/05/01
75/05/21
75/06/10
75/06/30
75/07/20
75/08/09
75/08/29
75/09/18
75/10/08
75/10/28
75/11/17
75/12/07
75/12/27
76/01/16
YR/M/D

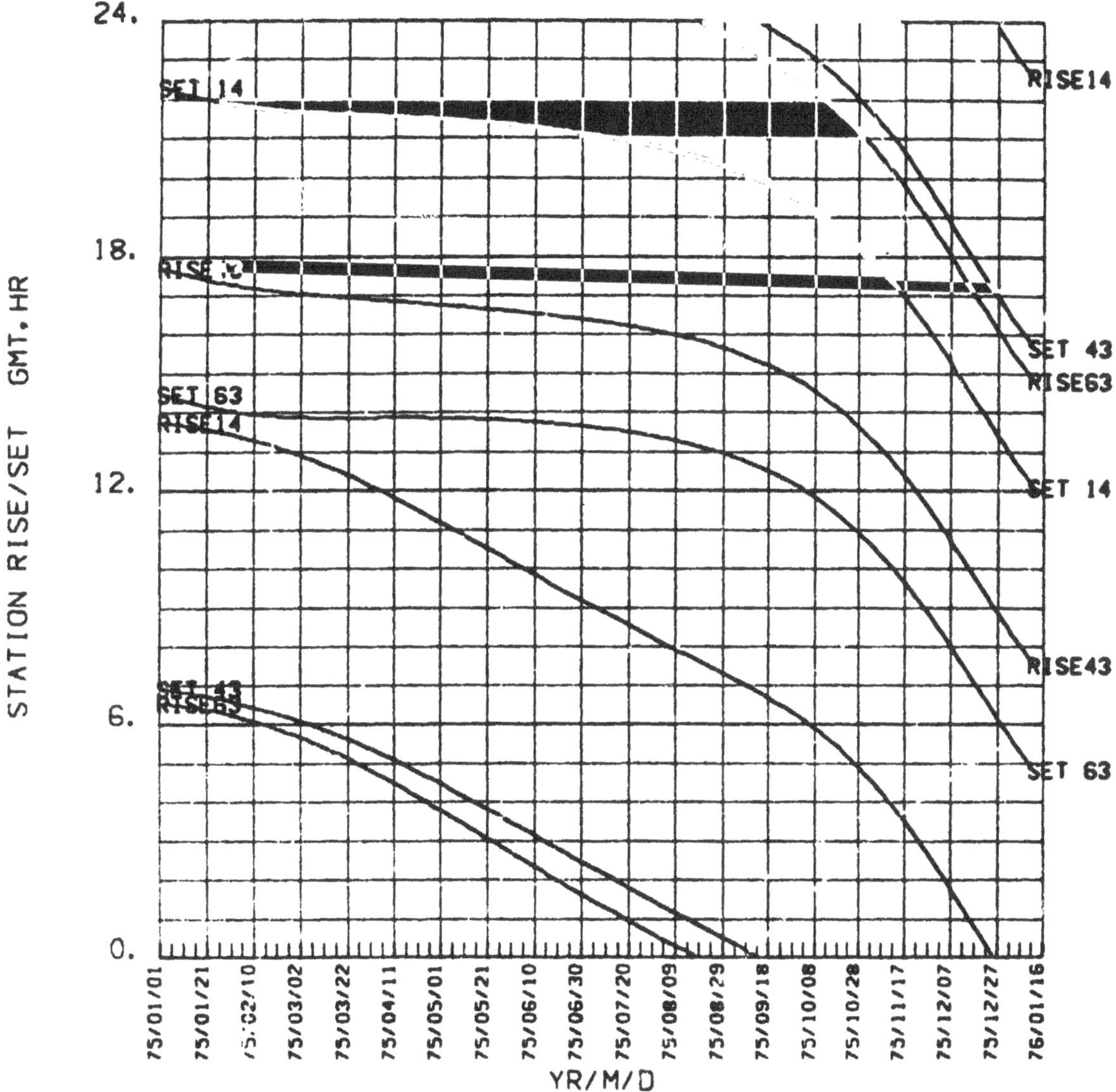

MARS 1975
24.
18.
12.
6.
0.
STATION RISE/SET GMT.HR
SET 14
RISE14
SET 63
SET 43
RISE63
SET 43
RISE63
SET 14
RISE43
SET 63
75/01/01
75/01/21
75/03/02
75/03/22
75/04/11
75/05/01
75/05/21
75/06/10
75/06/30
75/07/20
75/08/09
75/08/29
75/09/18
75/10/08
75/10/28
75/11/17
75/12/07
75/12/27
76/01/16
YR/M/D

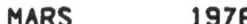

MARS 1976

DECLIN

40.
30.
20.
10.
0.
-10.
-20.
-30.
-40.

P S E

S E P

76/01/01 76/01/21 76/02/10 76/03/01 76/03/21 76/04/10 76/04/30 76/05/20 76/06/09 76/06/29 76/07/19 76/08/08 76/08/28 76/09/17 76/10/07 76/10/27 76/11/16 76/12/06 76/12/26 77/01/15

YR/M/D

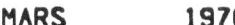

RT. ASC

360.
270.
180.
90.
0.

76/01/01 76/01/21 76/02/10 76/03/01 76/03/21 76/04/10 76/04/30 76/05/20 76/06/09 76/06/29 76/07/19 76/08/08 76/08/28 76/09/17 76/10/07 76/10/27 76/11/16 76/12/06 76/12/26 77/01/15

YR/M/D

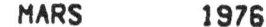
MARS 1976
360.
270.
180.
90.
0.
EC. LON
76/01/01
76/01/21
76/02/10
76/03/01
76/03/21
76/04/10
76/04/30
76/05/20
76/06/09
76/06/29
76/07/19
76/08/08
76/08/28
76/09/17
76/10/07
76/10/27
76/11/16
76/12/06
76/12/26
77/01/15
YR/M/D

CA, KA OF EARTH. CA CANOP

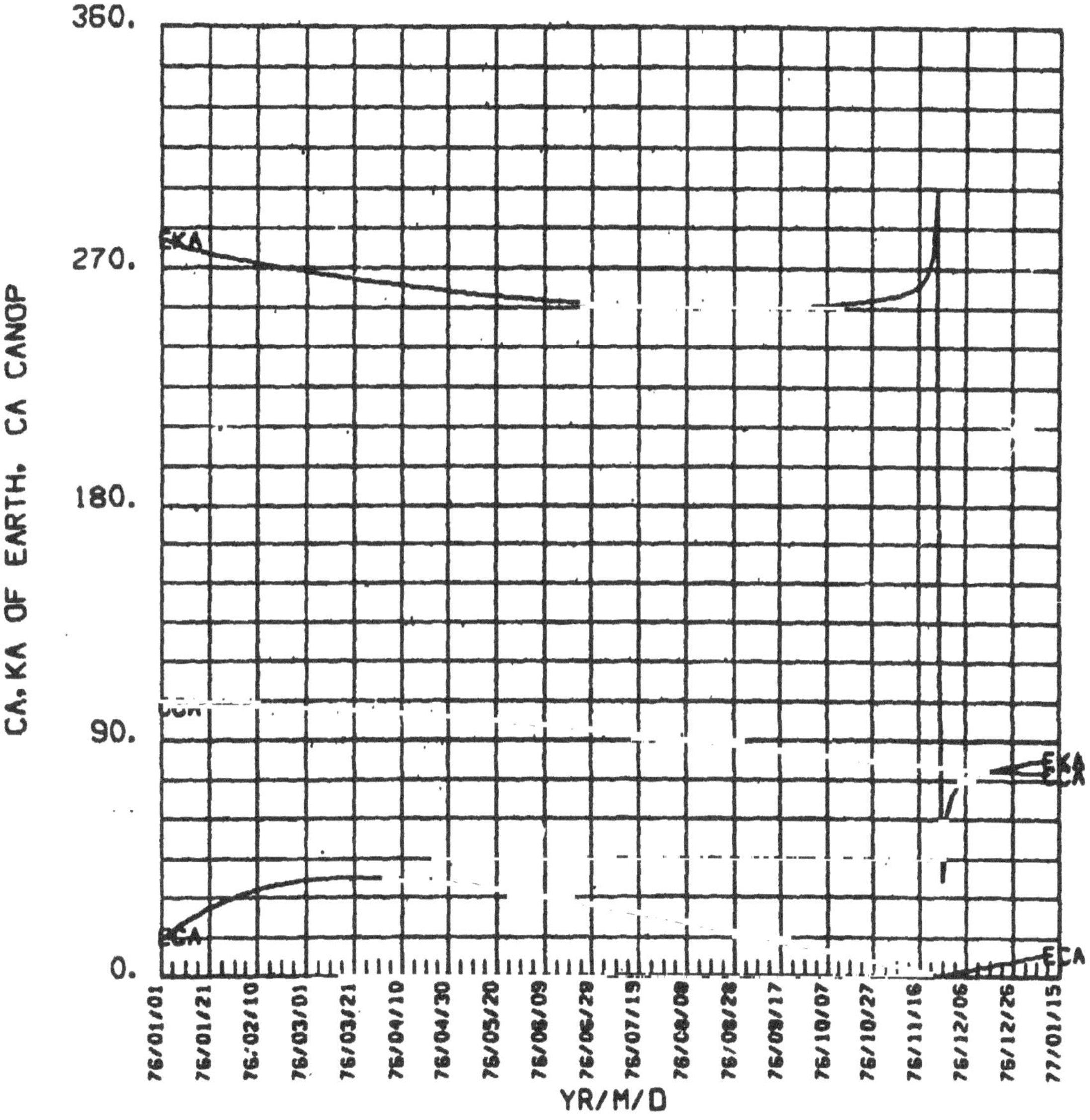

360.
270.
180.
90.
0.

EKA
ECA
CCA
EKA
CCA
ECA

76/01/01
76/01/21
76/02/10
76/03/01
76/03/21
76/04/10
76/04/30
76/05/20
76/06/09
76/06/29
76/07/19
76/08/08
76/08/28
76/09/17
76/10/07
76/10/27
76/11/16
76/12/06
76/12/26
77/01/15

YR/M/D

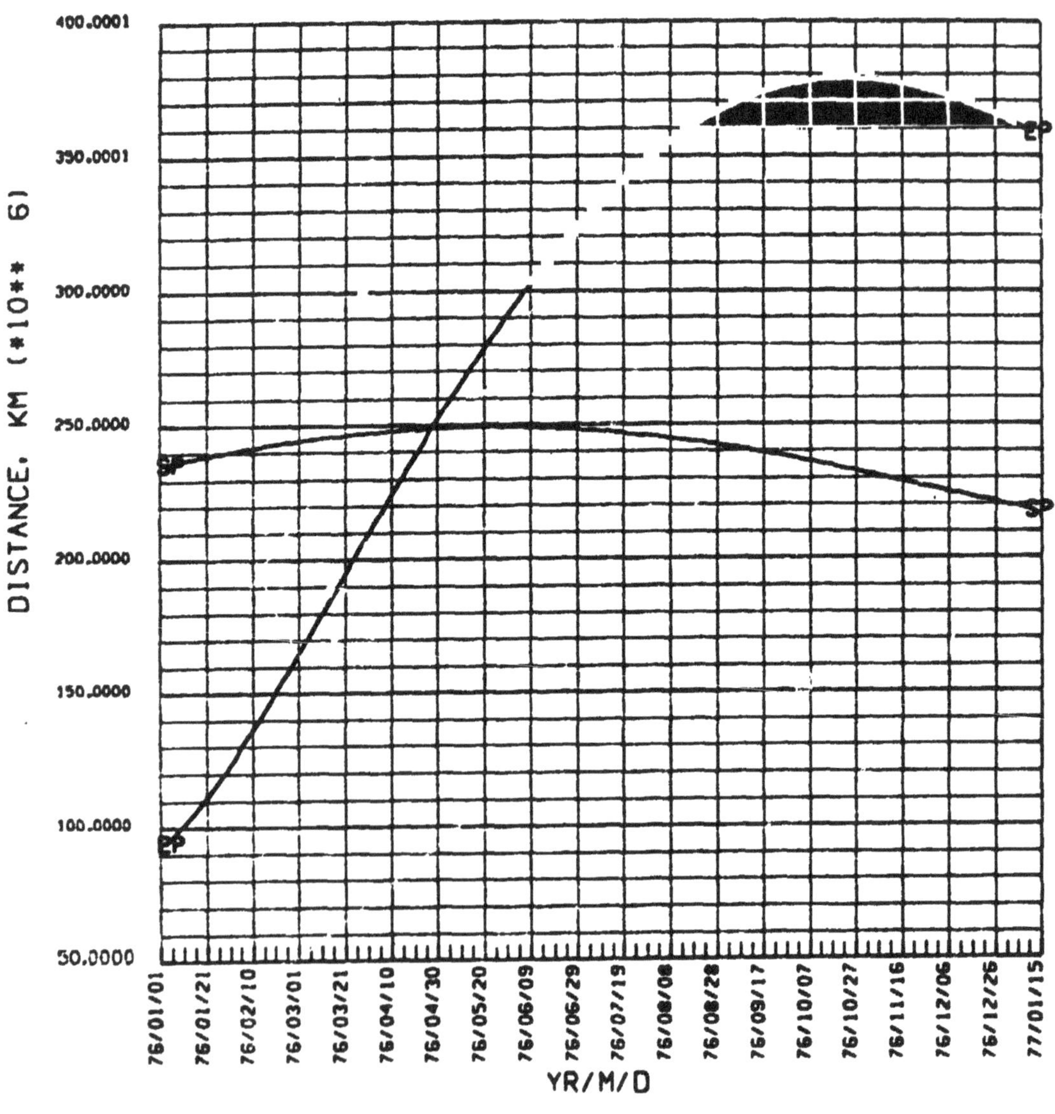
MARS 1976
DISTANCE, KM (*10** 6)
400.0001
350.0001
300.0000
250.0000
200.0000
150.0000
100.0000
50.0000
SP
EP
76/01/01
76/01/21
76/02/10
76/03/01
76/03/21
76/04/10
76/04/30
76/05/20
76/06/09
76/06/29
76/07/19
76/08/08
76/08/28
76/09/17
76/10/07
76/10/27
76/11/16
76/12/06
76/12/26
77/01/15
YR/M/D

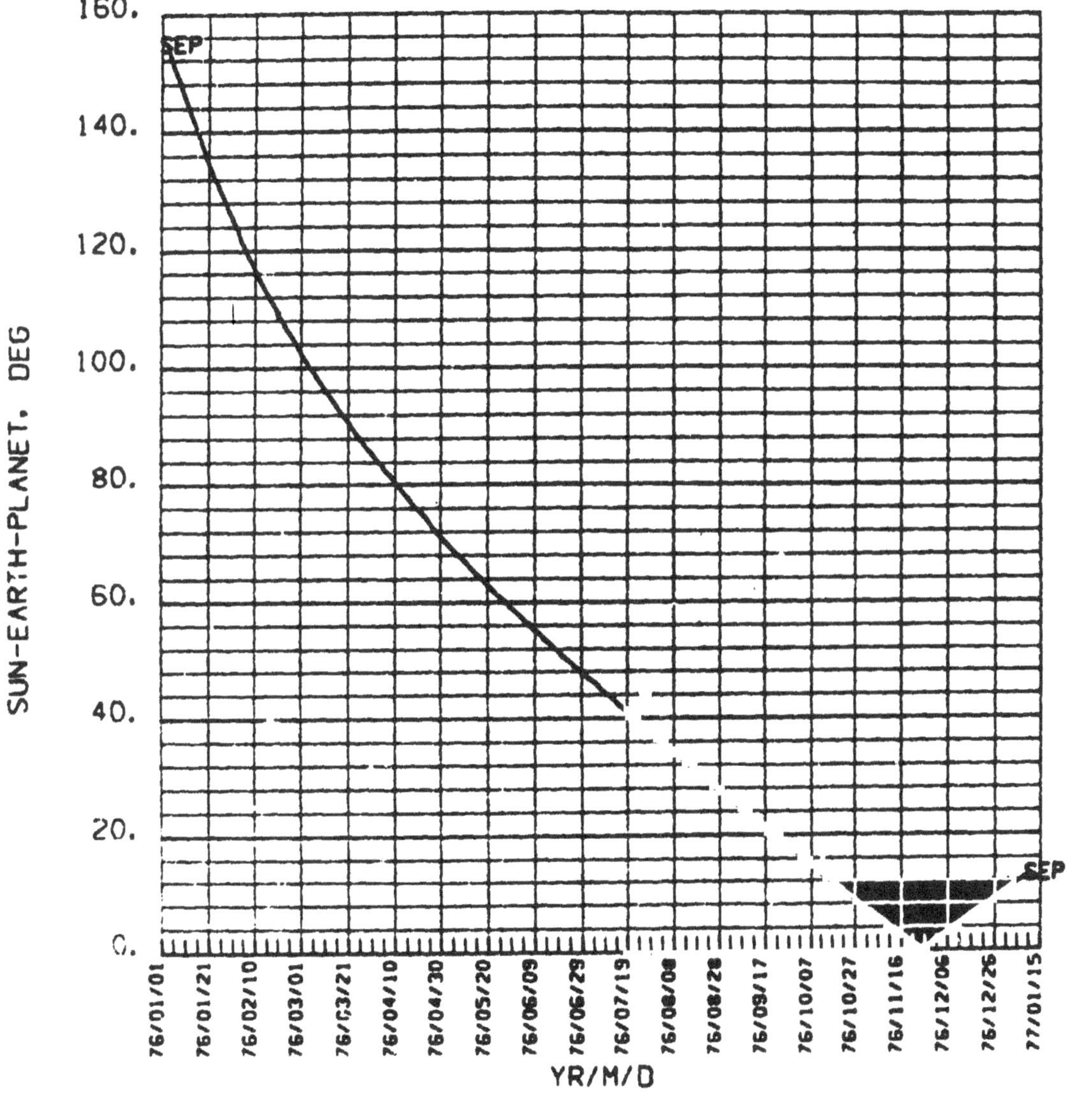
MARS 1976
SUN-EARTH-PLANET, DEG
160.
140.
120.
100.
80.
60.
40.
20.
0.
SEP
SEP
76/01/01
76/01/21
76/02/10
76/03/01
76/03/21
76/04/10
76/04/30
76/05/20
76/06/09
76/06/29
76/07/19
76/08/08
76/08/28
76/09/17
76/10/07
76/10/27
76/11/16
76/12/06
76/12/26
77/01/15
YR/M/D

MARS 1976

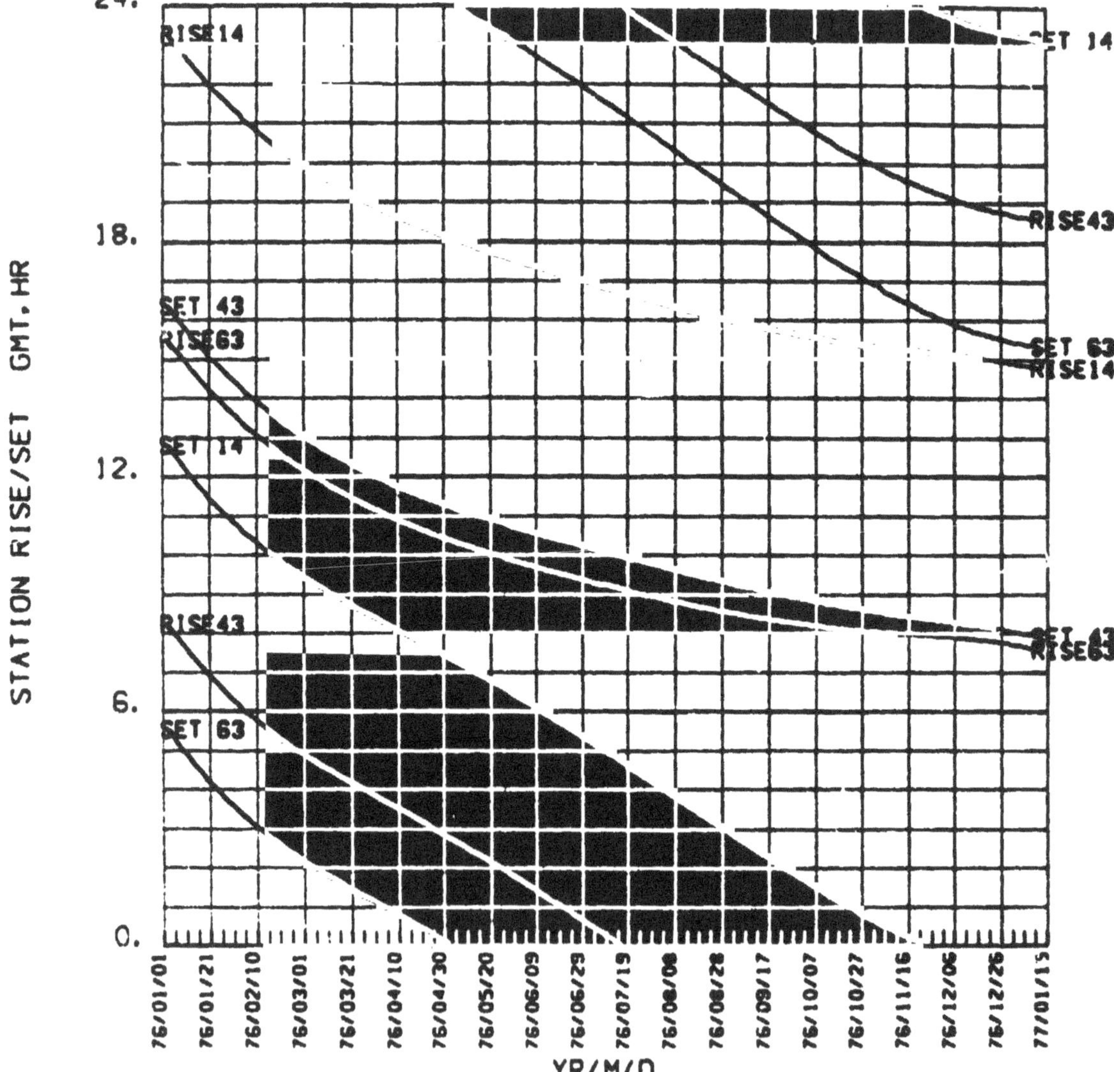

24.
18.
12.
6.
0.
STATION RISE/SET GMT.HR
RISE14
SET 43
RISE63
SET 14
RISE43
SET 63
SET 14
RISE43
SET 63
RISE14
SET 43
RISE63
76/01/01
76/01/21
76/02/10
76/03/01
76/03/21
76/04/10
76/04/30
76/05/20
76/06/09
76/06/29
76/07/19
76/08/08
76/08/28
76/09/17
76/10/07
76/10/27
76/11/16
76/12/06
76/12/26
77/01/15
YR/M/D

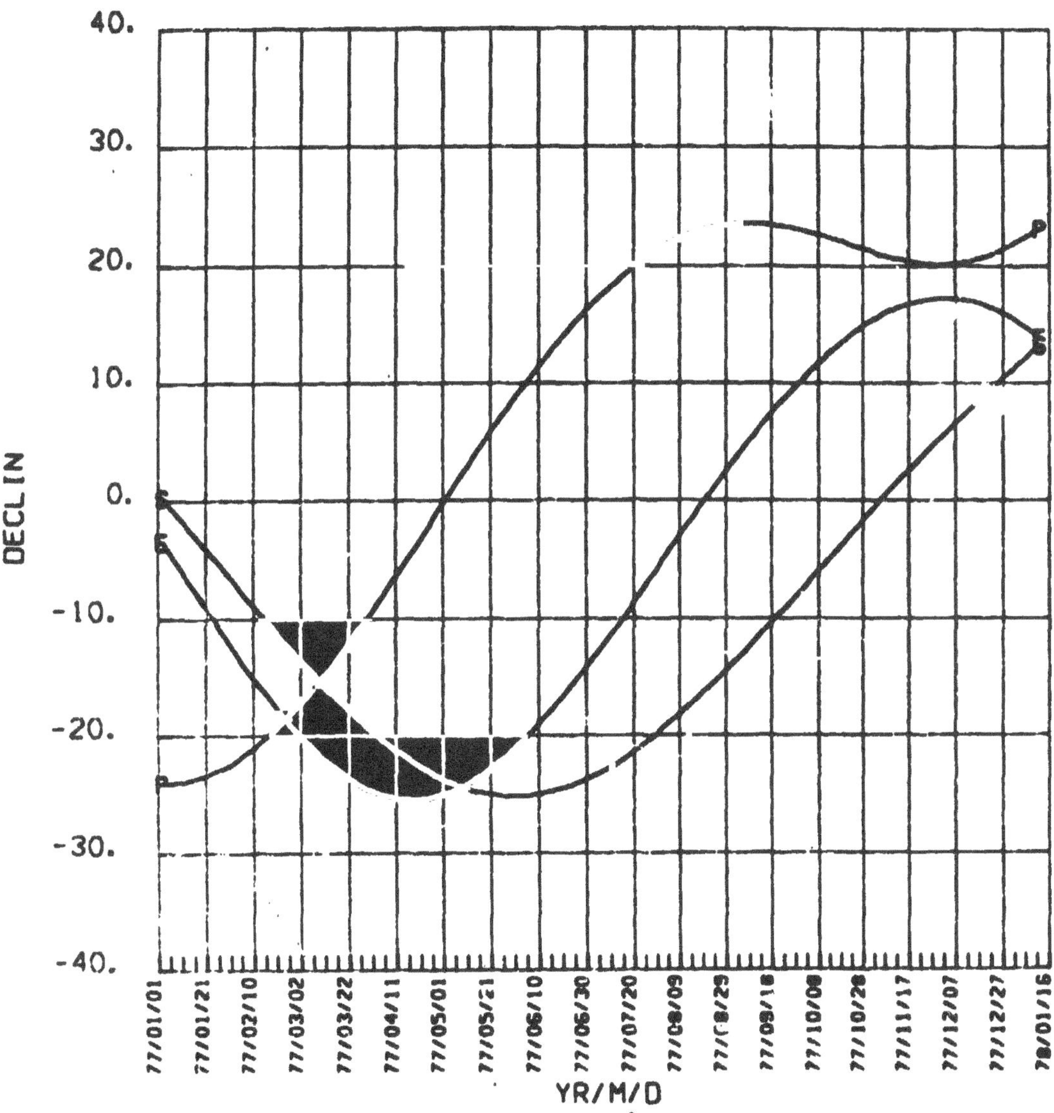
MARS 1977
40.
30.
20.
10.
0.
-10.
-20.
-30.
-40.
DECLIN
77/01/01
77/01/21
77/02/10
77/03/02
77/03/22
77/04/11
77/05/01
77/05/21
77/06/10
77/06/30
77/07/20
77/08/09
77/08/29
77/09/18
77/10/08
77/10/28
77/11/17
77/12/07
77/12/27
78/01/16
YR/M/D

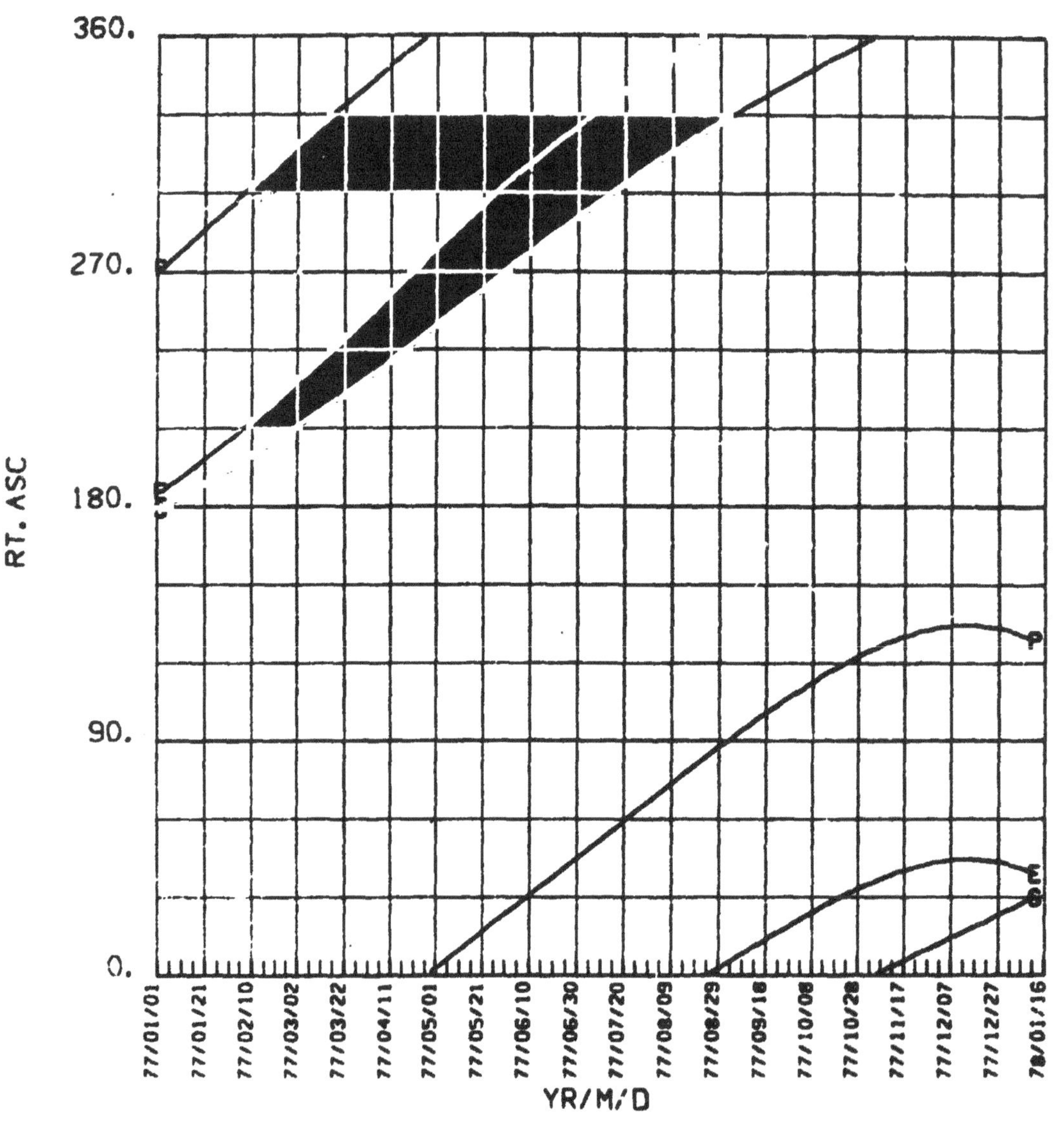

MARS 1977
360.
270.
180.
90.
0.
RT. ASC
77/01/01
77/01/21
77/02/10
77/03/02
77/03/22
77/04/11
77/05/01
77/05/21
77/06/10
77/06/30
77/07/20
77/08/09
77/08/29
77/09/18
77/10/08
77/10/28
77/11/17
77/12/07
77/12/27
78/01/16
YR/M/D

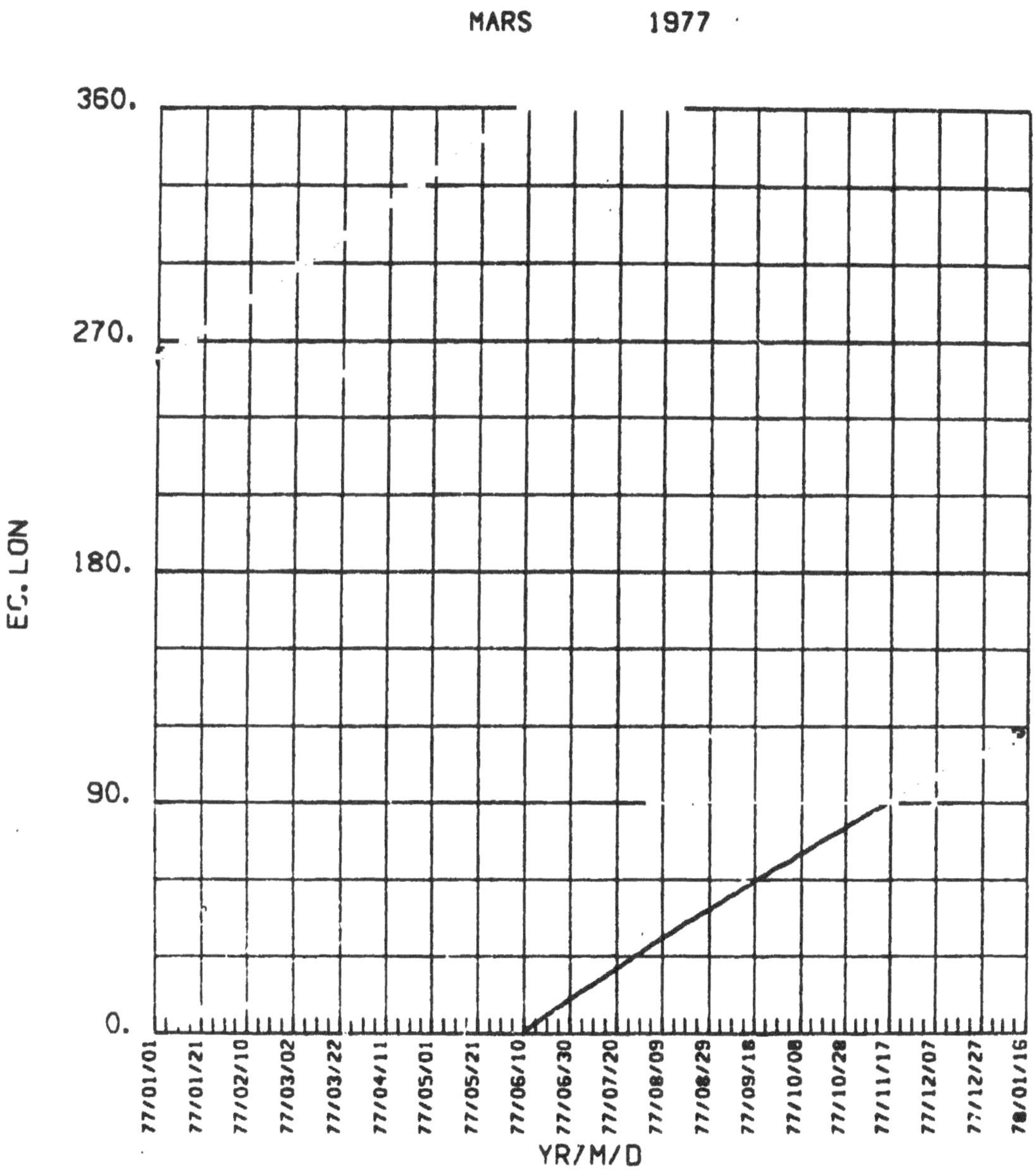

MARS 1977
EC. LON
360.
270.
180.
90.
0.
77/01/01
77/01/21
77/02/10
77/03/02
77/03/22
77/04/11
77/05/01
77/05/21
77/06/10
77/06/30
77/07/20
77/08/09
77/08/29
77/09/18
77/10/08
77/10/28
77/11/17
77/12/07
77/12/27
78/01/16
YR/M/D

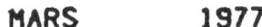

CA, KA OF EARTH. CA CANOP

360.
270.
180.
90.
0.

ECA
KA
ECA

77/01/01 77/01/21 77/02/10 77/03/02 77/03/22 77/04/11 77/05/01 77/05/21 77/06/10 77/06/30 77/07/20 77/08/09 77/08/29 77/09/18 77/10/08 77/10/28 77/11/17 77/12/07 77/12/27 78/01/16

YR/M/D

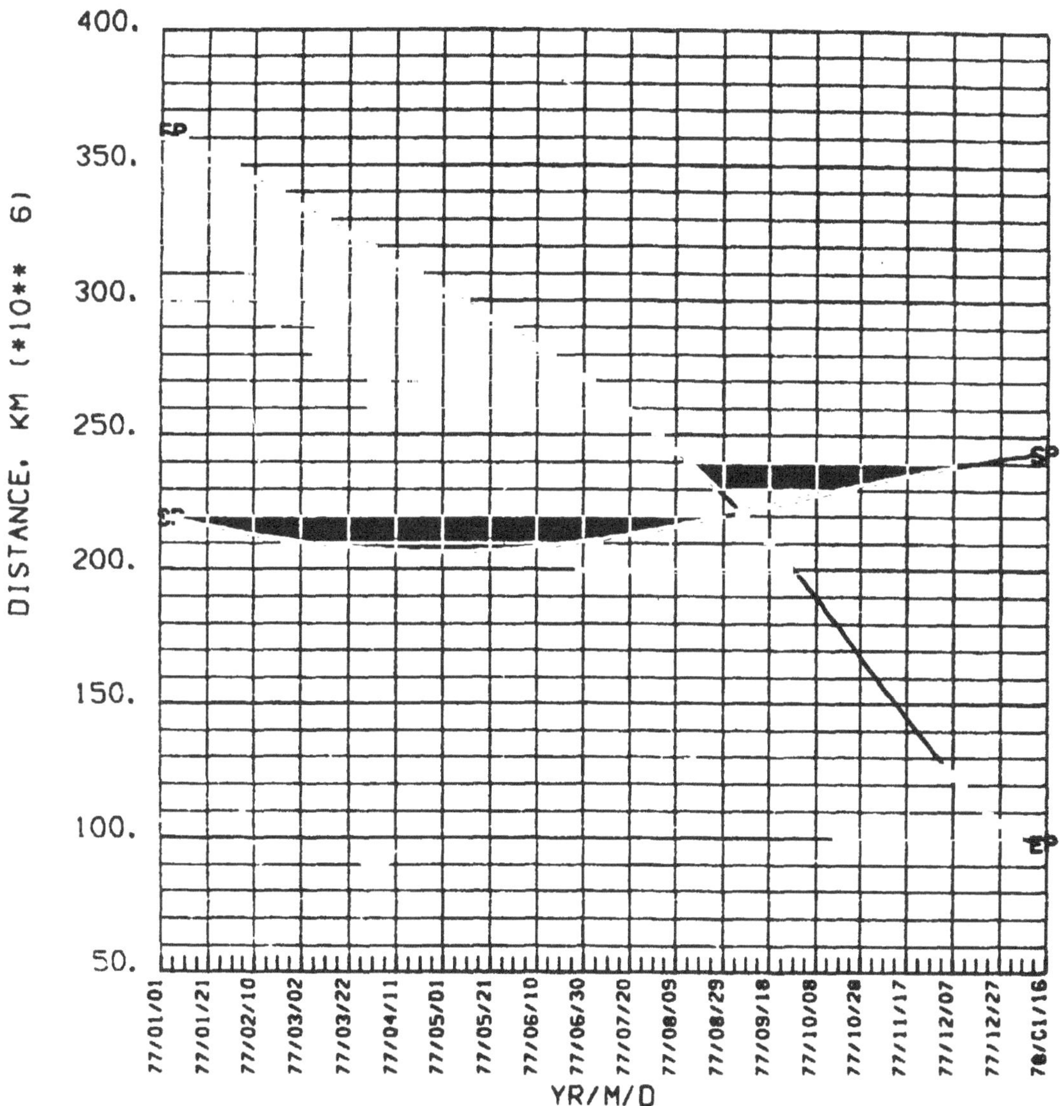

MARS 1977
400.
350.
300.
250.
200.
150.
100.
50.
DISTANCE, KM (*10** 6)
77/01/01
77/01/21
77/02/10
77/03/02
77/03/22
77/04/11
77/05/01
77/05/21
77/06/10
77/06/30
77/07/20
77/08/09
77/08/29
77/09/18
77/10/08
77/10/28
77/11/17
77/12/07
77/12/27
78/01/16
YR/M/D

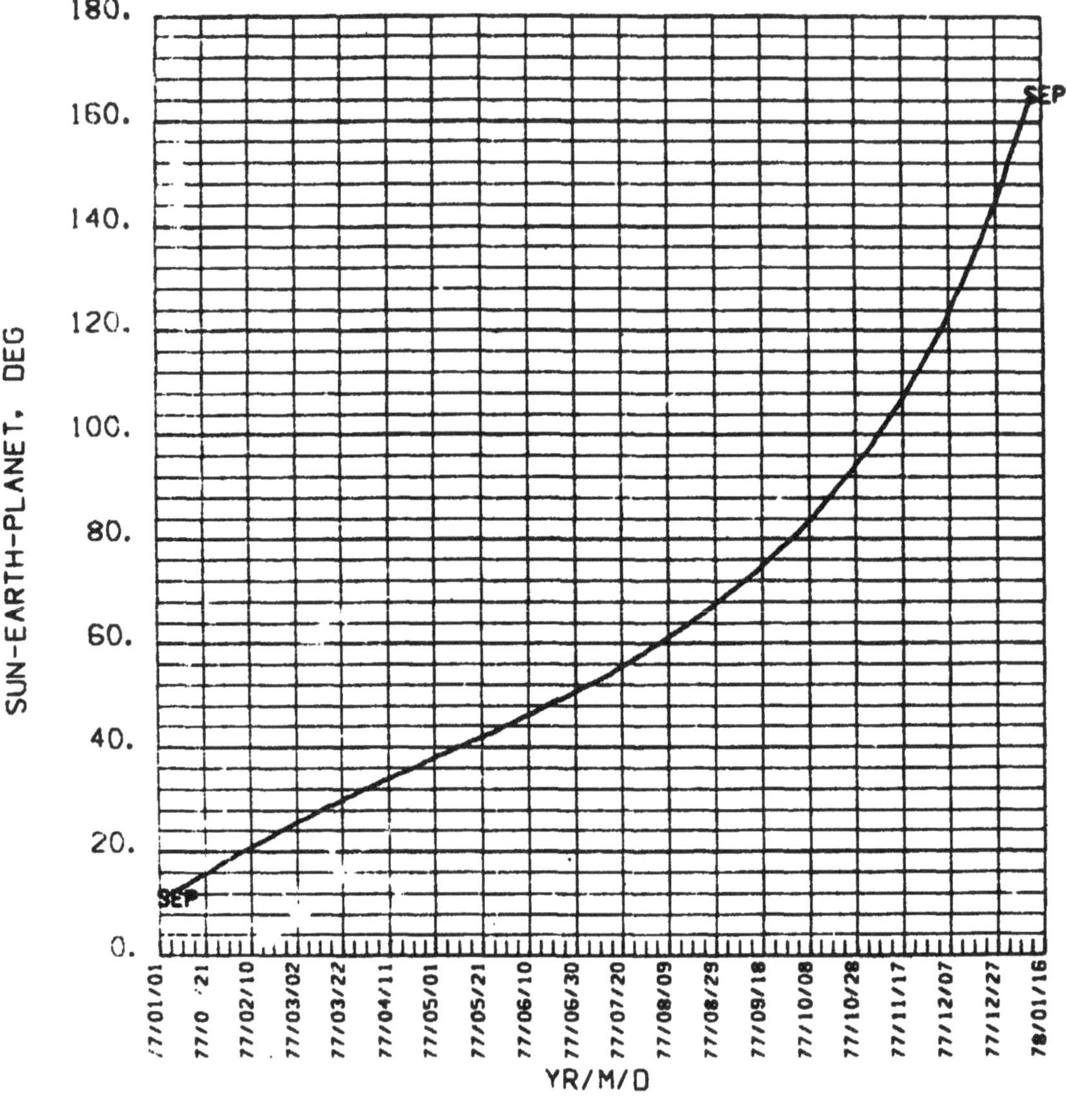

MARS 1977
SUN-EARTH-PLANET, DEG
180.
160.
140.
120.
100.
80.
60.
40.
20.
0.
SEP
SEP
77/01/01
77/01/21
77/02/10
77/03/02
77/03/22
77/04/11
77/05/01
77/05/21
77/06/10
77/06/30
77/07/20
77/08/09
77/08/29
77/09/18
77/10/08
77/10/28
77/11/17
77/12/07
77/12/27
78/01/16
YR/M/D

MARS 1977

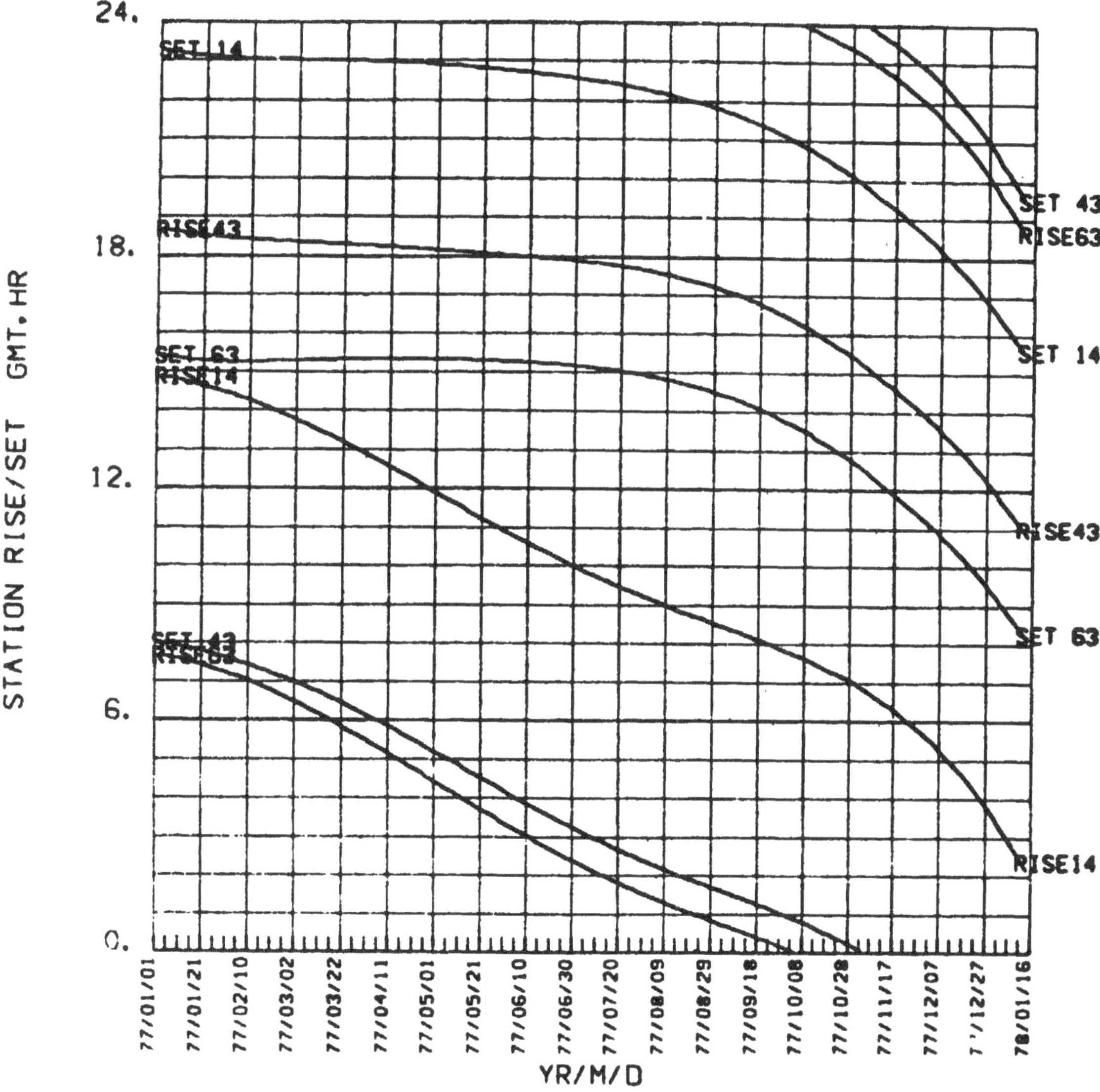

24.
18.
12.
6.
0.
STATION RISE/SET GMT.HR
SET 14
RISE43
SET 63
RISE14
SET 43
RISE63
SET 43
RISE63
SET 14
RISE43
SET 63
RISE14
77/01/01
77/01/21
77/02/10
77/03/02
77/03/22
77/04/11
77/05/01
77/05/21
77/06/10
77/06/30
77/07/20
77/08/09
77/08/29
77/09/18
77/10/08
77/10/28
77/11/17
77/12/07
78/01/16
YR/M/D

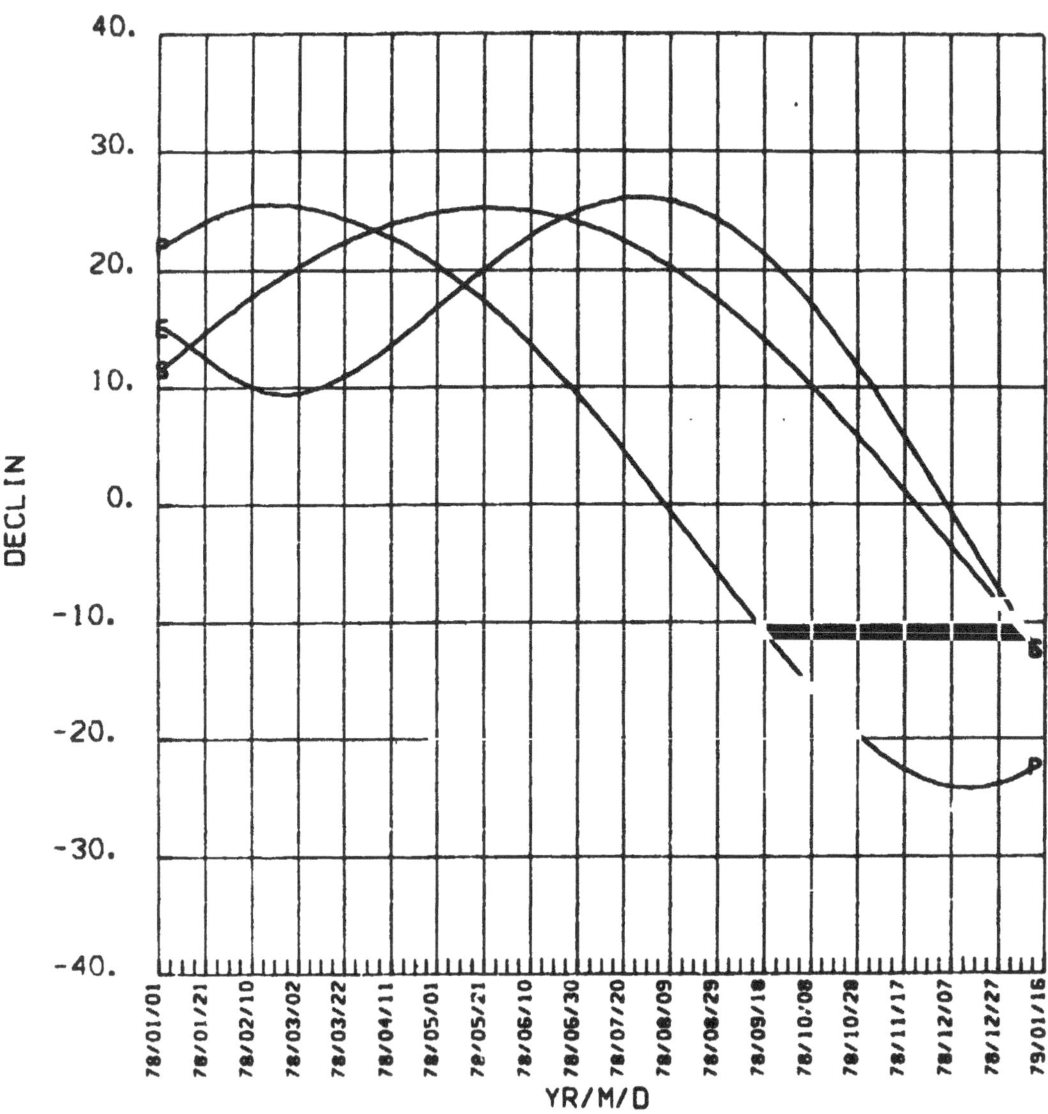
MARS 1978
40.
30.
20.
10.
0.
-10.
-20.
-30.
-40.
DECLIN
78/01/01
78/01/21
78/02/10
78/03/02
78/03/22
78/04/11
78/05/01
78/05/21
78/06/10
78/06/30
78/07/20
78/08/09
78/08/29
78/09/18
78/10/08
78/10/28
78/11/17
78/12/07
78/12/27
79/01/16
YR/M/D

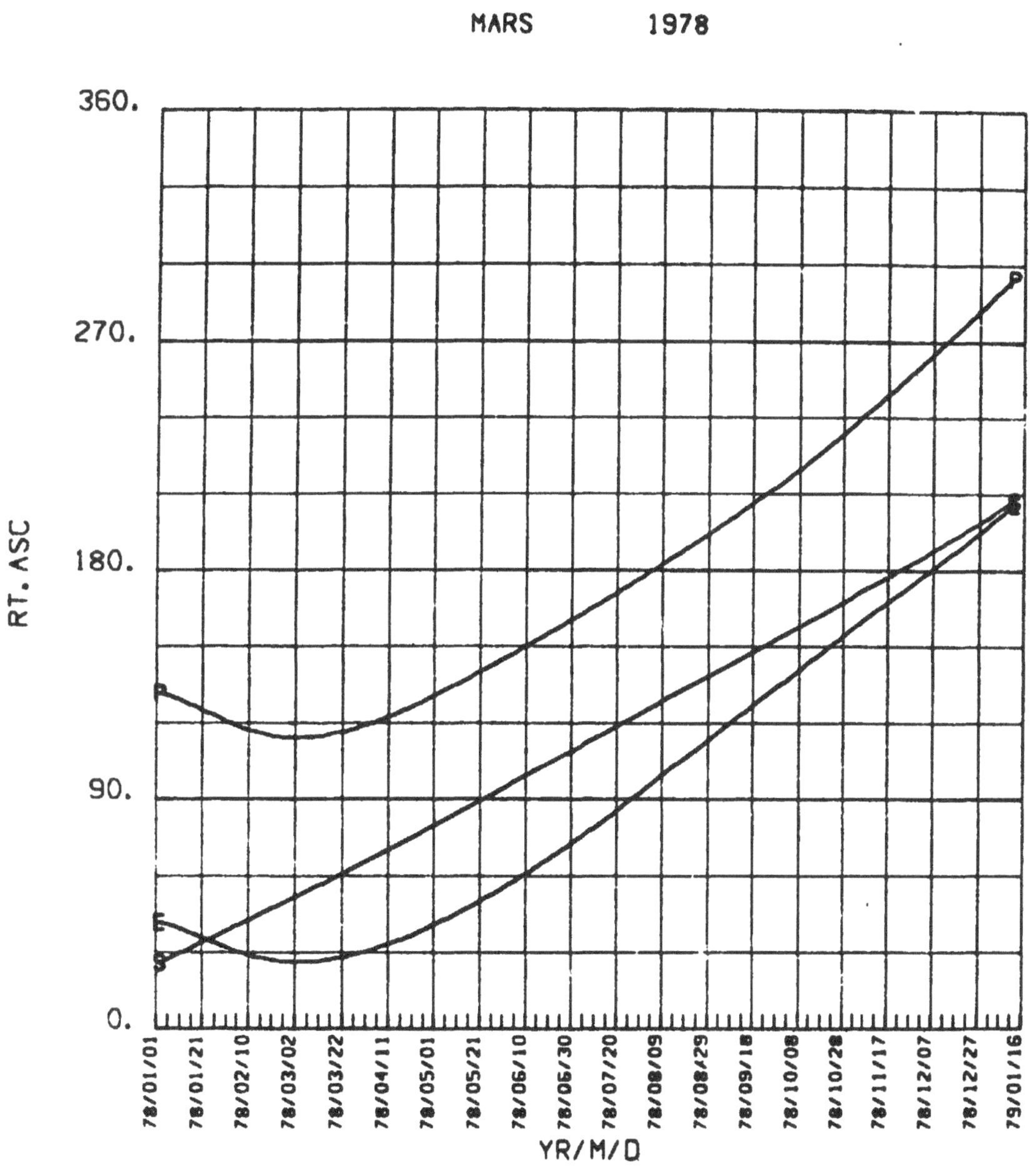

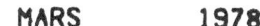

RT. ASC

360.
270.
180.
90.
0.

78/01/01
78/01/21
78/02/10
78/03/02
78/03/22
78/04/11
78/05/01
78/05/21
78/06/10
78/06/30
78/07/20
78/08/09
78/08/29
78/09/18
78/10/08
78/10/28
78/11/17
78/12/07
78/12/27
79/01/16

YR/M/D

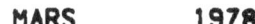
MARS 1978
360.
270.
180.
90.
0.
EC. LON
78/01/01
78/01/21
78/02/10
78/03/02
78/03/22
78/04/11
78/05/01
78/05/21
78/06/10
78/06/30
78/07/20
78/08/09
78/08/29
78/09/18
78/10/08
78/10/28
78/11/17
78/12/07
78/12/27
79/01/16
YR/M/D

CA, KA OF EARTH, CA CANOP

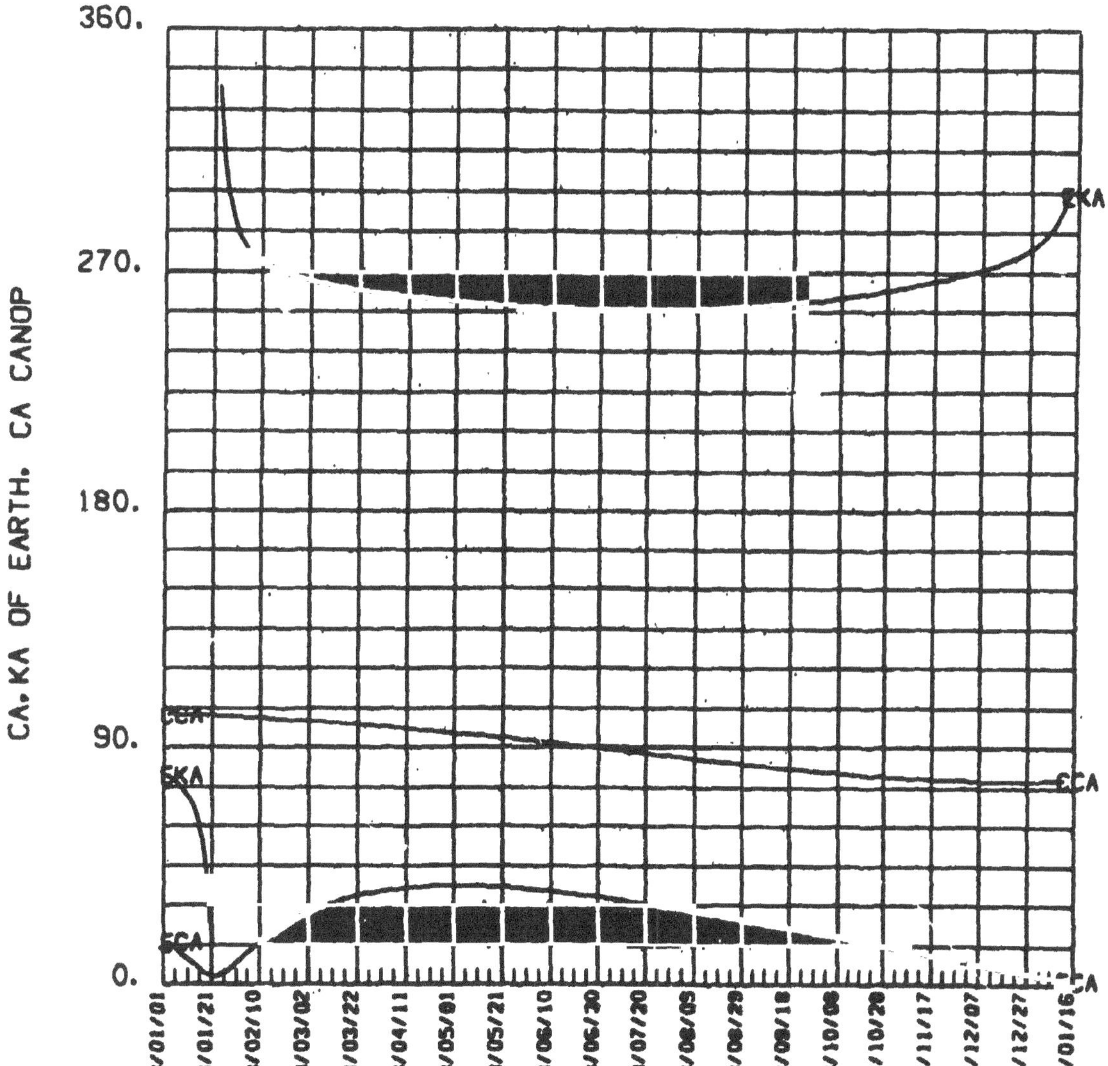

360.
270.
180.
90.
0.

EKA
ECA
CCA
ECA

78/01/01
78/01/21
78/02/10
78/03/02
78/03/22
78/04/11
78/05/01
78/05/21
78/06/10
78/06/30
78/07/20
78/08/09
78/08/29
78/09/18
78/10/08
78/10/28
78/11/17
78/12/07
78/12/27
79/01/16

YR/M/D

MARS 1978

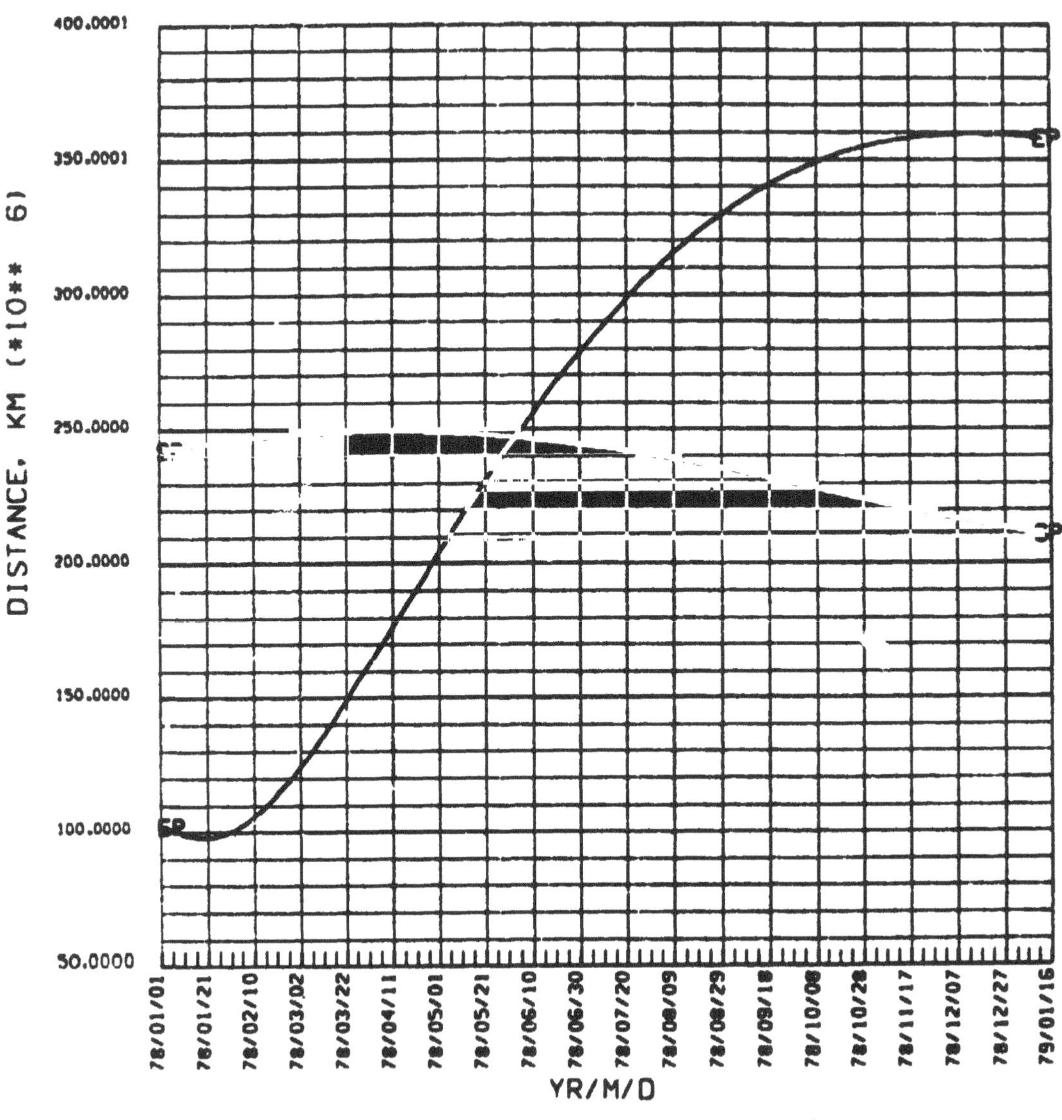

400.0001
350.0001
300.0000
250.0000
200.0000
150.0000
100.0000
50.0000
DISTANCE, KM (*10** 6)
78/01/01
78/01/21
78/02/10
78/03/02
78/03/22
78/04/11
78/05/01
78/05/21
78/06/10
78/06/30
78/07/20
78/08/09
78/08/29
78/09/18
78/10/08
78/10/28
78/11/17
78/12/07
78/12/27
79/01/16
YR/M/D

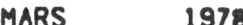
MARS 1978
SUN-EARTH-PLANET, DEG
180.
160.
140.
120.
100.
80.
60.
40.
20.
0.
SEP
SEP
78/01/01
78/01/21
78/02/10
78/03/02
78/03/22
78/04/11
78/05/01
78/05/21
78/06/10
78/06/30
78/07/20
78/08/09
78/08/29
78/09/18
78/10/08
78/10/28
78/11/17
78/12/07
78/12/27
79/01/16
YR/M/D

MARS 1978

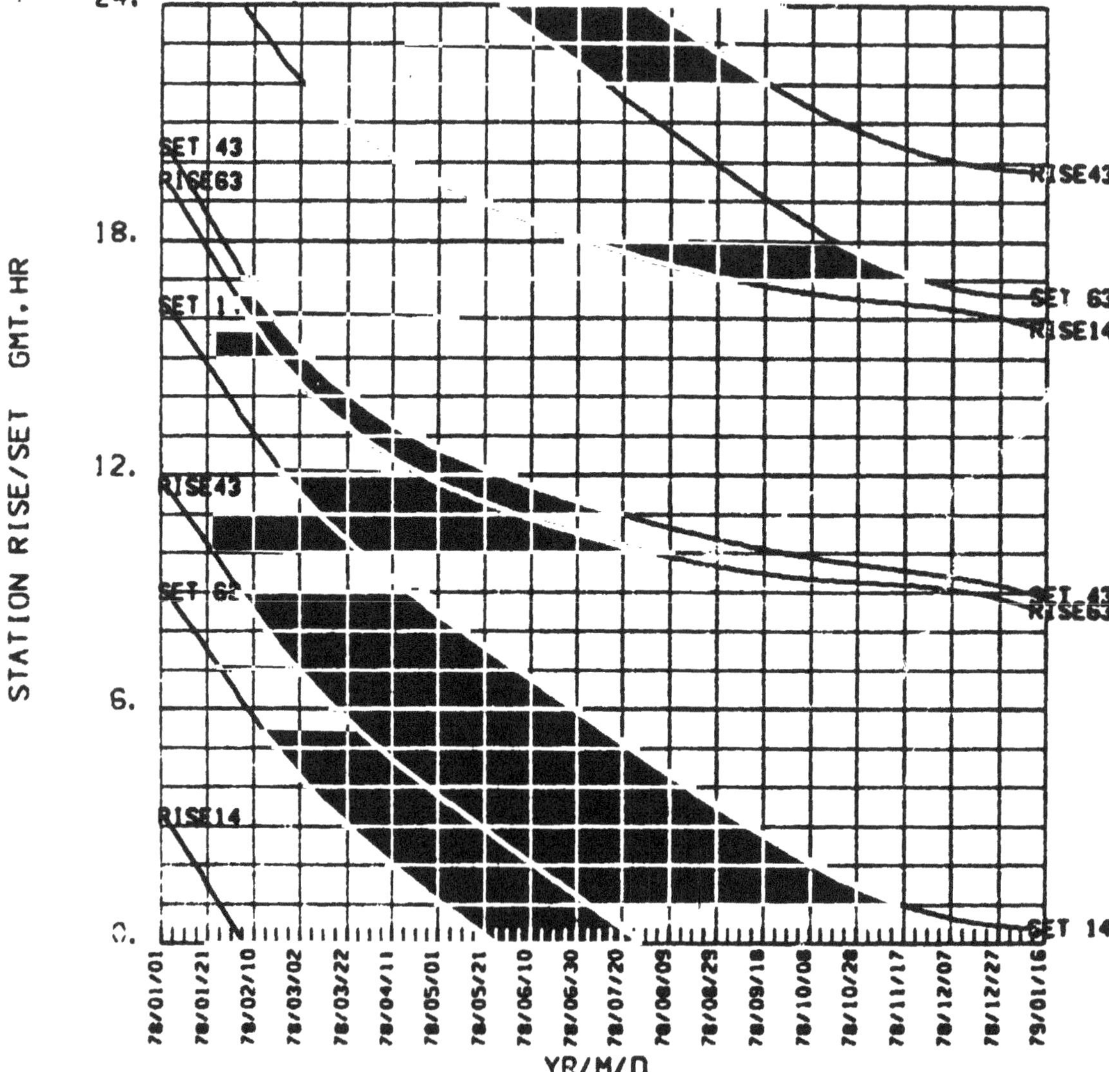

24.
18.
12.
6.
0.
STATION RISE/SET GMT. HR
SET 43
RISE63
SET 1
RISE43
SET 62
RISE14
RISE43
SET 63
RISE14
SET 43
RISE63
SET 14
78/01/01
78/01/21
78/02/10
78/03/02
78/03/22
78/04/11
78/05/01
78/05/21
78/06/10
78/06/30
78/07/20
78/08/09
78/08/29
78/09/18
78/10/08
78/10/28
78/11/17
78/12/07
78/12/27
79/01/16
YR/M/D

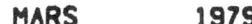
MARS 1979
40.
30.
20.
10.
0.
-10.
-20.
-30.
-40.
DECLIN
79/01/01
79/01/21
79/02/10
79/03/02
79/03/22
79/04/11
79/05/01
79/05/21
79/06/10
79/06/30
79/07/20
79/08/09
79/08/29
79/09/18
79/10/08
79/10/28
79/11/17
79/12/07
79/12/27
YR/M/D

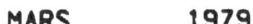

RT. ASC

360.
270.
180.
90.
0.

79/01/01 79/01/21 79/02/10 79/03/02 79/03/22 79/04/11 79/05/01 79/05/21 79/06/10 79/06/30 79/07/20 79/08/09 79/08/29 79/09/18 79/10/08 79/10/28 79/11/17 79/12/07 79/12/27 80/01/16

YR/M/D

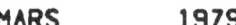

EC. LON

360.
270.
180.
90.
0.

79/01/01
79/01/21
79/02/10
79/03/02
79/03/22
79/04/11
79/05/01
79/05/21
79/06/10
79/06/30
79/07/20
79/08/09
79/08/29
79/09/18
79/10/08
79/10/28
79/11/17
79/12/07
79/12/27
80/01/16

YR/M/D

MARS 1979

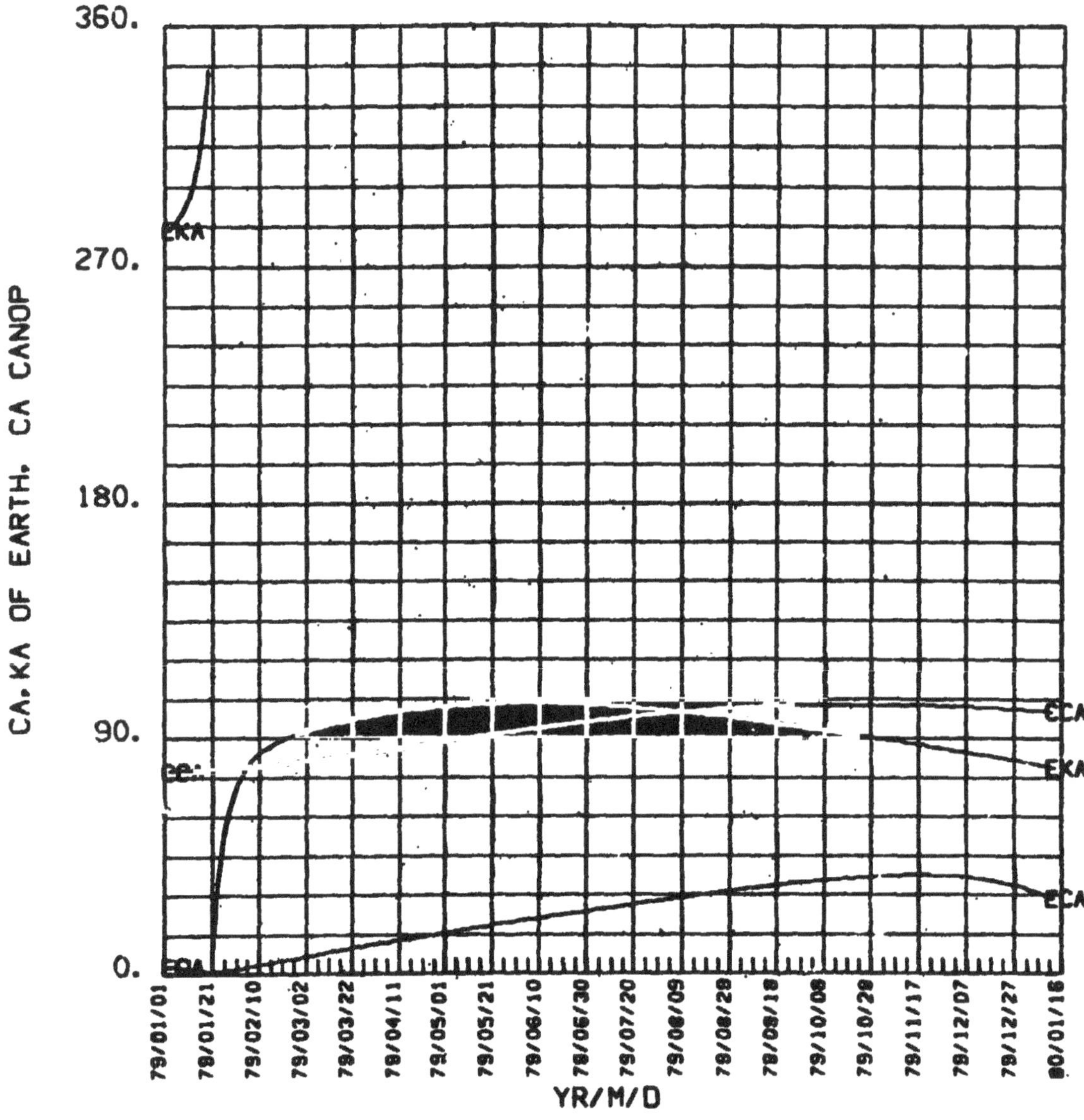

360.
270.
180.
90.
0.
CA.KA OF EARTH. CA CANOP
EKA
ECA
EKA
ECA
79/01/01
79/01/21
79/02/10
79/03/02
79/03/22
79/04/11
79/05/01
79/05/21
79/06/10
79/06/30
79/07/20
79/08/09
79/08/29
79/09/18
79/10/08
79/10/28
79/11/17
79/12/07
79/12/27
80/01/16
YR/M/D

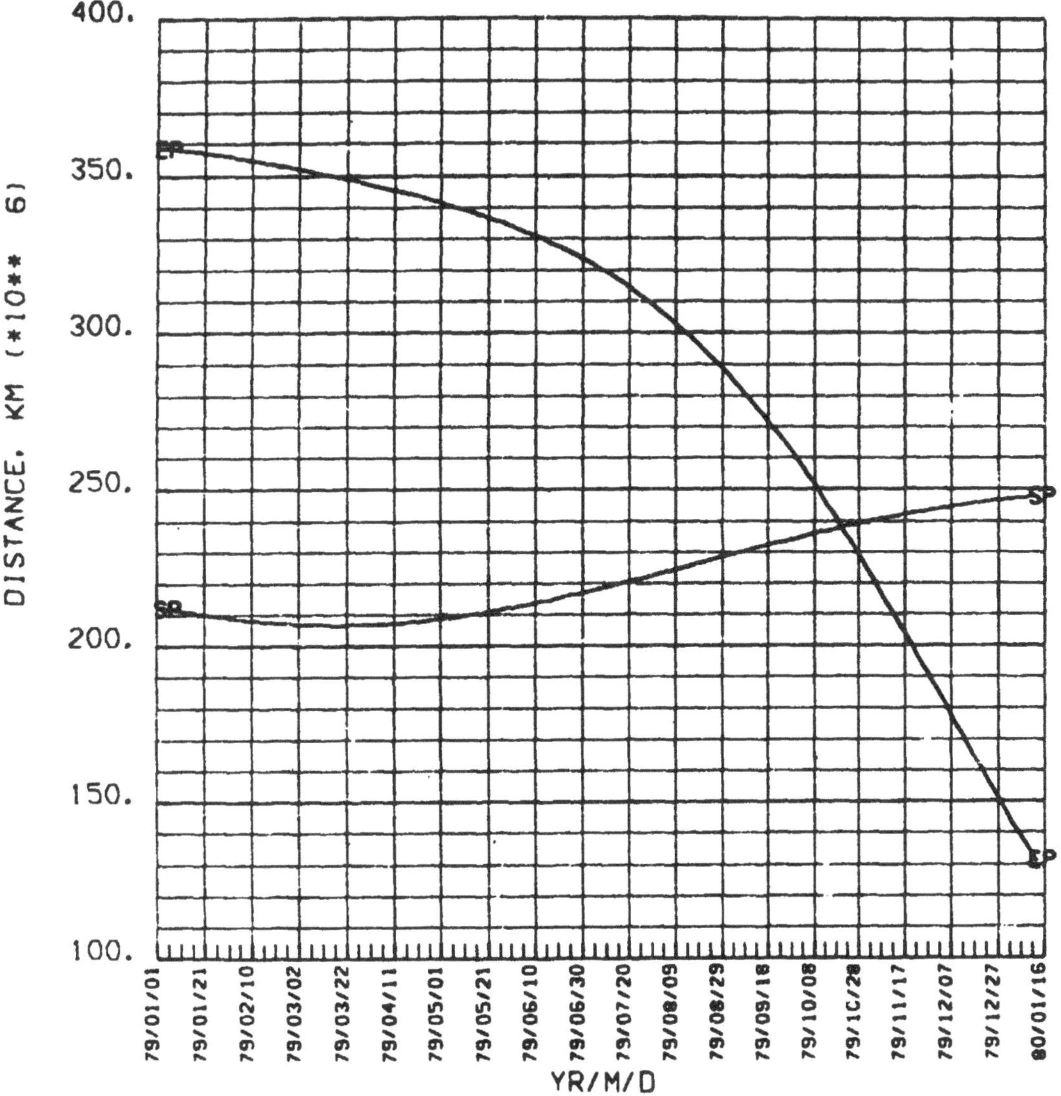

MARS 1979
400.
350.
300.
250.
200.
150.
100.
DISTANCE, KM (*10** 6)
EP
SP
SP
EP
79/01/01
79/01/21
79/02/10
79/03/02
79/03/22
79/04/11
79/05/01
79/05/21
79/06/10
79/06/30
79/07/20
79/08/09
79/08/29
79/09/18
79/10/08
79/10/28
79/11/17
79/12/07
79/12/27
80/01/16
YR/M/D

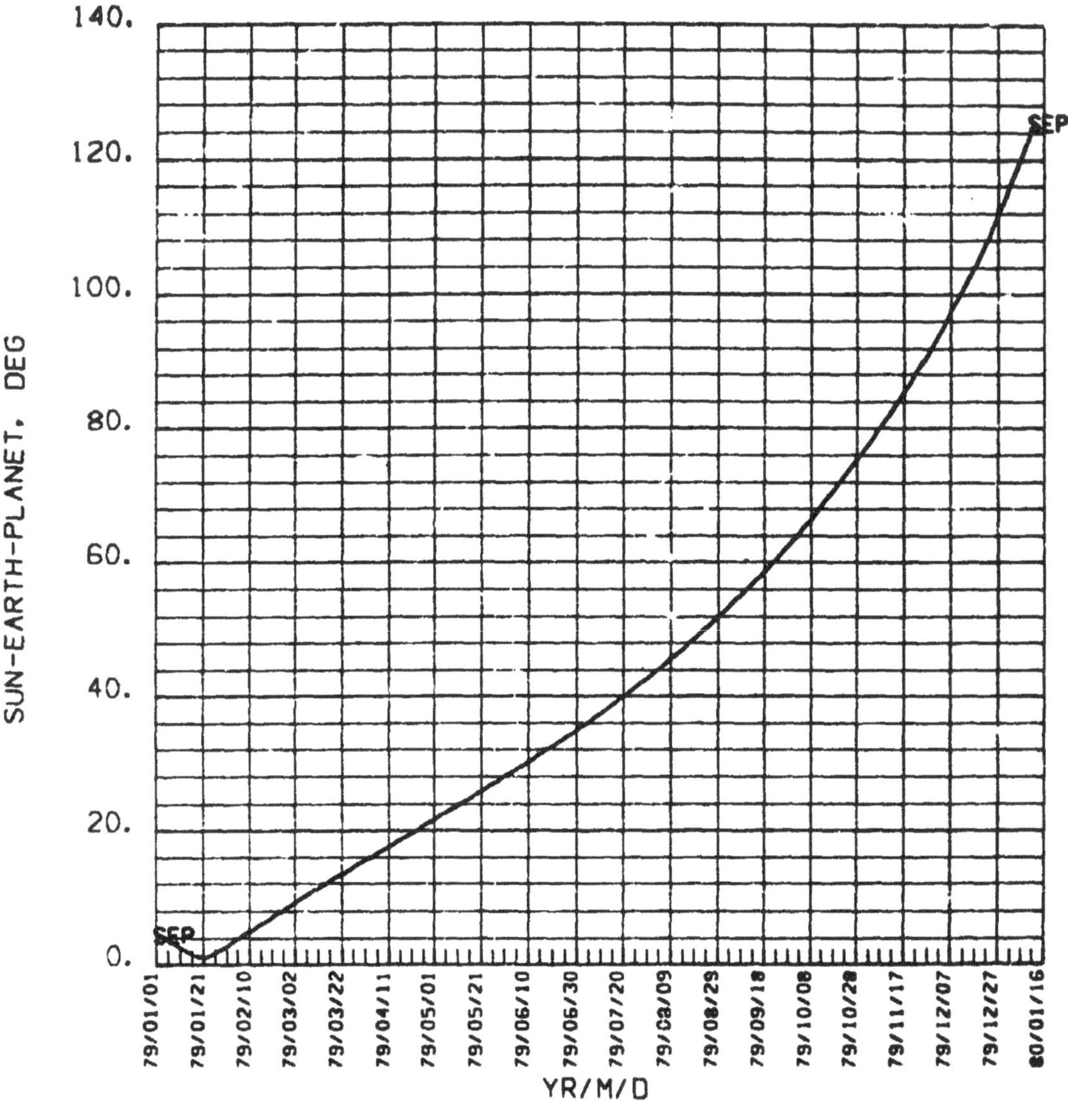
MARS 1979
SUN-EARTH-PLANET, DEG
140.
120.
100.
80.
60.
40.
20.
0.
SEP
SEP
79/01/01
79/01/21
79/02/10
79/03/02
79/03/22
79/04/11
79/05/01
79/05/21
79/06/10
79/06/30
79/07/20
79/08/09
79/08/29
79/09/18
79/10/08
79/10/28
79/11/17
79/12/07
79/12/27
80/01/16
YR/M/D

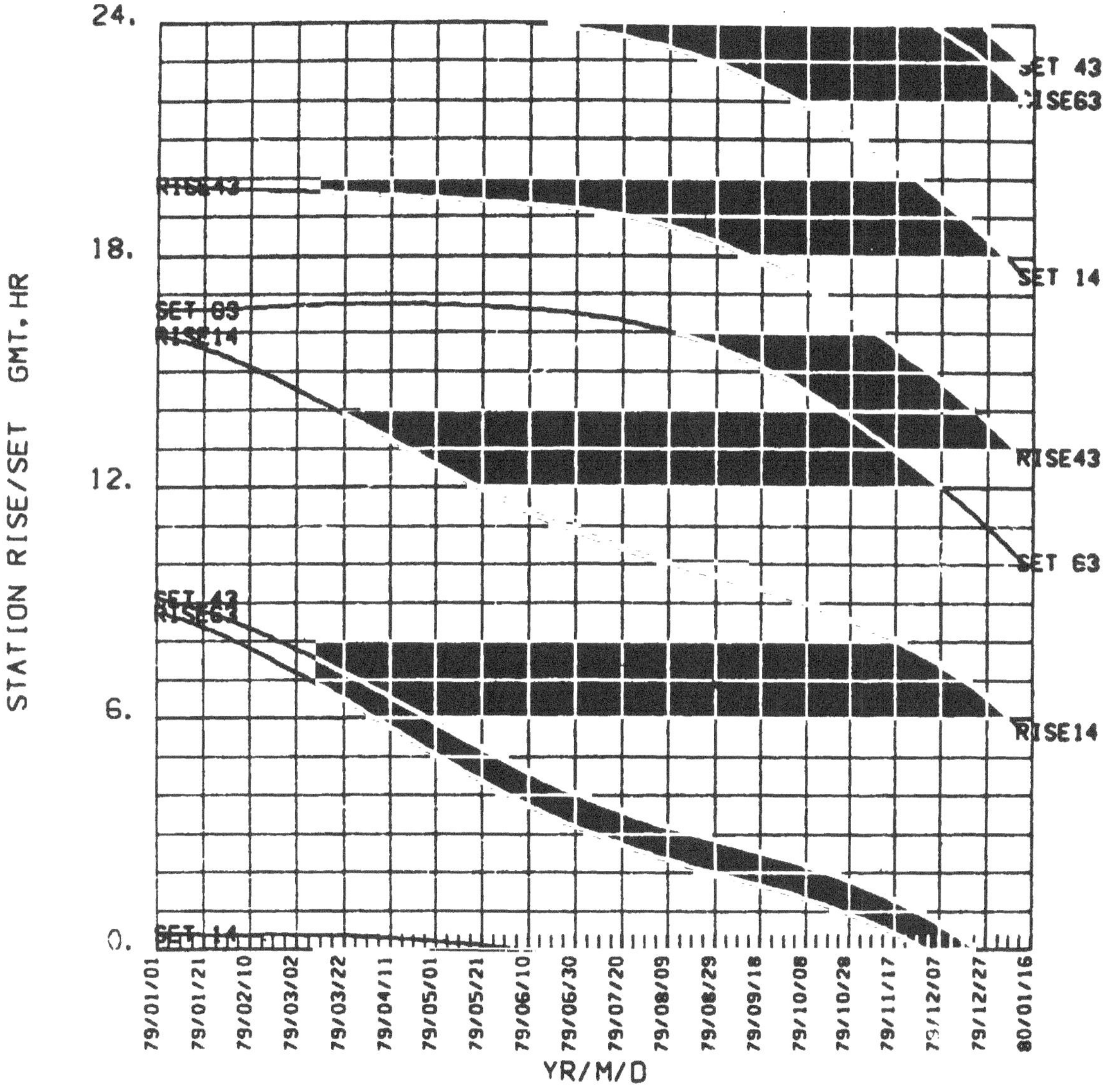

STATION RISE/SET GMT, HR

24.
18.
12.
6.
0.

SET 43
RISE63
RISE43
SET 14
SET 63
RISE14
RISE43
SET 63
SET 43
RISE63
RISE14
SET 14

79/01/01
79/01/21
79/02/10
79/03/02
79/03/22
79/04/11
79/05/01
79/05/21
79/06/10
79/06/30
79/07/20
79/08/09
79/08/29
79/09/18
79/10/08
79/10/28
79/11/17
79/12/07
79/12/27
80/01/16

YR/M/D

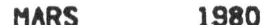
MARS 1980
40.
30.
20.
10.
0.
-10.
-20.
-30.
-40.
DECLIN
80/01/01
80/01/21
80/02/10
80/03/01
80/03/21
80/04/10
80/04/30
80/05/20
80/06/09
80/06/29
80/07/19
80/08/08
80/08/28
80/09/17
80/10/07
80/10/27
80/11/16
80/12/06
80/12/26
81/01/15
YR/M/D

MARS 1980 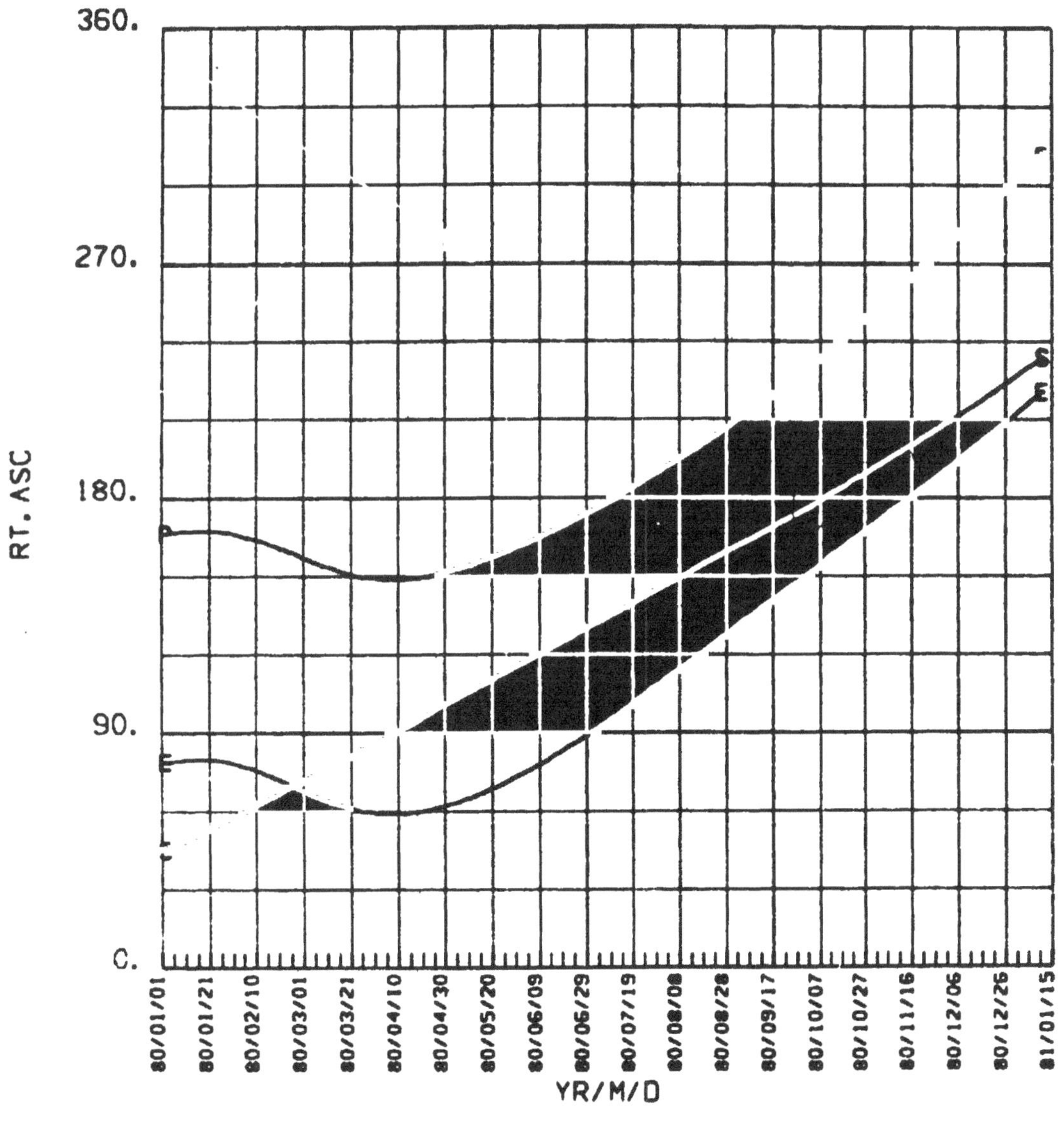
RT. ASC
360.
270.
180.
90.
0.
80/01/01
80/01/21
80/02/10
80/03/01
80/03/21
80/04/10
80/04/30
80/05/20
80/06/09
80/06/29
80/07/19
80/08/08
80/08/28
80/09/17
80/10/07
80/10/27
80/11/16
80/12/06
80/12/26
81/01/15
YR/M/D

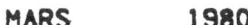

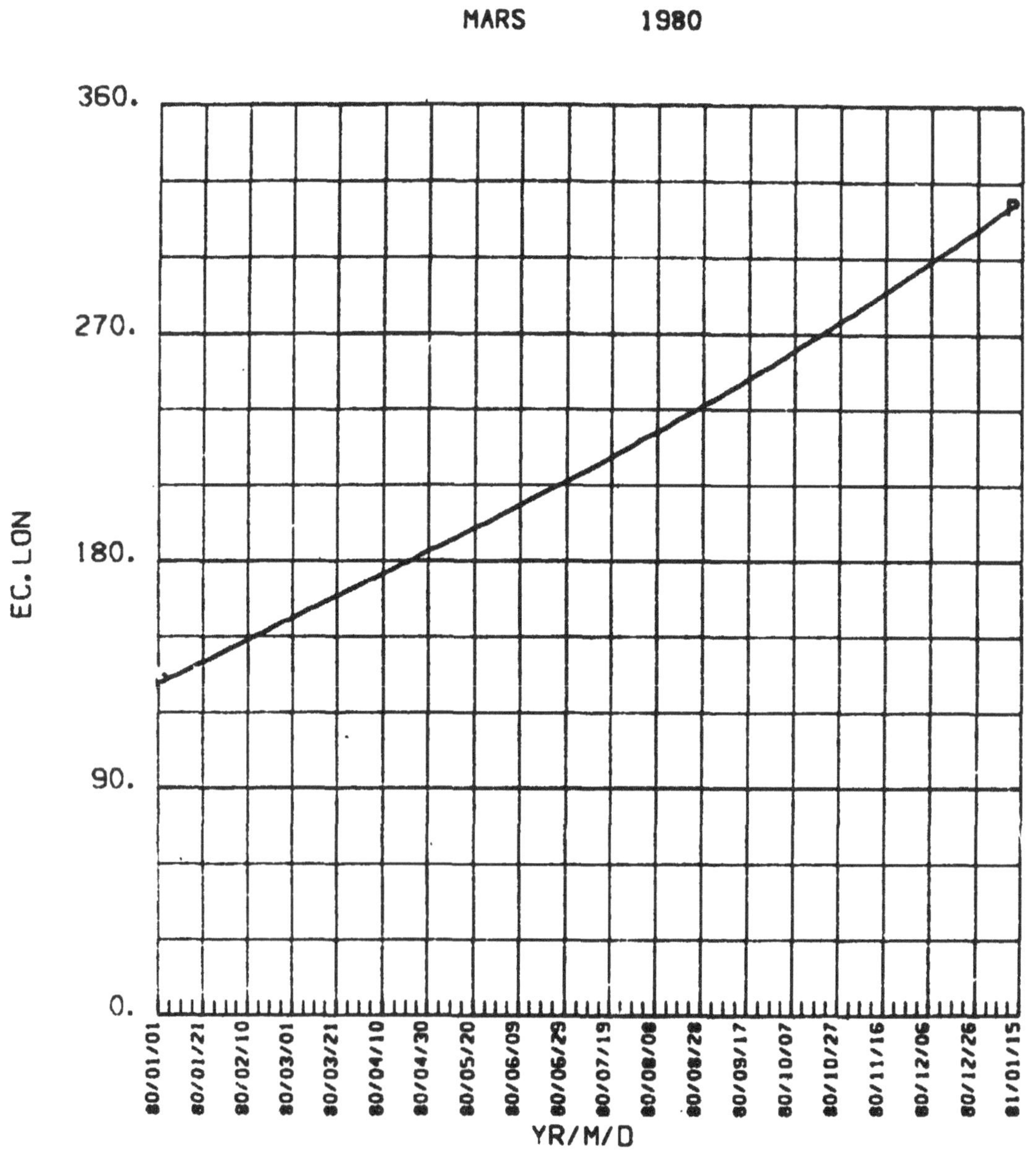

360.
270.
180.
90.
0.

EC. LON

80/01/01 80/01/21 80/02/10 80/03/01 80/03/21 80/04/10 80/04/30 80/05/20 80/06/09 80/06/29 80/07/19 80/08/08 80/08/28 80/09/17 80/10/07 80/10/27 80/11/16 80/12/06 80/12/26 81/01/15

YR/M/D

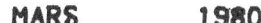
MARS 1980
360.
270.
180.
90.
0.
CA, KA OF EARTH, CA CANOP
EKA
CCA
EKA
ECA
CCA
ECA
80/01/01
80/01/21
80/02/10
80/03/01
80/03/21
80/04/10
80/04/30
80/05/20
80/06/09
80/06/29
80/07/19
80/08/08
80/08/28
80/09/17
80/10/07
80/10/27
80/11/16
80/12/06
80/12/26
81/01/15
YR/M/D

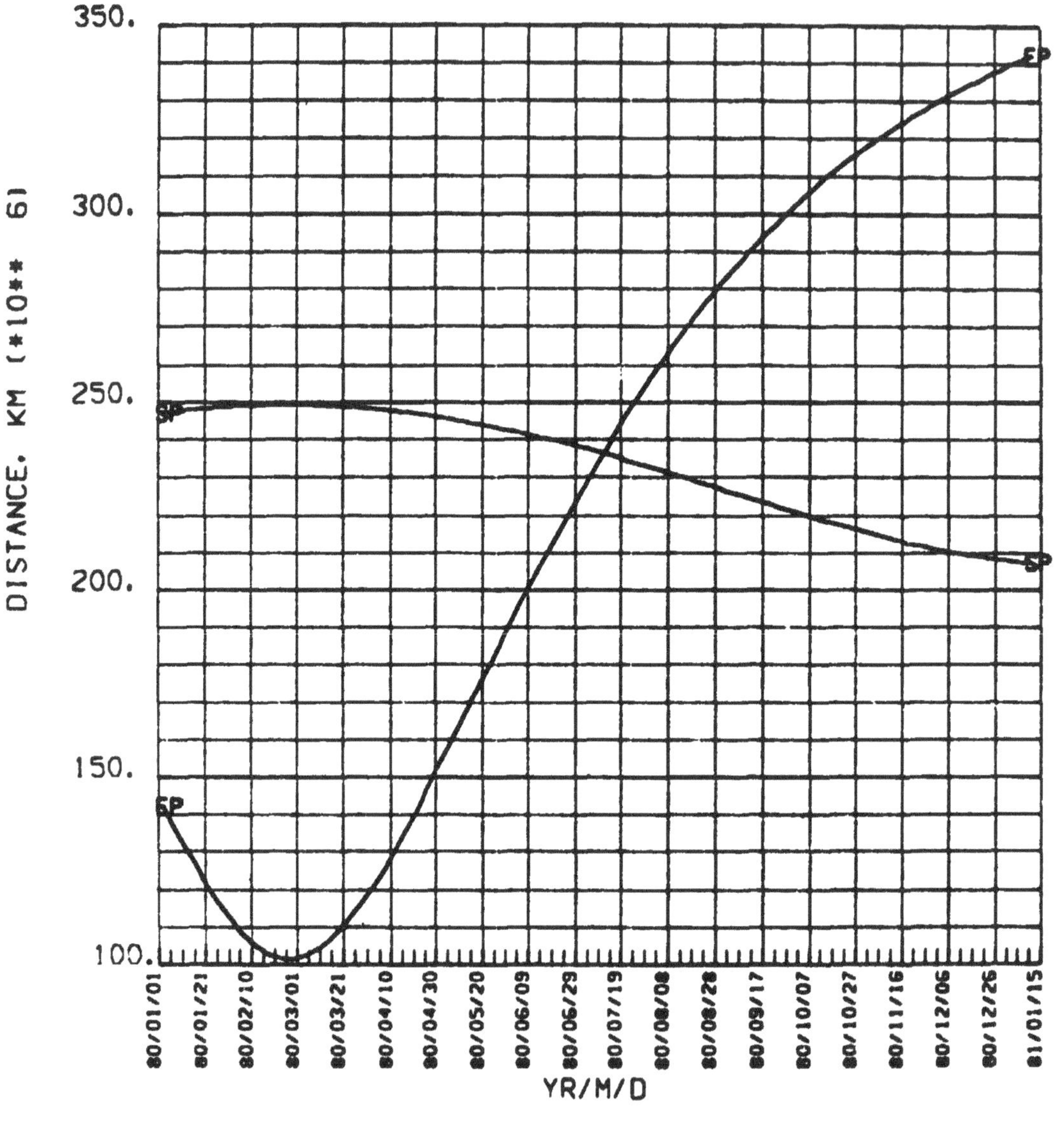
MARS 1980
DISTANCE, KM (*10** 6)
350.
300.
250.
200.
150.
100.
EP
SP
YR/M/D
80/01/01
80/01/21
80/02/10
80/03/01
80/03/21
80/04/10
80/04/30
80/05/20
80/06/09
80/06/29
80/07/19
80/08/08
80/08/28
80/09/17
80/10/07
80/10/27
80/11/16
80/12/06
80/12/26
81/01/15

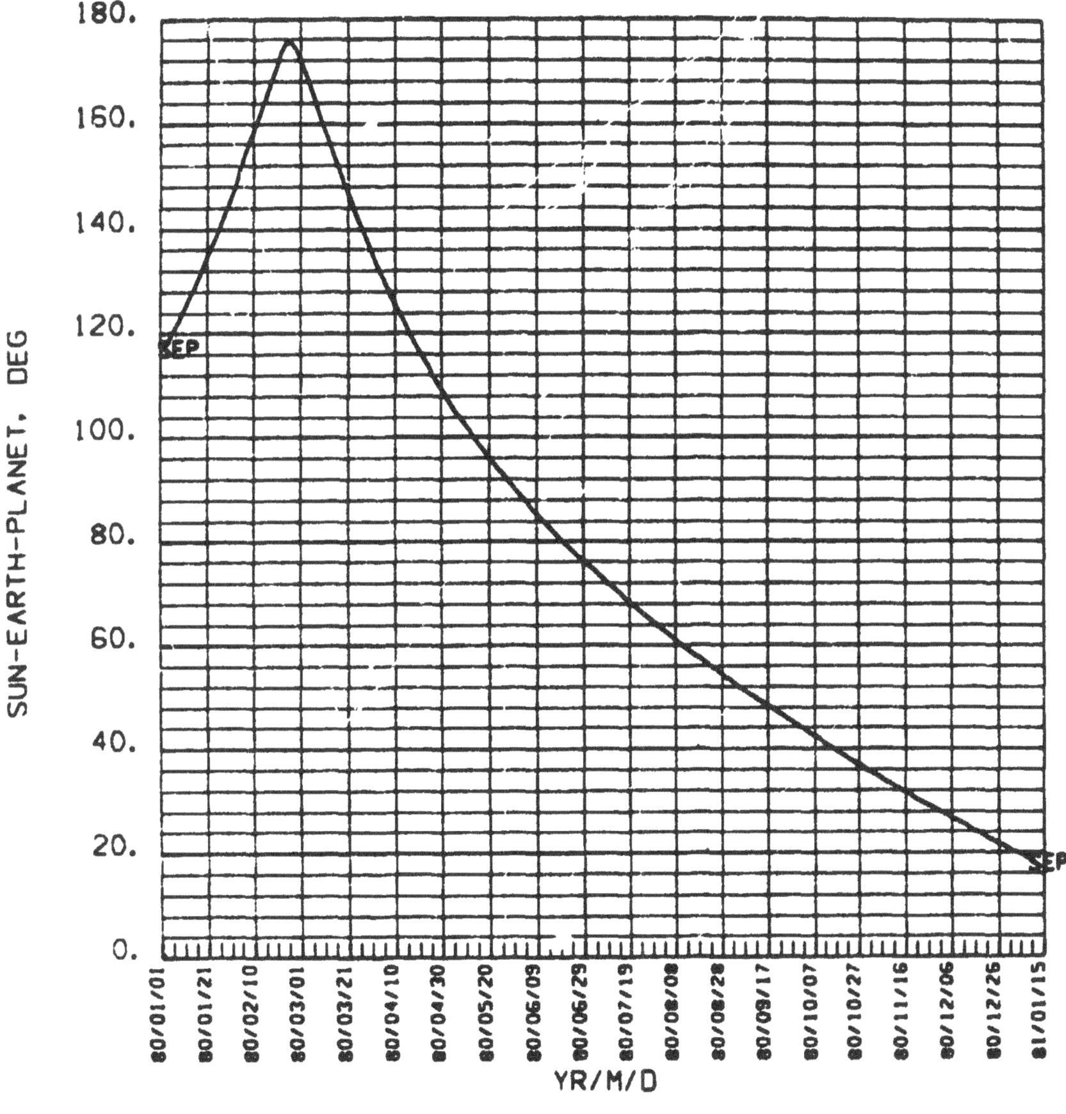

MARS 1980
SUN-EARTH-PLANET, DEG
180.
160.
140.
120.
100.
80.
60.
40.
20.
0.
XEP
XEP
80/01/01
80/01/21
80/02/10
80/03/01
80/03/21
80/04/10
80/04/30
80/05/20
80/06/09
80/06/29
80/07/19
80/08/08
80/08/28
80/09/17
80/10/07
80/10/27
80/11/16
80/12/06
80/12/26
81/01/15
YR/M/D

MARS 1980

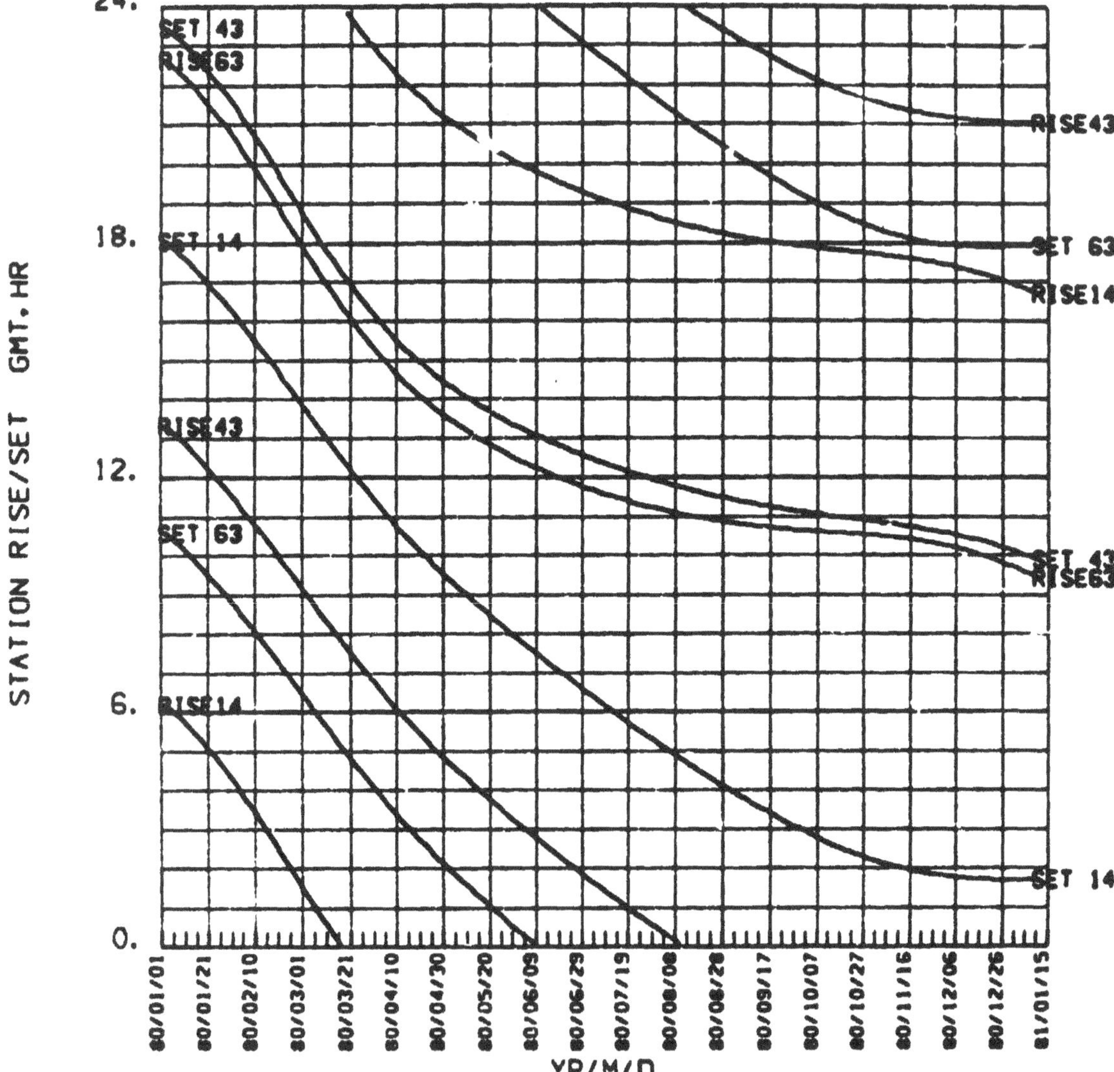

STATION RISE/SET GMT.HR
24.
18.
12.
6.
0.
SET 43
RISE63
SET 14
RISE43
SET 63
RISE14
RISE43
SET 63
RISE14
SET 43
RISE63
SET 14
80/01/01
80/01/21
80/02/10
80/03/01
80/03/21
80/04/10
80/04/30
80/05/20
80/06/09
80/06/29
80/07/19
80/08/08
80/08/28
80/09/17
80/10/07
80/10/27
80/11/16
80/12/06
80/12/26
81/01/15
YR/M/D

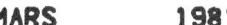
MARS 1981
DECLIN
40.
30.
20.
10.
0.
-10.
-20.
-30.
-40.
81/01/01
81/01/21
81/02/10
81/03/02
81/03/22
81/04/11
81/05/01
81/05/21
81/06/10
81/06/30
81/07/20
81/08/09
81/08/29
81/09/18
81/10/08
81/10/28
81/11/17
81/12/07
81/12/27
82/01/16
YR/M/D

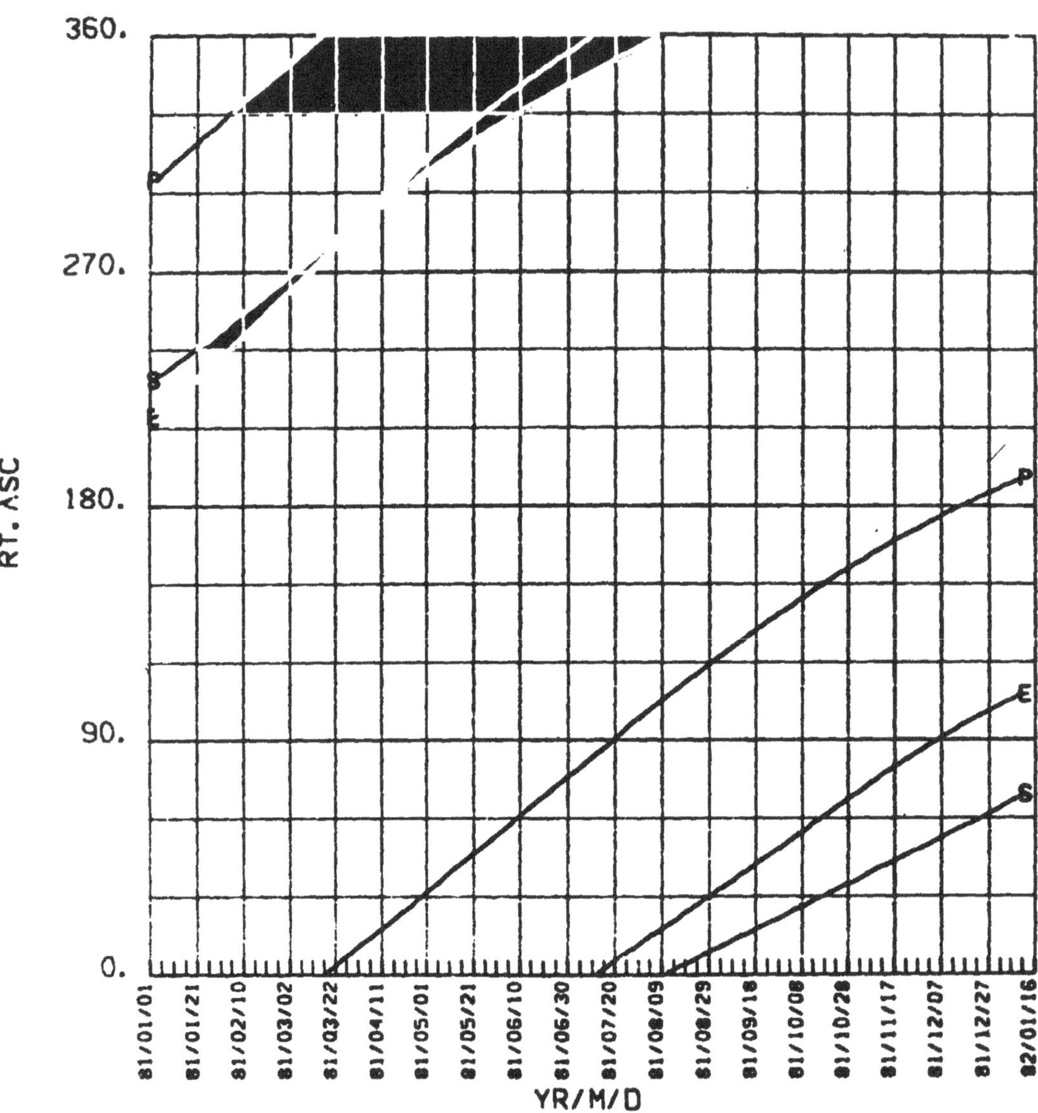

MARS 1981
360.
270.
180.
90.
0.
RT. ASC
P
S
E
P
E
S
81/01/01
81/01/21
81/02/10
81/03/02
81/03/22
81/04/11
81/05/01
81/05/21
81/06/10
81/06/30
81/07/20
81/08/09
81/08/29
81/09/18
81/10/08
81/10/28
81/11/17
81/12/07
81/12/27
82/01/16
YR/M/D

MARS 1981 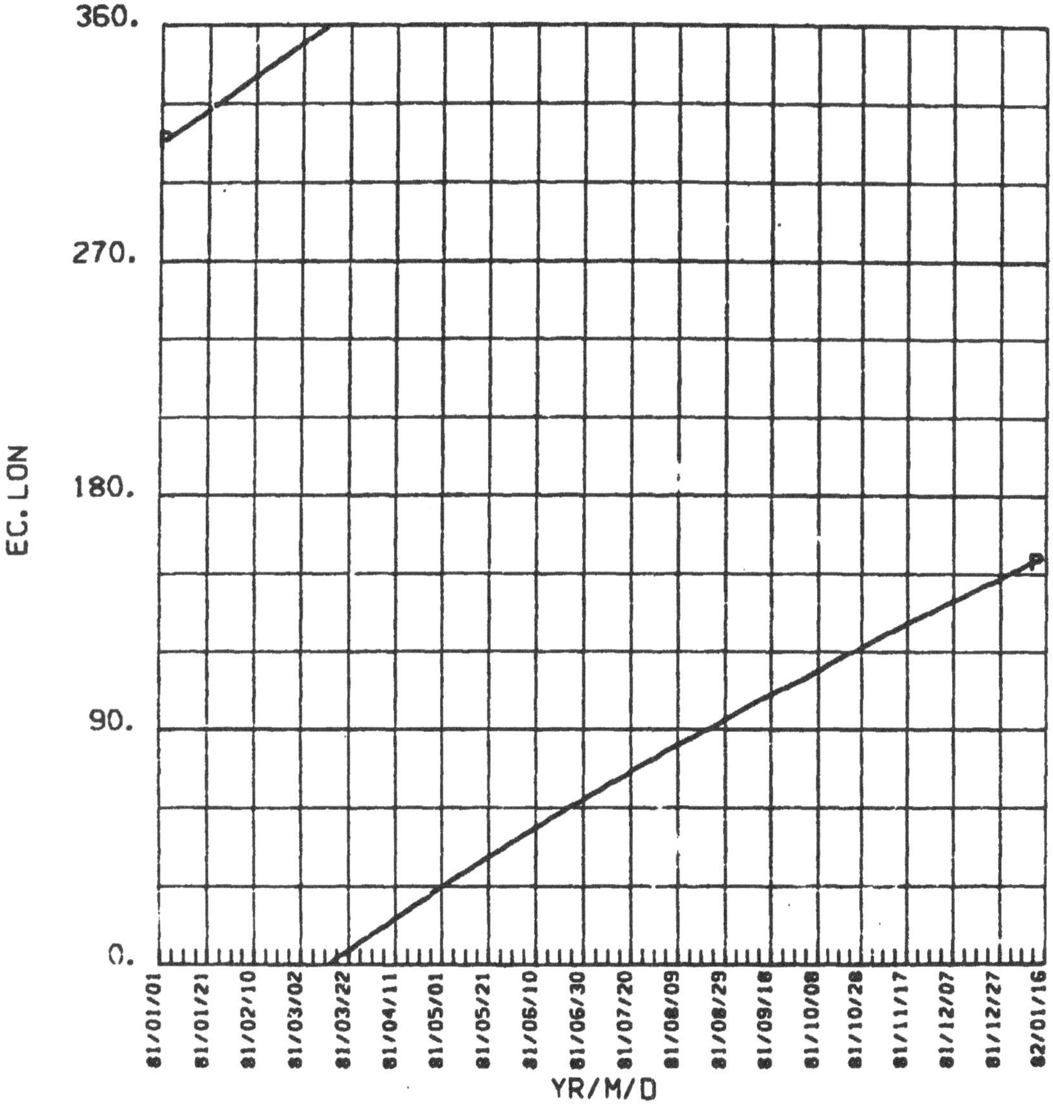
EC.LON
360.
270.
180.
90.
0.
P
P
81/01/01
81/01/21
81/02/10
81/03/02
81/03/22
81/04/11
81/05/01
81/05/21
81/06/10
81/06/30
81/07/20
81/08/09
81/08/29
81/09/18
81/10/08
81/10/28
81/11/17
81/12/07
81/12/27
82/01/16
YR/M/D

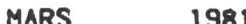

MARS 1981

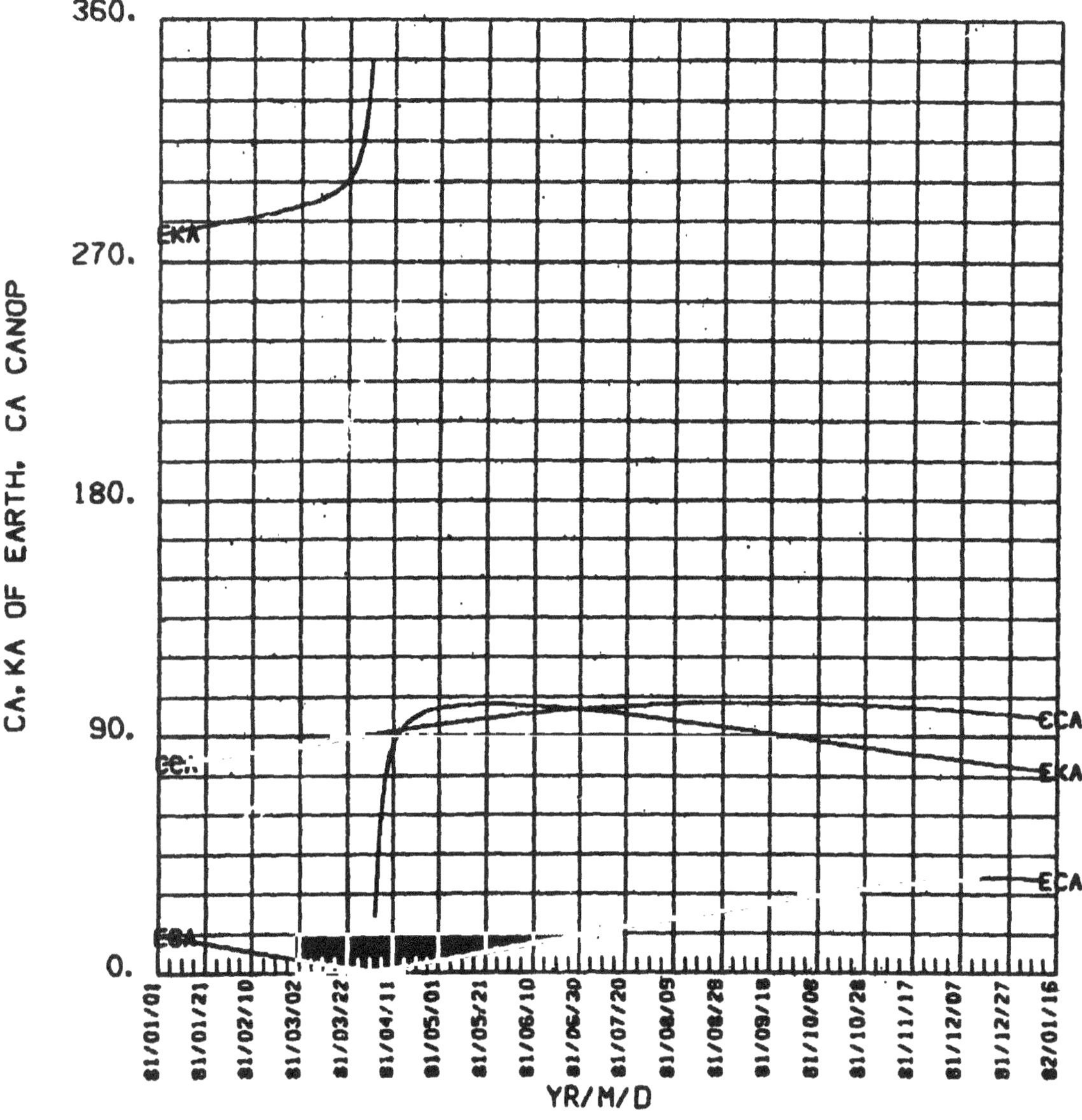

360.
270.
180.
90.
0.
CA, KA OF EARTH, CA CANOP
EKA
ECA
EKA
ECA
81/01/01
81/01/21
81/02/10
81/03/02
81/03/22
81/04/11
81/05/01
81/05/21
81/06/10
81/06/30
81/07/20
81/08/09
81/08/29
81/09/18
81/10/08
81/10/28
81/11/17
81/12/07
81/12/27
82/01/16
YR/M/D

MARS 1981 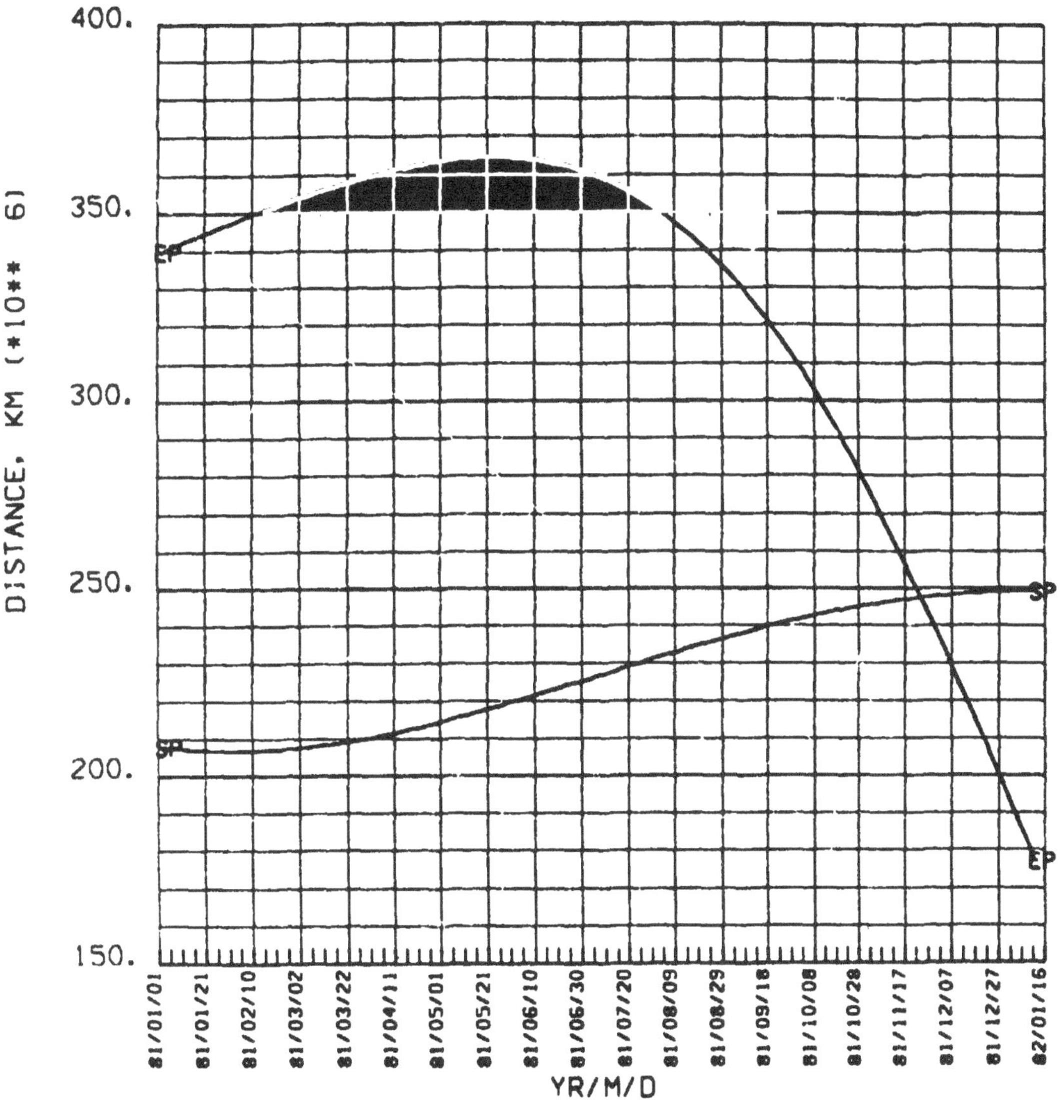
DISTANCE, KM (*10** 6)
400.
350.
300.
250.
200.
150.
EP
SP
SP
EP
81/01/01
81/01/21
81/02/10
81/03/02
81/03/22
81/04/11
81/05/01
81/05/21
81/06/10
81/06/30
81/07/20
81/08/09
81/08/29
81/09/18
81/10/08
81/10/28
81/11/17
81/12/07
81/12/27
82/01/16
YR/M/D

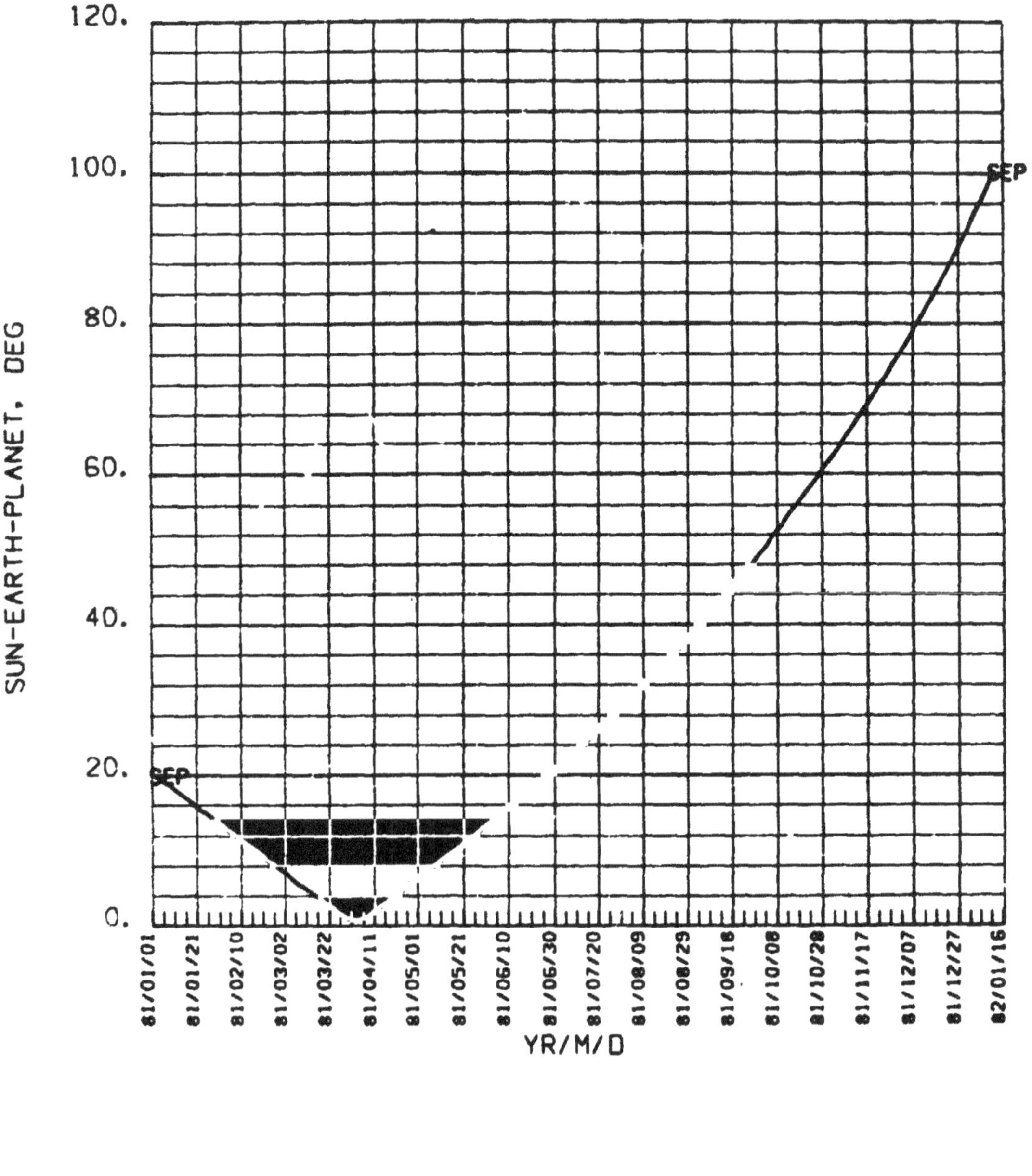

SUN-EARTH-PLANET, DEG

120.
100.
80.
60.
40.
20.
0.

SEP
SEP

81/01/01
81/01/21
81/02/10
81/03/02
81/03/22
81/04/11
81/05/01
81/05/21
81/06/10
81/06/30
81/07/20
81/08/09
81/08/29
81/09/18
81/10/08
81/10/28
81/11/17
81/12/07
81/12/27
82/01/16

YR/M/D

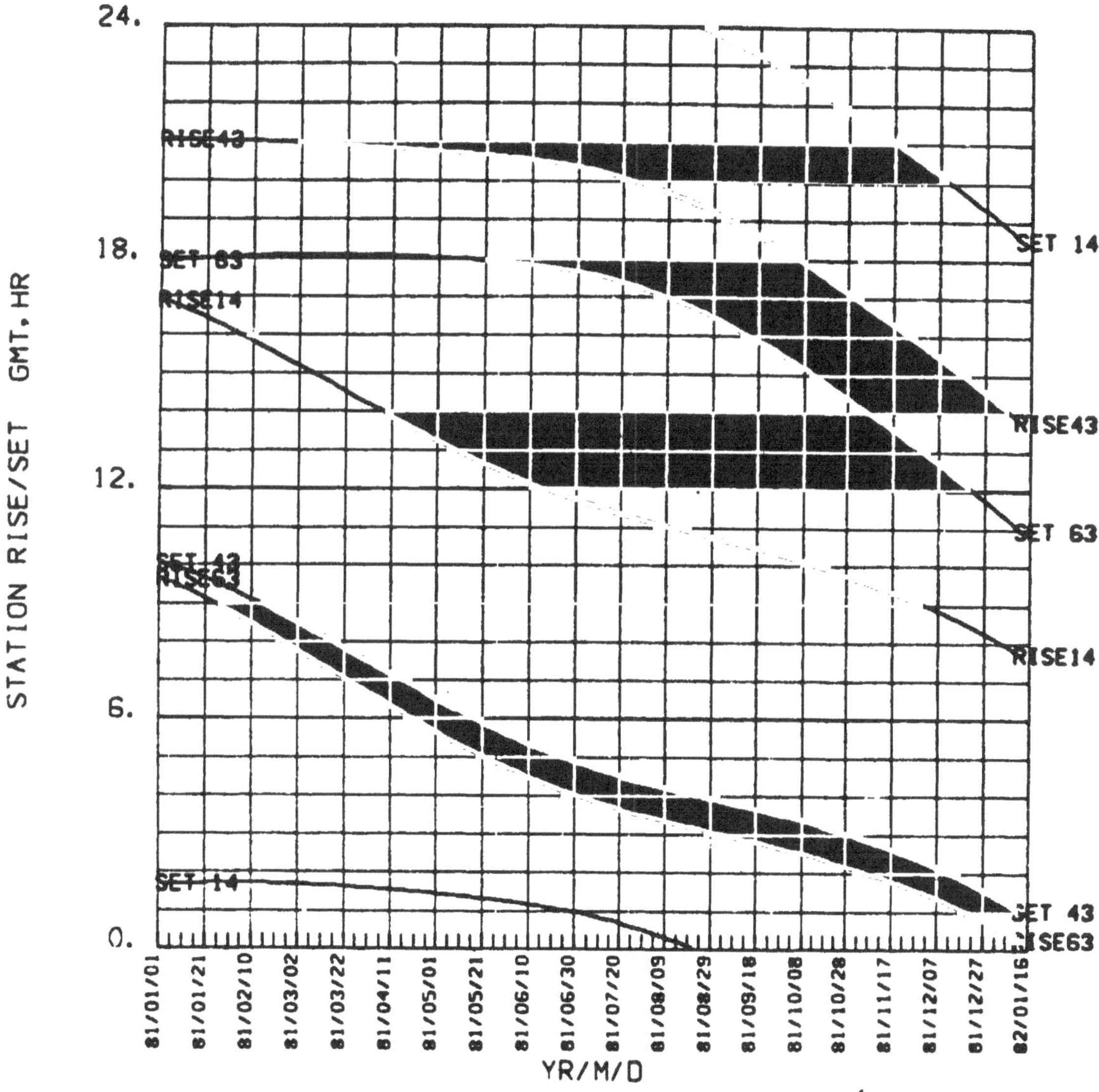

MARS 1981
24.
18.
12.
6.
0.
STATION RISE/SET GMT, HR
RISE43
SET 63
RISE14
SET 43
RISE63
SET 14
SET 14
RISE43
SET 63
RISE14
SET 43
RISE63
81/01/01
81/01/21
81/02/10
81/03/02
81/03/22
81/04/11
81/05/01
81/05/21
81/06/10
81/06/30
81/07/20
81/08/09
81/08/29
81/09/18
81/10/08
81/10/28
81/11/17
81/12/07
81/12/27
82/01/16
YR/M/D

MARS 1982

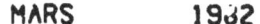
DECLIN
40.
30.
20.
10.
0.
-10.
-20.
-30.
-40.
82/01/01
82/01/21
82/02/10
82/03/02
82/03/22
82/04/11
82/05/01
82/05/21
82/06/10
82/06/30
82/07/20
82/08/09
82/08/29
82/09/18
82/10/08
82/10/28
82/11/17
82/12/07
82/12/27
83/01/16
YR/M/D
D
E
S

RT. ASC

360.
270.
180.
90.
0.

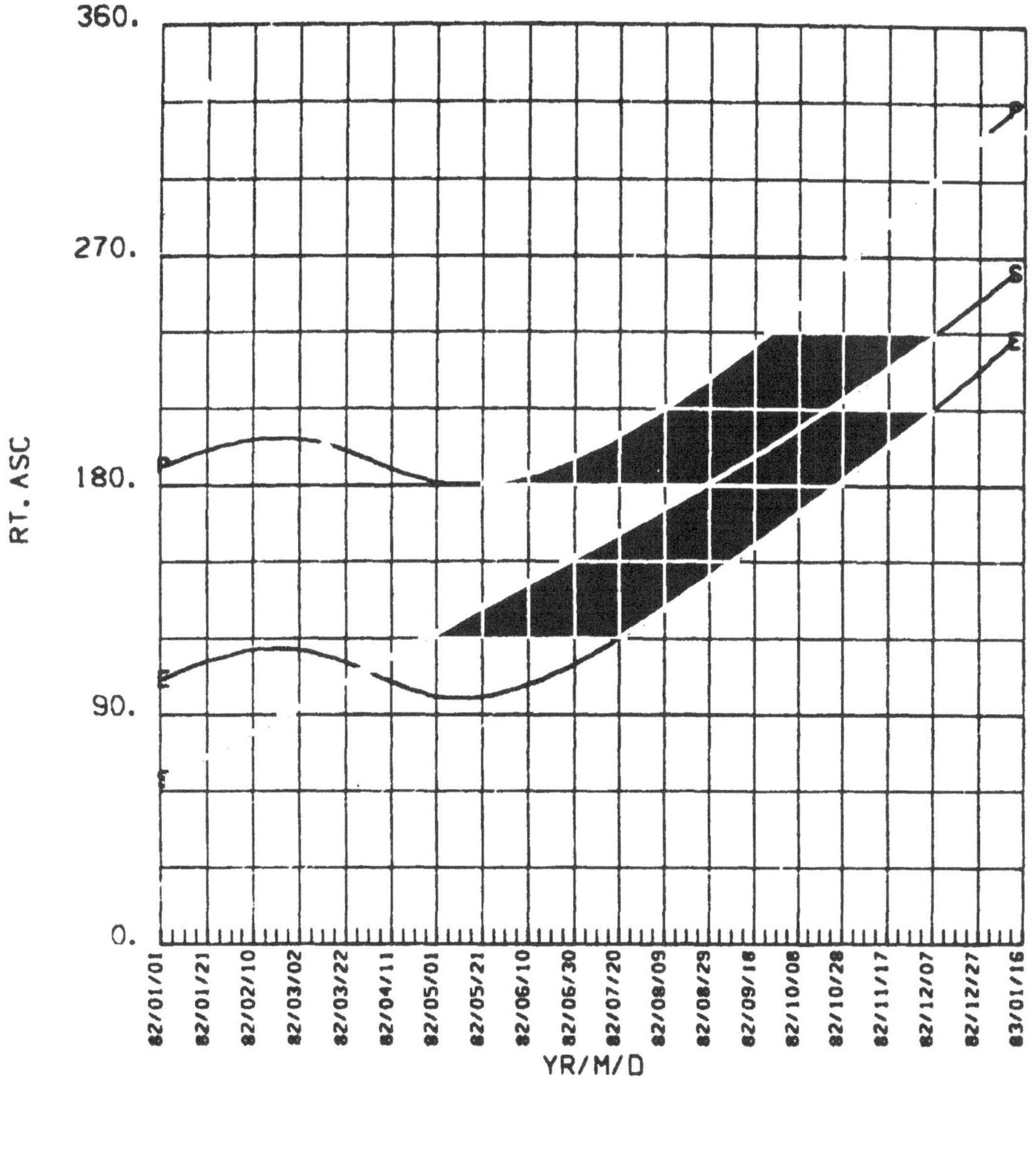

82/01/01 82/01/21 82/02/10 82/03/02 82/03/22 82/04/11 82/05/01 82/05/21 82/06/10 82/06/30 82/07/20 82/08/09 82/08/29 82/09/18 82/10/08 82/10/28 82/11/17 82/12/07 82/12/27 83/01/16

YR/M/D

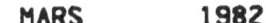

EC. LON

360.
270.
180.
90.
0.

82/01/01
82/01/21
82/02/10
82/03/02
82/03/22
82/04/11
82/05/01
82/05/21
82/06/10
82/06/30
82/07/20
82/08/09
82/08/29
82/09/18
82/10/08
82/10/28
82/11/17
82/12/07
82/12/27
83/01/16

YR/M/D

MARS 1982

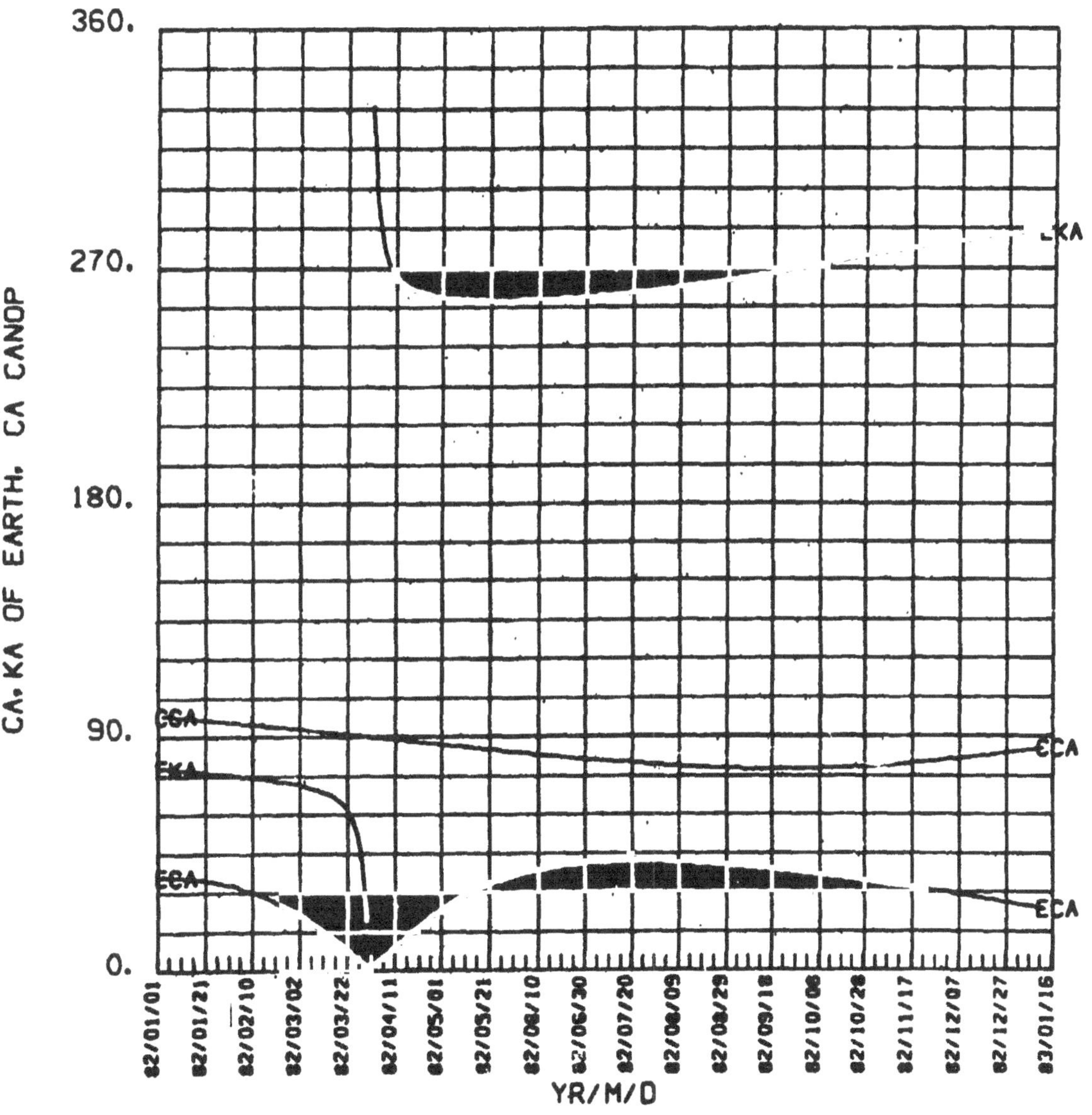

360.
270.
180.
90.
0.
CA, KA OF EARTH, CA CANOP
EKA
ECA
CCA
YR/M/D
82/01/01
82/01/21
82/02/10
82/03/02
82/03/22
82/04/11
82/05/01
82/05/21
82/06/10
82/06/30
82/07/20
82/08/09
82/08/29
82/09/18
82/10/08
82/10/28
82/11/17
82/12/07
82/12/27
83/01/16

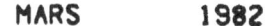

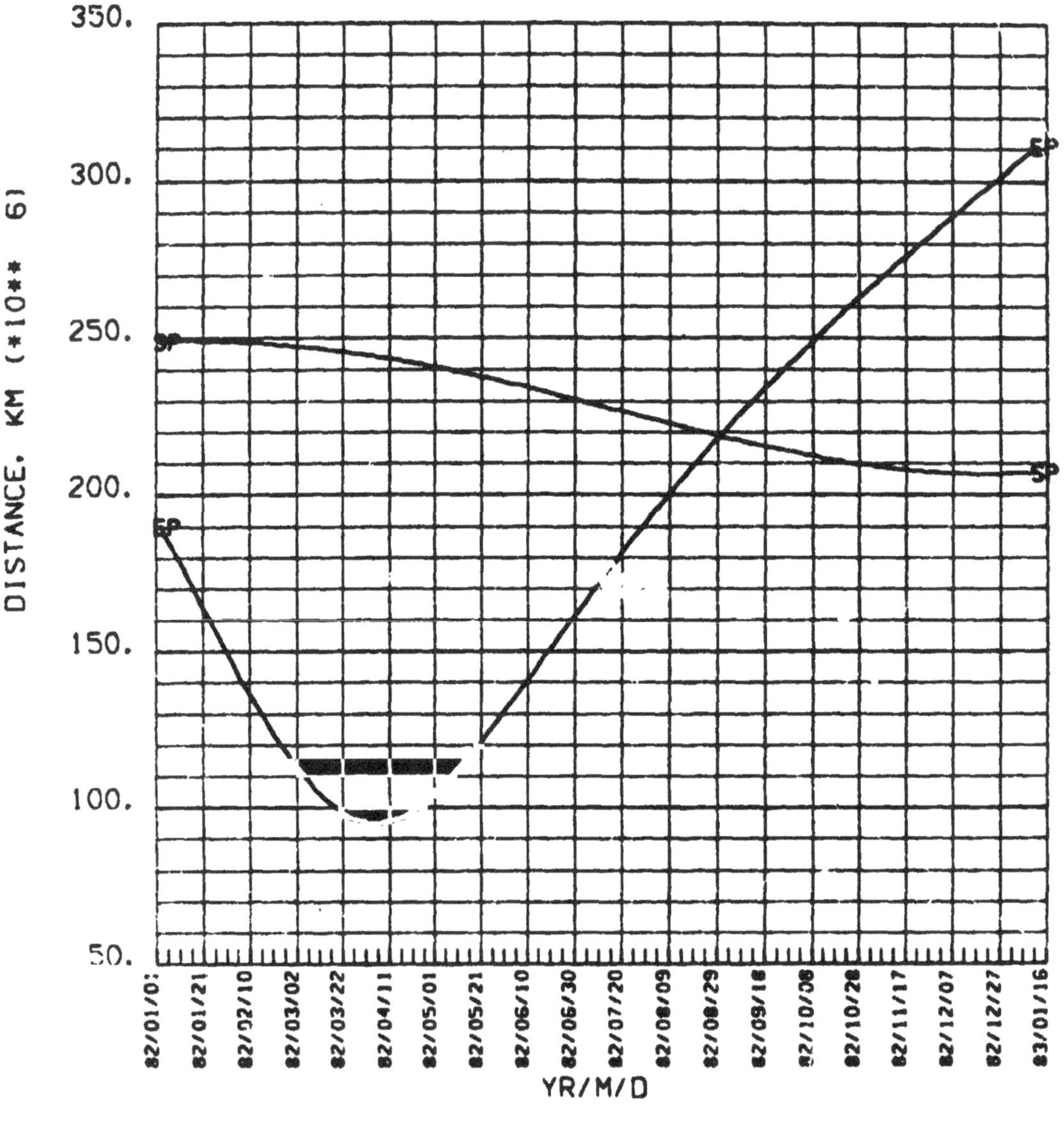

350.
300.
250.
200.
150.
100.
50.

DISTANCE, KM (*10** 6)

SP
EP
EP
SP

82/01/01
82/01/21
82/02/10
82/03/02
82/03/22
82/04/11
82/05/01
82/05/21
82/06/10
82/06/30
82/07/20
82/08/09
82/08/29
82/09/18
82/10/08
82/10/28
82/11/17
82/12/07
82/12/27
83/01/16

YR/M/D

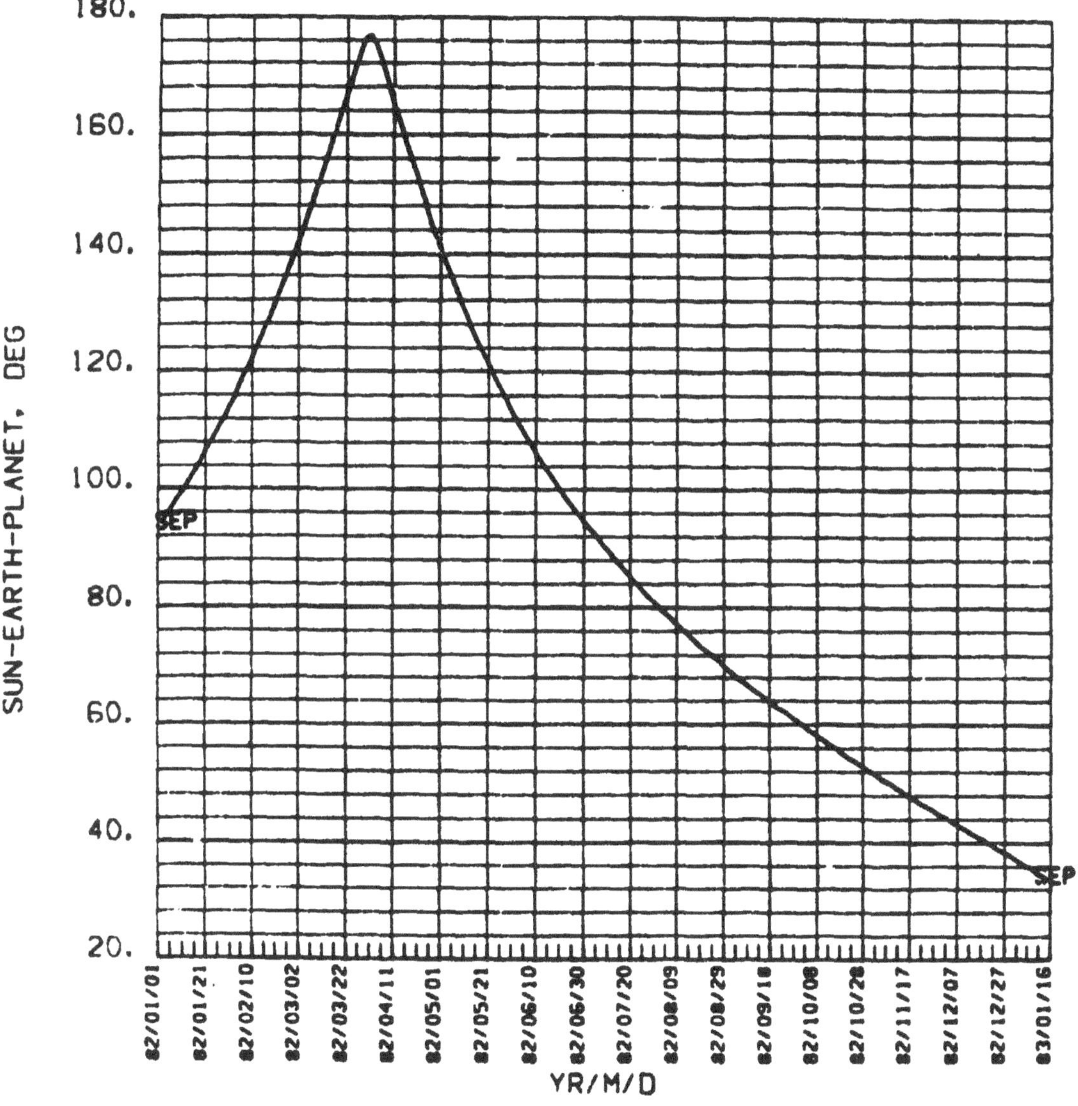

MARS 1982
SUN-EARTH-PLANET, DEG
180.
160.
140.
120.
100.
80.
60.
40.
20.
SEP
SEP
82/01/01
82/01/21
82/02/10
82/03/02
82/03/22
82/04/11
82/05/01
82/05/21
82/06/10
82/06/30
82/07/20
82/08/09
82/08/29
82/09/18
82/10/08
82/10/28
82/11/17
82/12/07
82/12/27
83/01/16
YR/M/D

MARS 1982

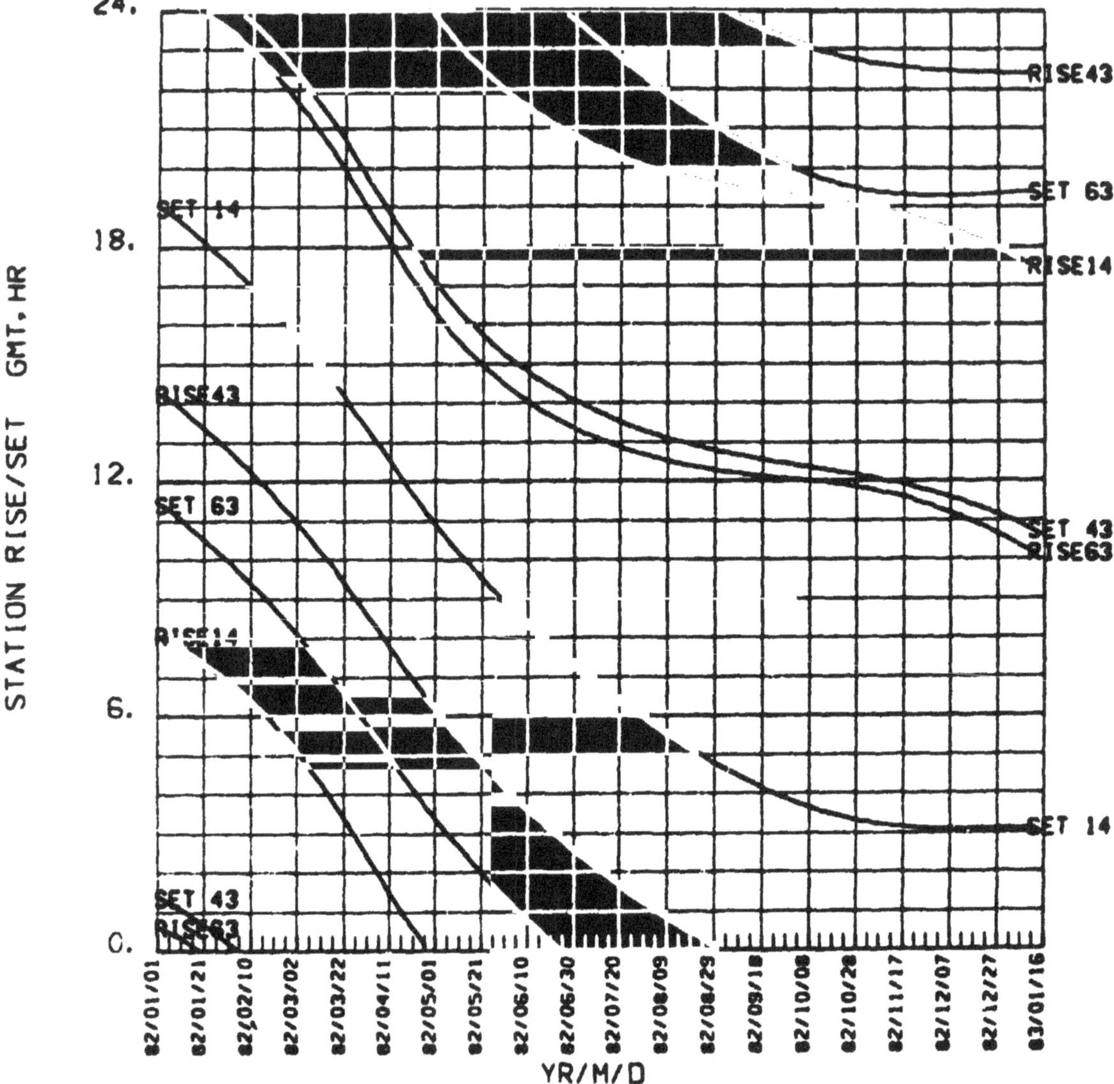

STATION RISE/SET GMT.HR
24.
18.
12.
6.
0.
SET 14
RISE43
SET 63
RISE14
SET 43
RISE63
RISE43
SET 63
RISE14
SET 43
RISE63
SET 14
82/01/01
82/01/21
82/02/10
82/03/02
82/03/22
82/04/11
82/05/01
82/05/21
82/06/10
82/06/30
82/07/20
82/08/09
82/08/29
82/09/18
82/10/08
82/10/28
82/11/17
82/12/07
82/12/27
83/01/16
YR/M/D

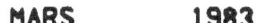

DECLIN

40.
30.
20.
10.
0.
-10.
-20.
-30.
-40.

83/01/01 83/01/21 83/02/10 83/03/02 83/03/22 83/04/11 83/05/01 83/05/21 83/06/10 83/06/30 83/07/20 83/08/09 83/08/29 83/09/18 83/10/08 83/10/28 83/11/17 83/12/07 83/12/27 84/01/16

YR/M/D

MARS 1983

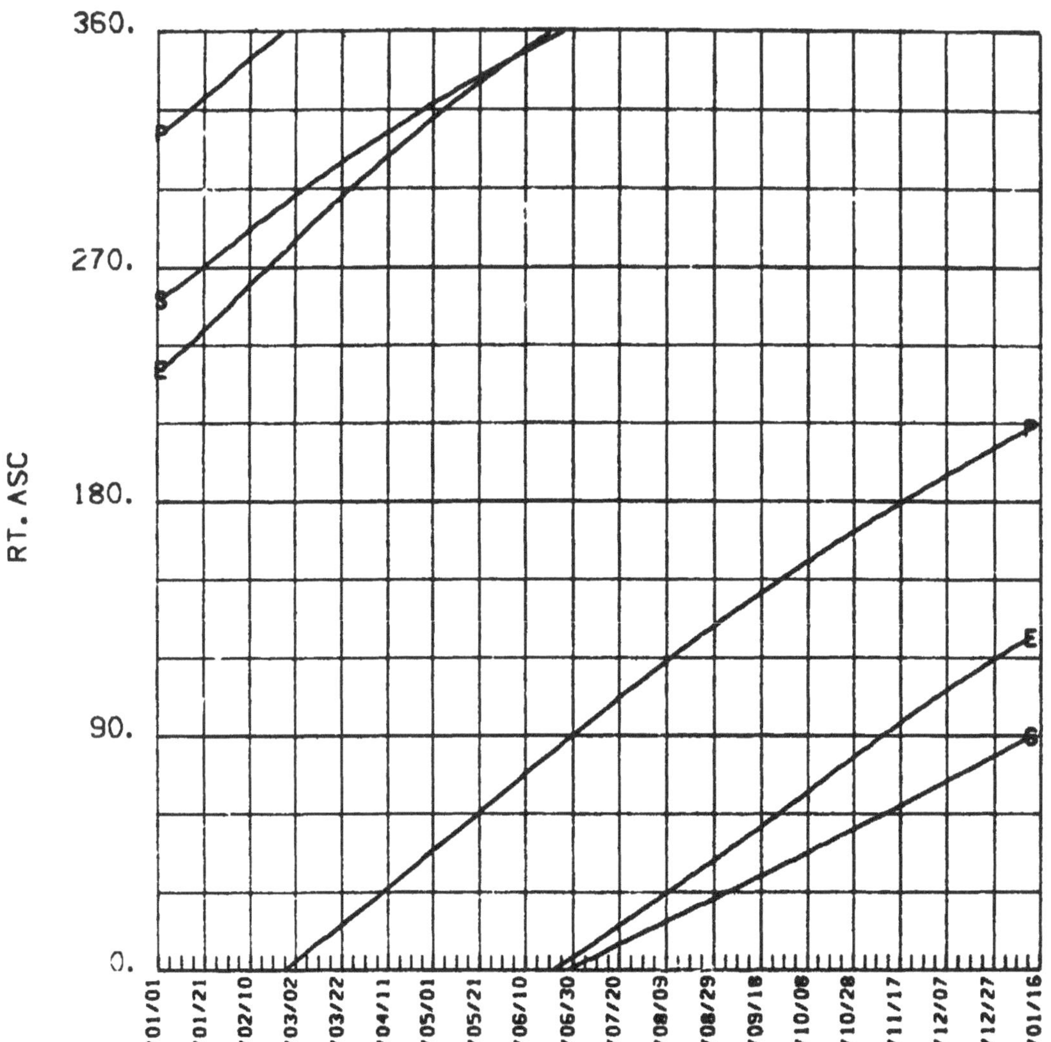

360.
270.
180.
90.
0.
RT. ASC
83/01/01
83/01/21
83/02/10
83/03/02
83/03/22
83/04/11
83/05/01
83/05/21
83/06/10
83/06/30
83/07/20
83/08/09
83/08/29
83/09/18
83/10/08
83/10/28
83/11/17
83/12/07
83/12/27
84/01/16
YR/M/D
P
S
E

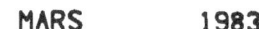
MARS 1983
360.
270.
180.
90.
0.
EC. LON
83/01/01
83/01/21
83/02/10
83/03/02
83/03/22
83/04/11
83/05/01
83/05/21
83/06/10
83/06/30
83/07/20
83/08/09
83/08/23
83/09/18
83/10/08
83/10/28
83/11/17
83/12/07
83/12/27
84/01/16
YR/M/D

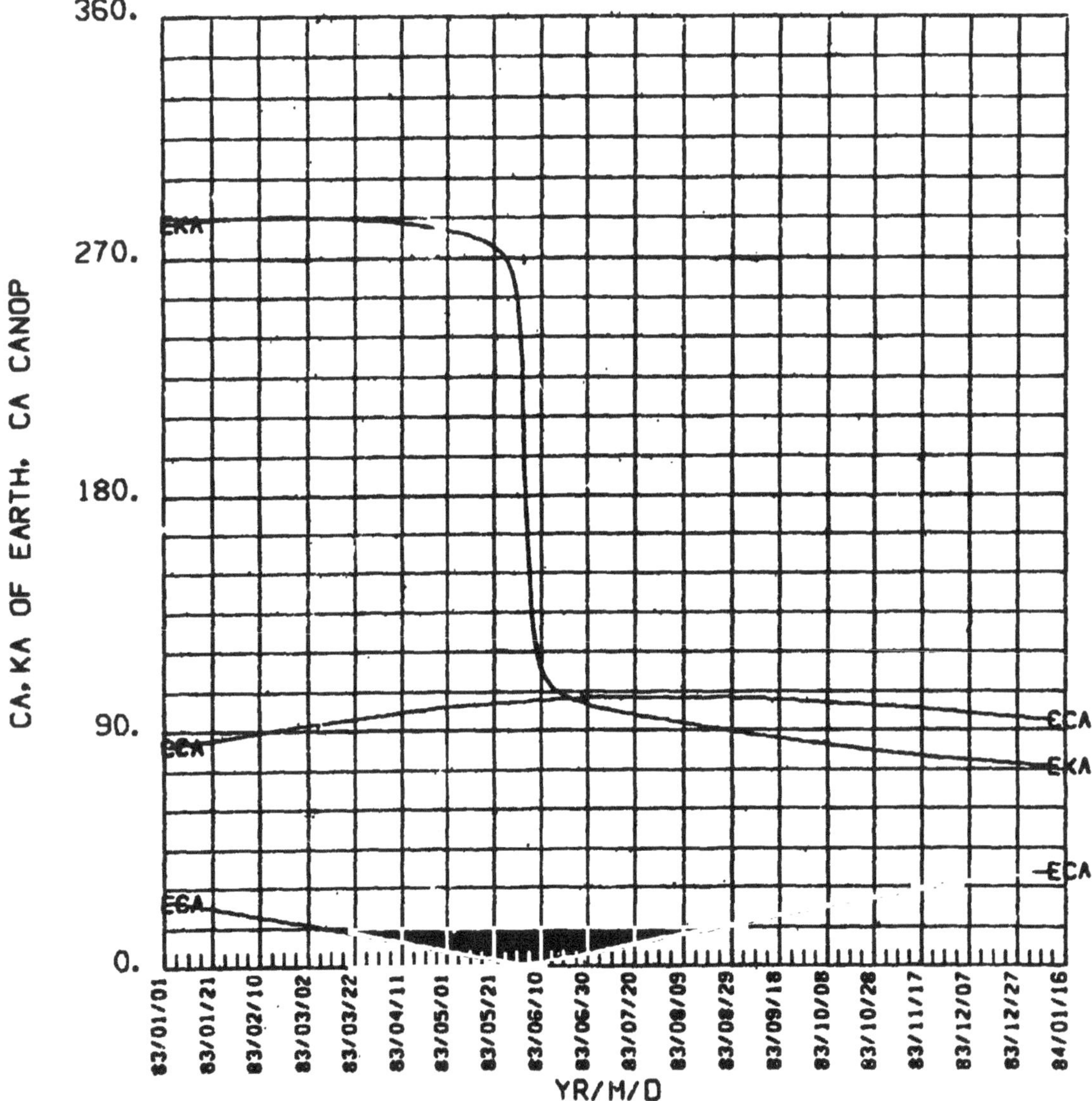

MARS 1983
360.
270.
180.
90.
0.
CA, KA OF EARTH, CA CANOP
EKA
ECA
ECA
ECA
EKA
ECA
83/01/01
83/01/21
83/02/10
83/03/02
83/03/22
83/04/11
83/05/01
83/05/21
83/06/10
83/06/30
83/07/20
83/08/09
83/08/29
83/09/18
83/10/08
83/10/28
83/11/17
83/12/07
83/12/27
84/01/16
YR/M/D

MARS 1983

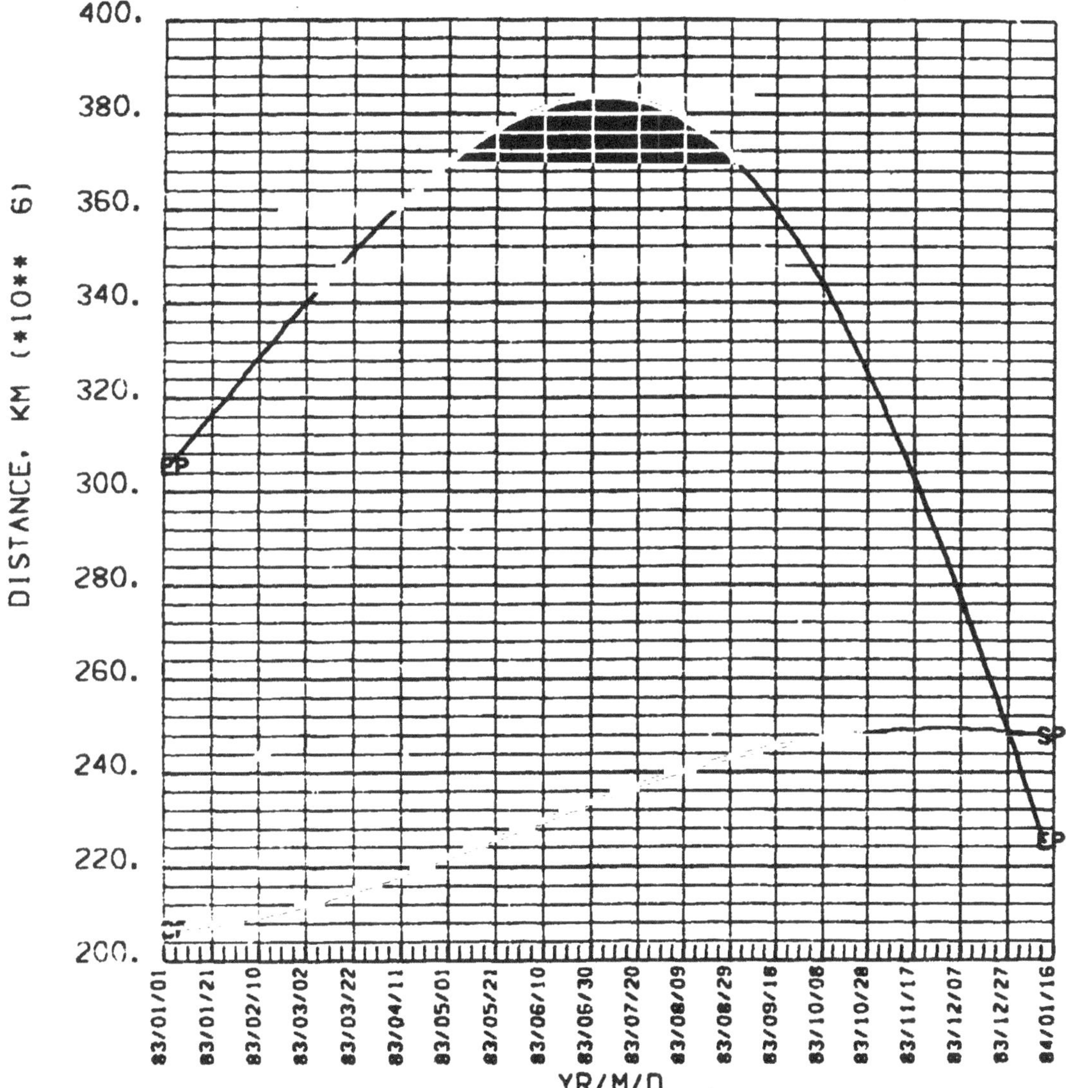

400.
380.
360.
340.
320.
300.
280.
260.
240.
220.
200.
DISTANCE, KM (*10** 6)
EP
SP
EP
83/01/01
83/01/21
83/02/10
83/03/02
83/03/22
83/04/11
83/05/01
83/05/21
83/06/10
83/06/30
83/07/20
83/08/09
83/08/29
83/09/18
83/10/08
83/10/28
83/11/17
83/12/07
83/12/27
84/01/16
YR/M/D

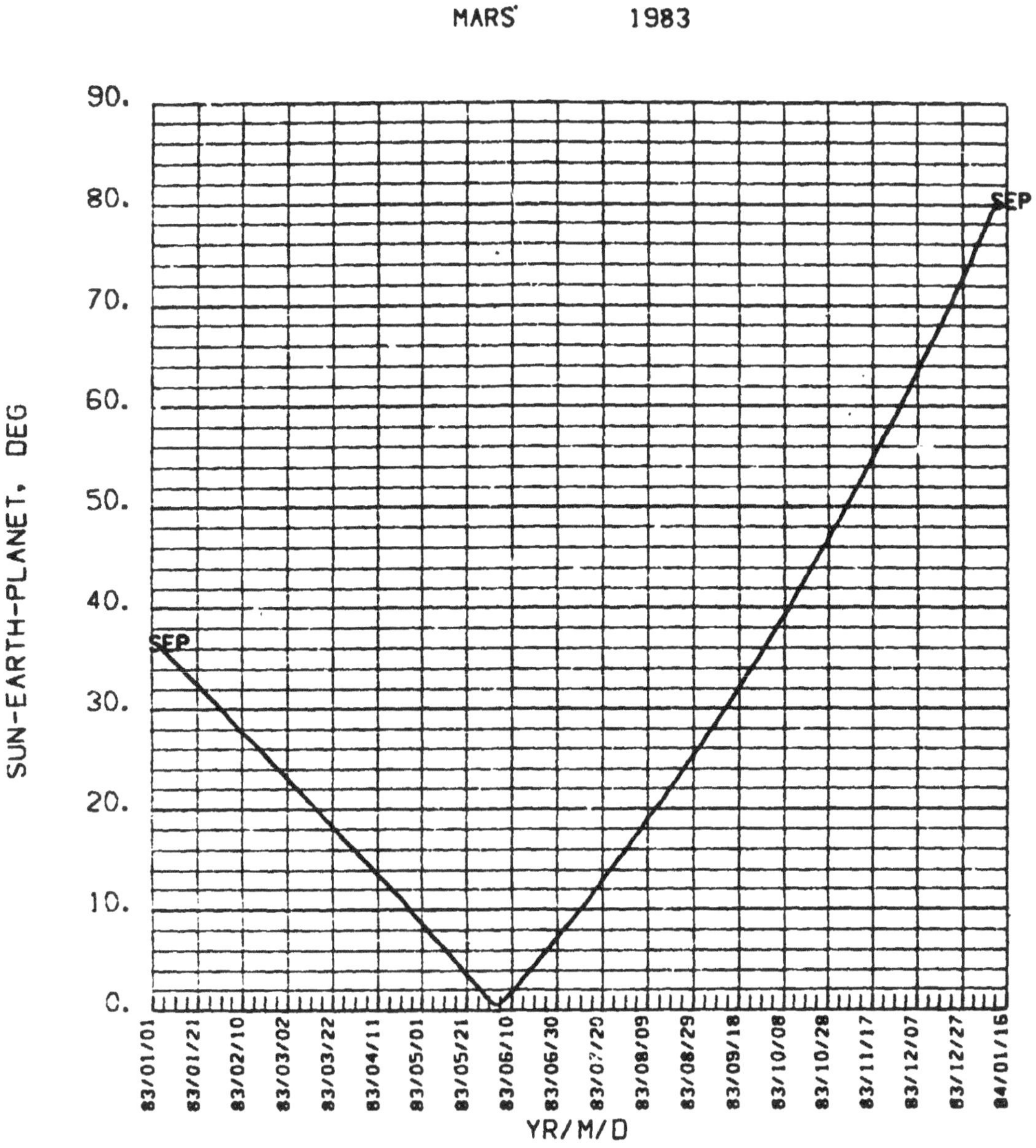

MARS 1983
90.
80.
70.
60.
50.
40.
30.
20.
10.
0.
SUN-EARTH-PLANET, DEG
SEP
SEP
83/01/01
83/01/21
83/02/10
83/03/02
83/03/22
83/04/11
83/05/01
83/05/21
83/06/10
83/06/30
83/07/20
83/08/09
83/08/29
83/09/18
83/10/08
83/10/28
83/11/17
83/12/07
83/12/27
84/01/16
YR/M/D

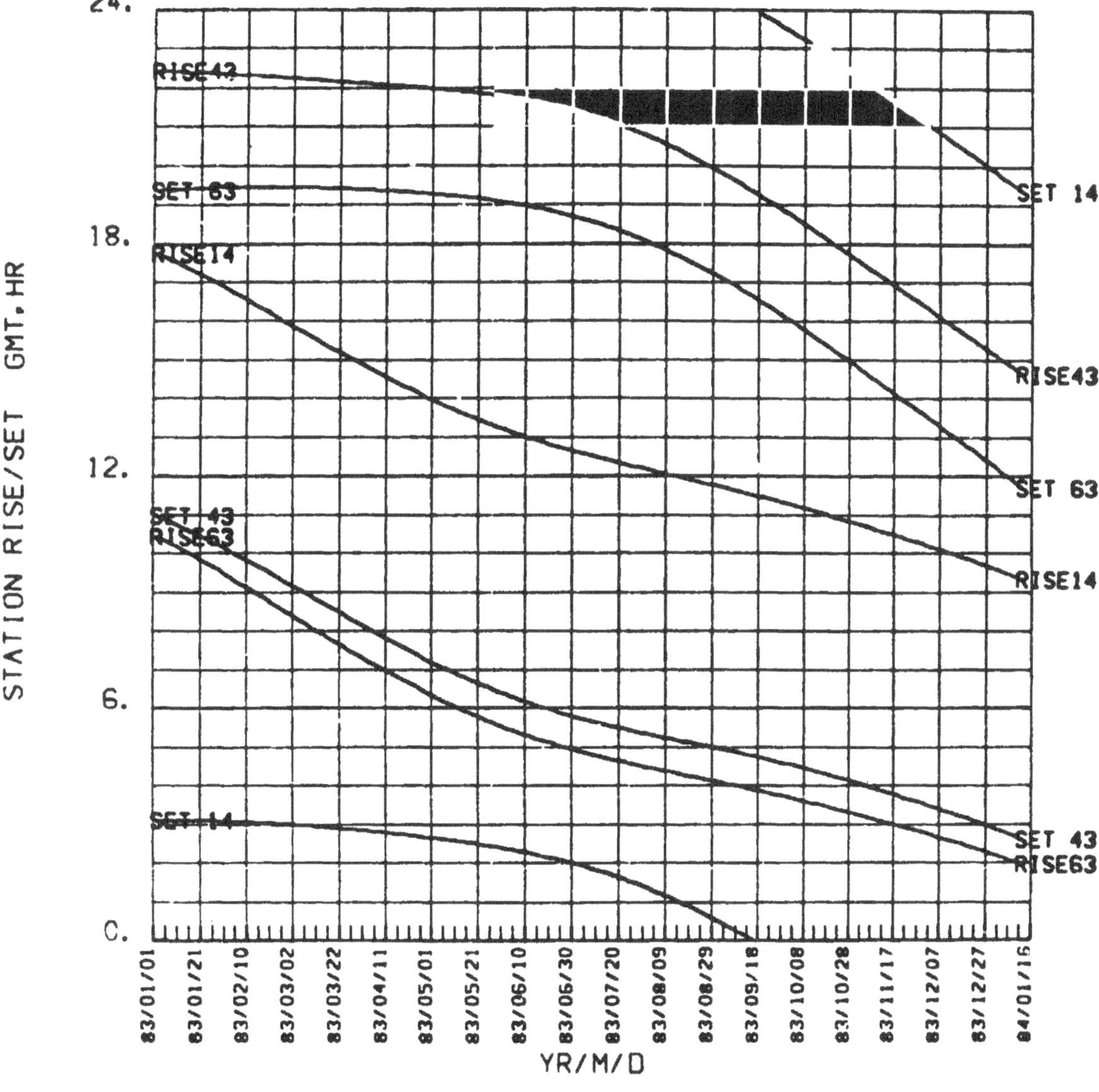

MARS 1983
24.
18.
12.
6.
0.
STATION RISE/SET GMT.HR
RISE43
SET 63
RISE14
SET 43
RISE63
SET 14
SET 14
RISE43
SET 63
RISE14
SET 43
RISE63
83/01/01
83/01/21
83/02/10
83/03/02
83/03/22
83/04/11
83/05/01
83/05/21
83/06/10
83/06/30
83/07/20
83/08/09
83/08/29
83/09/18
83/10/08
83/10/28
83/11/17
83/12/07
83/12/27
84/01/16
YR/M/D

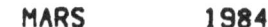

MARS 1984

40.
30.
20.
10.
0.
-10.
-20.
-30.
-40.
DECLIN
S
E
P
84/01/01
84/01/21
84/02/10
84/03/01
84/03/21
84/04/10
84/04/30
84/05/20
84/06/09
84/06/29
84/07/19
84/08/08
84/08/28
84/09/17
84/10/07
84/10/27
84/11/16
84/12/06
84/12/26
85/01/15
YR/M/D

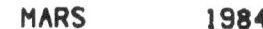

RT. ASC

360.
270.
180.
90.
0.

P
S
E

84/01/01
84/01/21
84/02/10
84/03/01
84/03/21
84/04/10
84/04/30
84/05/20
84/06/09
84/06/29
84/07/19
84/08/08
84/08/28
84/09/17
84/10/07
84/10/27
84/11/16
84/12/06
84/12/26
85/01/15

YR/M/D

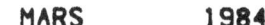
MARS 1984

360.
270.
180.
90.
0.
EC. LON
84/01/01
84/01/21
84/02/10
84/03/01
84/03/21
84/04/10
84/04/30
84/05/20
84/06/09
84/06/29
84/07/19
84/08/08
84/08/28
84/09/17
84/10/07
84/10/27
84/11/16
84/12/06
84/12/26
85/01/15
YR/M/D

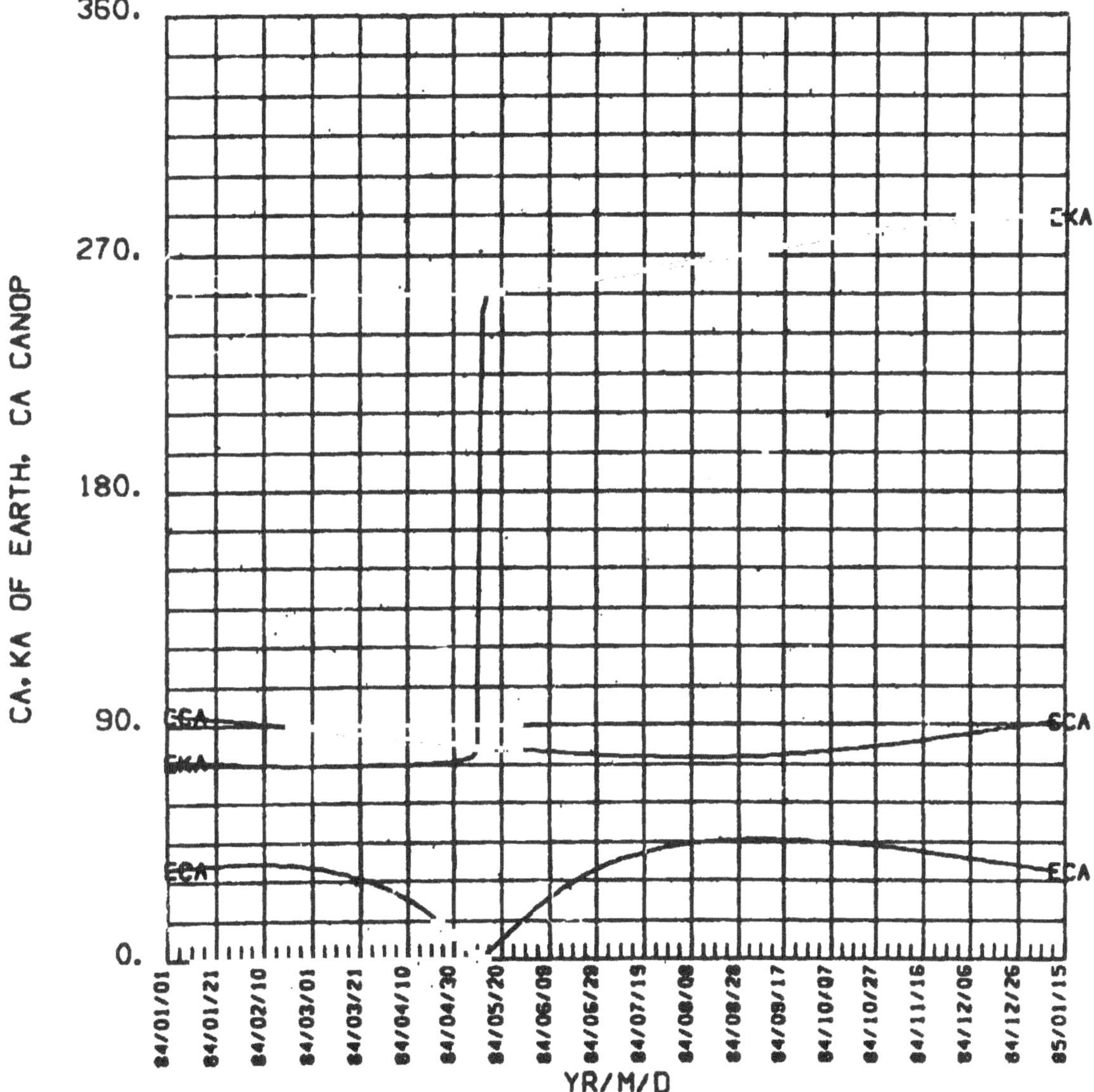
MARS 1984
360.
270.
180.
90.
0.
CA, KA OF EARTH, CA CANOP
EKA
ECA
ECA
EKA
ECA
ECA
84/01/01
84/01/21
84/02/10
84/03/01
84/03/21
84/04/10
84/04/30
84/05/20
84/06/09
84/06/29
84/07/19
84/08/08
84/08/28
84/09/17
84/10/07
84/10/27
84/11/16
84/12/06
84/12/26
85/01/15
YR/M/D

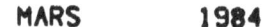
MARS 1984
300.
250.
200.
150.
100.
50.
DISTANCE. KM (*10** 6)
SP
EP
EP
SP
84/01/01
84/01/21
84/02/10
84/03/01
84/03/21
84/04/10
84/04/30
84/05/20
84/06/09
84/06/29
84/07/19
84/08/08
84/08/28
84/09/17
84/10/07
84/10/27
84/11/16
84/12/06
84/12/26
85/01/15
YR/M/D

MARS 1984 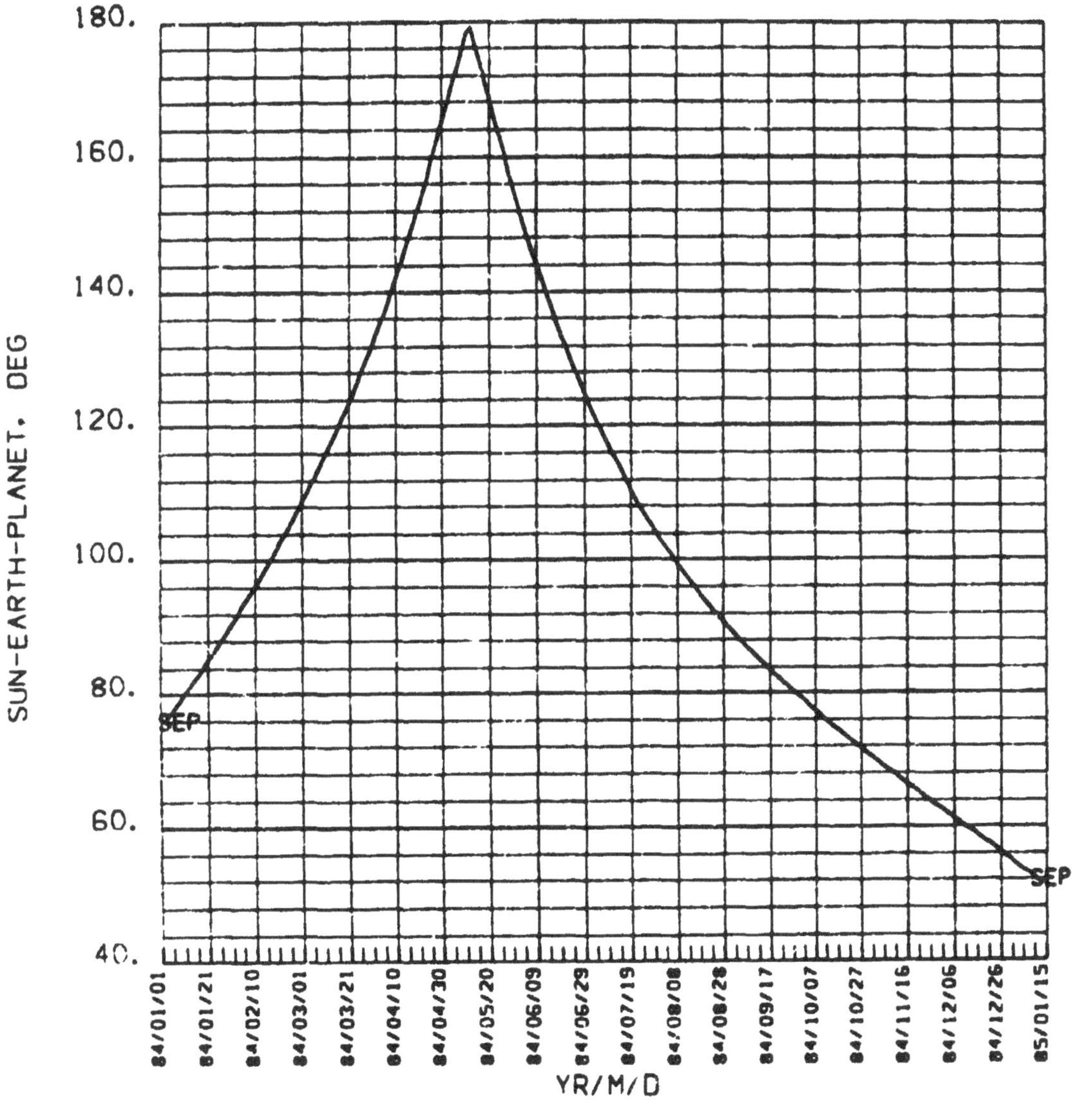
SUN-EARTH-PLANET, DEG
180.
160.
140.
120.
100.
80.
60.
40.
SEP
SEP
84/01/01
84/01/21
84/02/10
84/03/01
84/03/21
84/04/10
84/04/30
84/05/20
84/06/09
84/06/29
84/07/19
84/08/08
84/08/28
84/09/17
84/10/07
84/10/27
84/11/16
84/12/06
84/12/26
85/01/15
YR/M/D

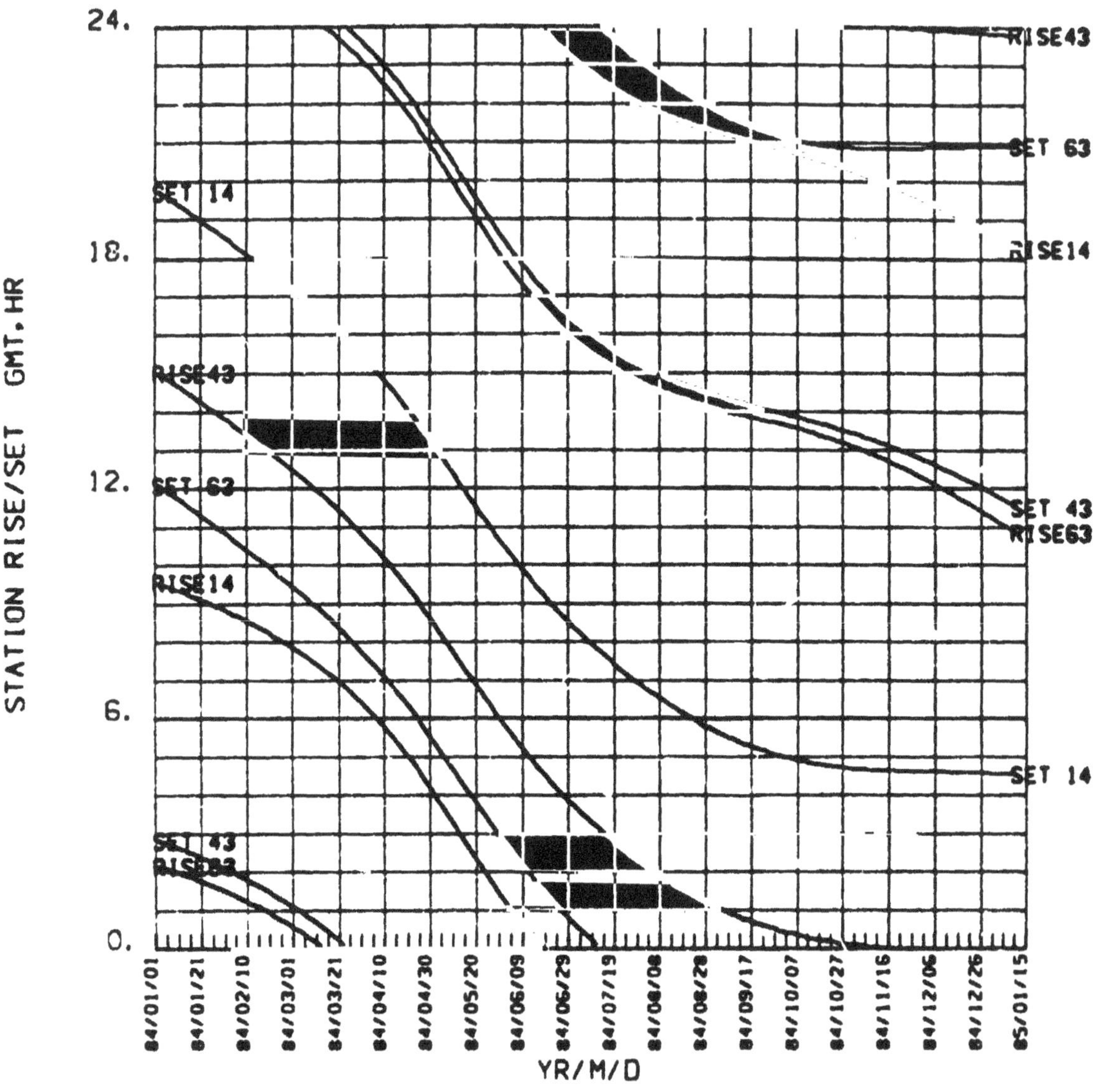

MARS 1984
STATION RISE/SET GMT.HR
24.
18.
12.
6.
0.
SET 14
RISE43
SET 63
RISE14
SET 43
RISE63
RISE43
SET 63
RISE14
SET 43
RISE63
SET 14
84/01/01
84/01/21
84/02/10
84/03/01
84/03/21
84/04/10
84/04/30
84/05/20
84/06/09
84/06/29
84/07/19
84/08/08
84/08/28
84/09/17
84/10/07
84/10/27
84/11/16
84/12/06
84/12/26
85/01/15
YR/M/D

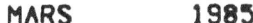
MARS 1985

40.
30.
20.
10.
0.
-10.
-20.
-30.
-40.
DECLIN
S
E
P
85/01/01
85/01/21
85/02/10
85/03/02
85/03/22
85/04/11
85/05/01
85/05/21
85/06/10
85/06/30
85/07/20
85/08/09
85/08/29
85/09/18
85/10/08
85/10/28
85/11/17
85/12/07
85/12/27
86/01/16
YR/M/D

360.
270.
180.
90.
0.

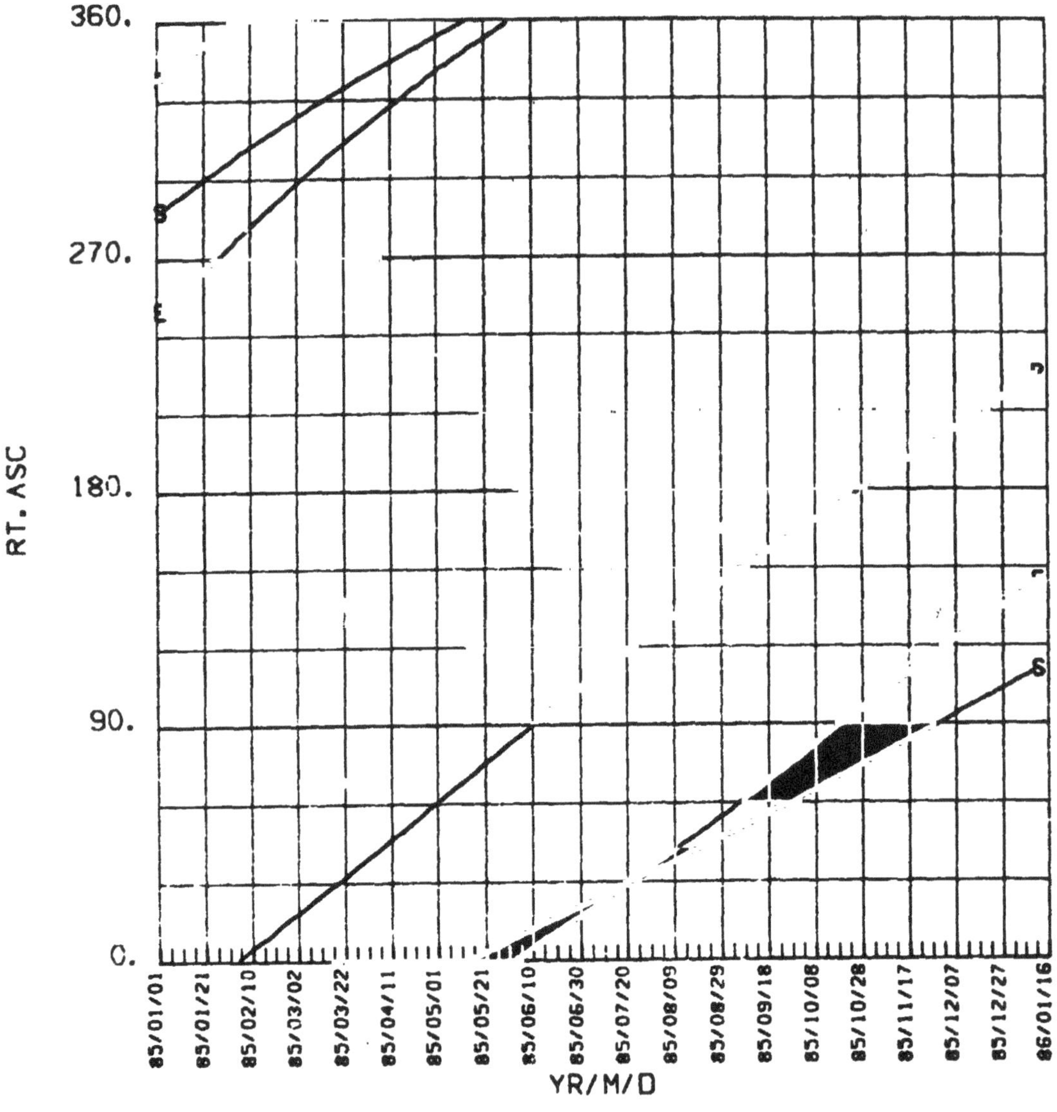

RT. ASC

85/01/01 85/01/21 85/02/10 85/03/02 85/03/22 85/04/11 85/05/01 85/05/21 85/06/10 85/06/30 85/07/20 85/08/09 85/08/29 85/09/18 85/10/08 85/10/28 85/11/17 85/12/07 85/12/27 86/01/16

YR/M/D

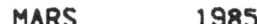

MARS 1985
360.
270.
180.
90.
0.
EC. LON
85/01/01
85/01/21
85/02/10
85/03/02
85/03/22
85/04/11
85/05/01
85/05/21
85/06/10
85/06/30
85/07/20
85/08/09
85/08/29
85/09/18
85/10/08
85/10/28
85/11/17
85/12/07
85/12/27
86/01/16
YR/M/D

MARS 1985

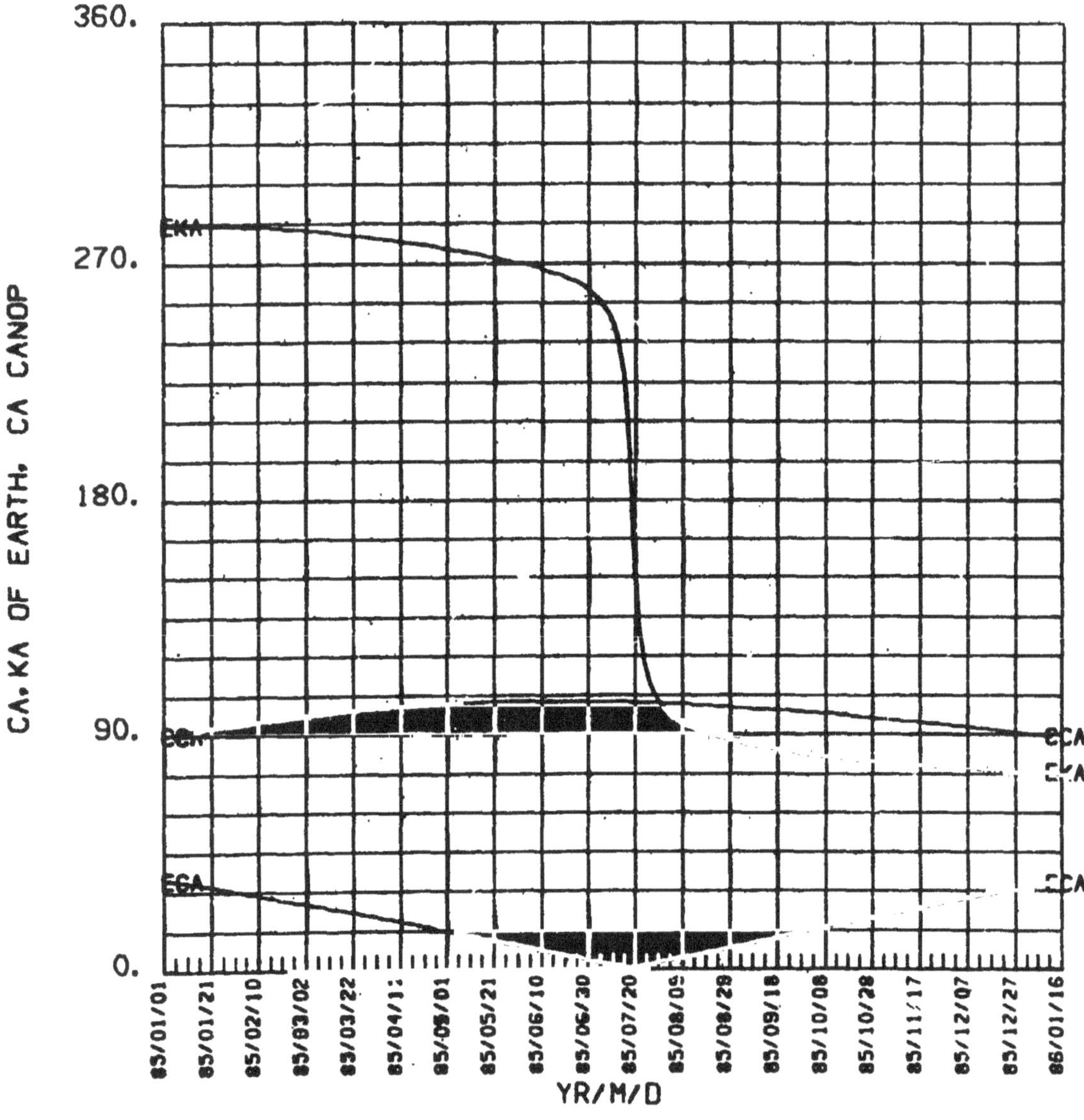

360.
270.
180.
90.
0.
CA, KA OF EARTH, CA CANOP
EKA
ECA
YR/M/D
85/01/01
85/01/21
85/02/10
85/03/02
85/03/22
85/05/01
85/05/21
85/06/10
85/06/30
85/07/20
85/08/29
85/09/18
85/10/08
85/10/28
85/12/07
85/12/27
86/01/16

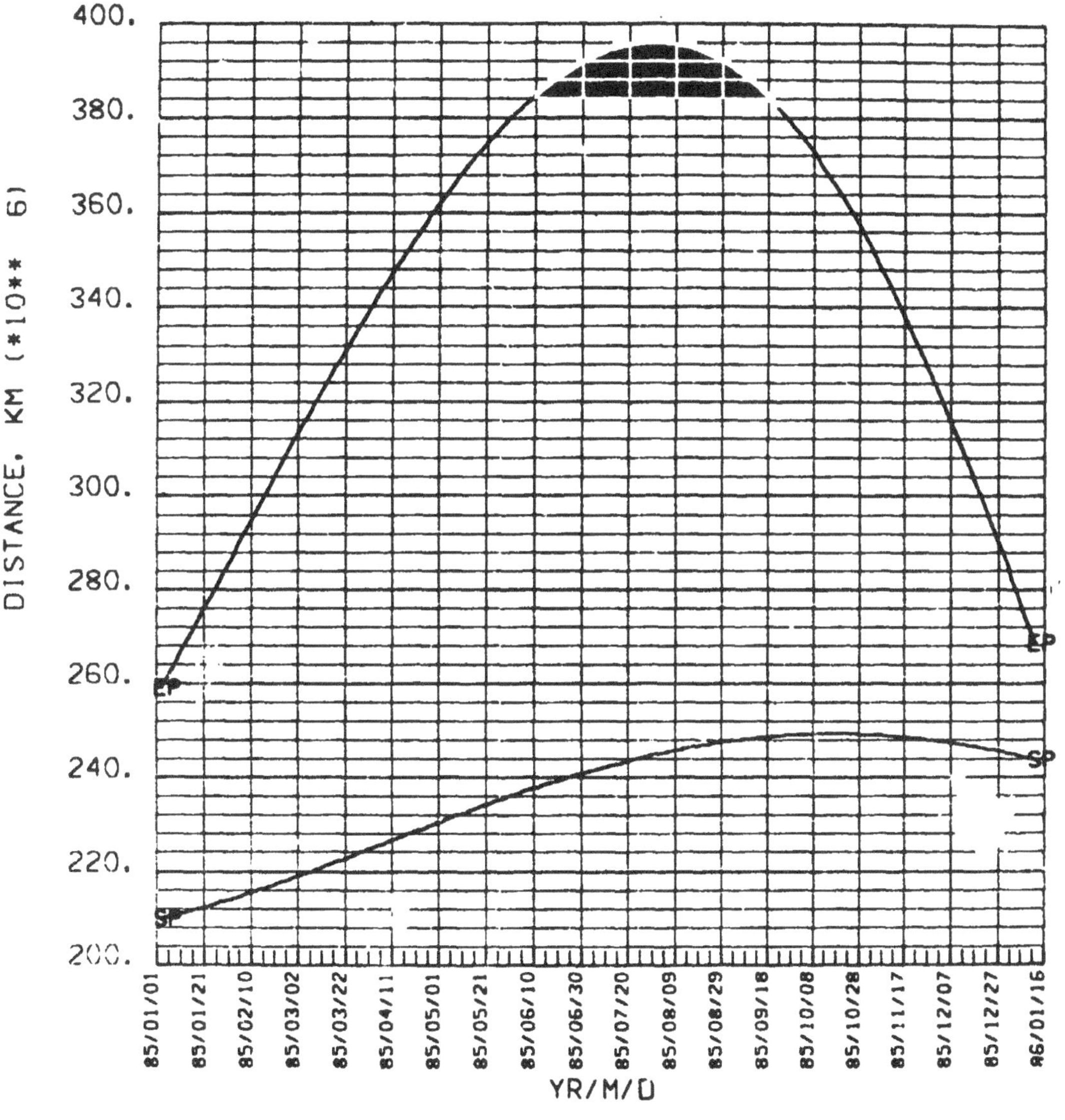
MARS 1985
DISTANCE, KM (*10** 6)
400.
380.
360.
340.
320.
300.
280.
260.
240.
220.
200.
EP
SP
85/01/01
85/01/21
85/02/10
85/03/02
85/03/22
85/04/11
85/05/01
85/05/21
85/06/10
85/06/30
85/07/20
85/08/09
85/08/29
85/09/18
85/10/08
85/10/28
85/11/17
85/12/07
85/12/27
86/01/16
YR/M/D

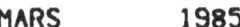

SUN-EARTH-PLANET. DEG

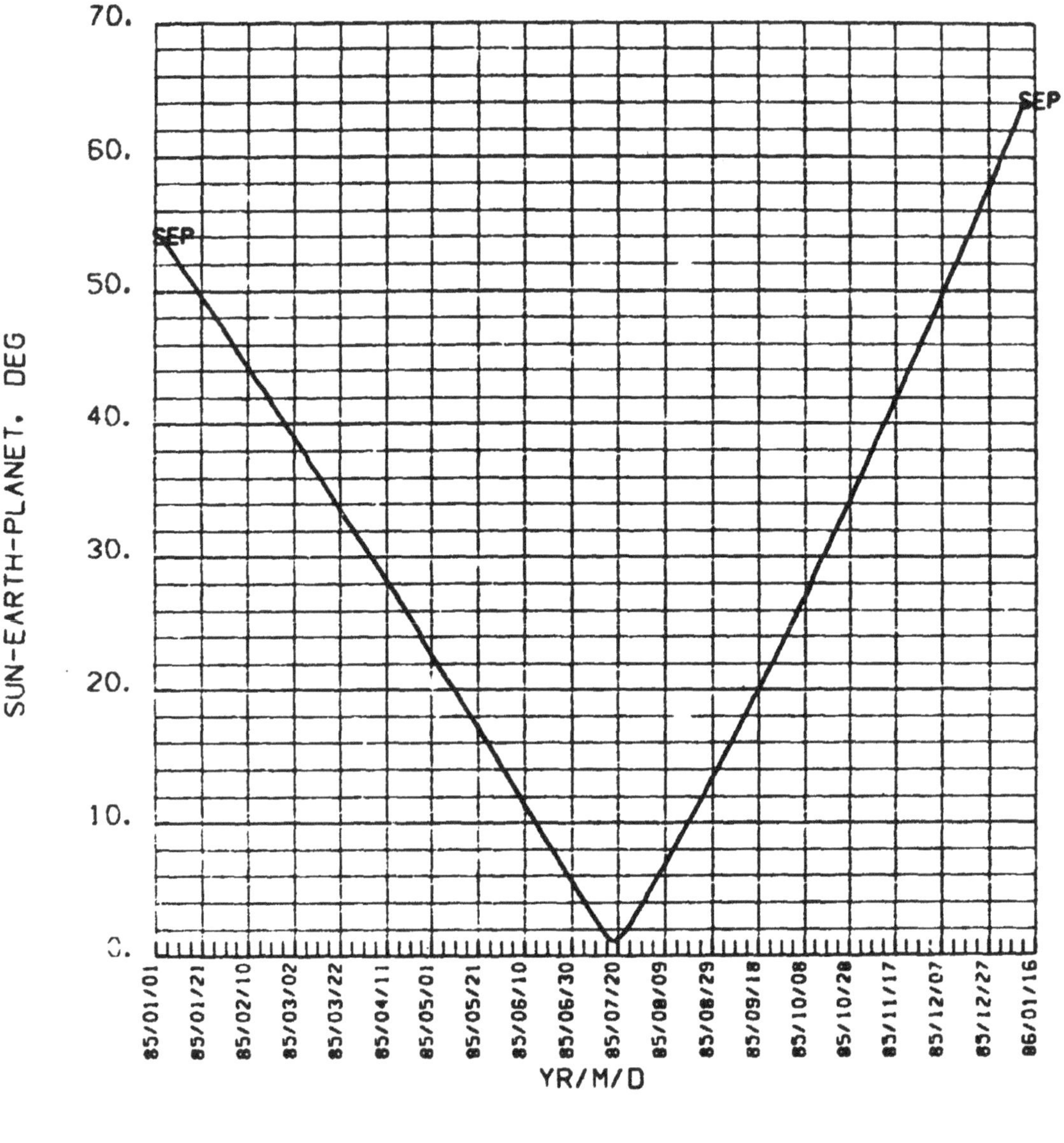

70.
60.
50.
40.
30.
20.
10.
0.

SEP
SEP

85/01/01 85/01/21 85/02/10 85/03/02 85/03/22 85/04/11 85/05/01 85/05/21 85/06/10 85/06/30 85/07/20 85/08/09 85/08/29 85/09/18 85/10/08 85/10/28 85/11/17 85/12/07 85/12/27 86/01/16

YR/M/D

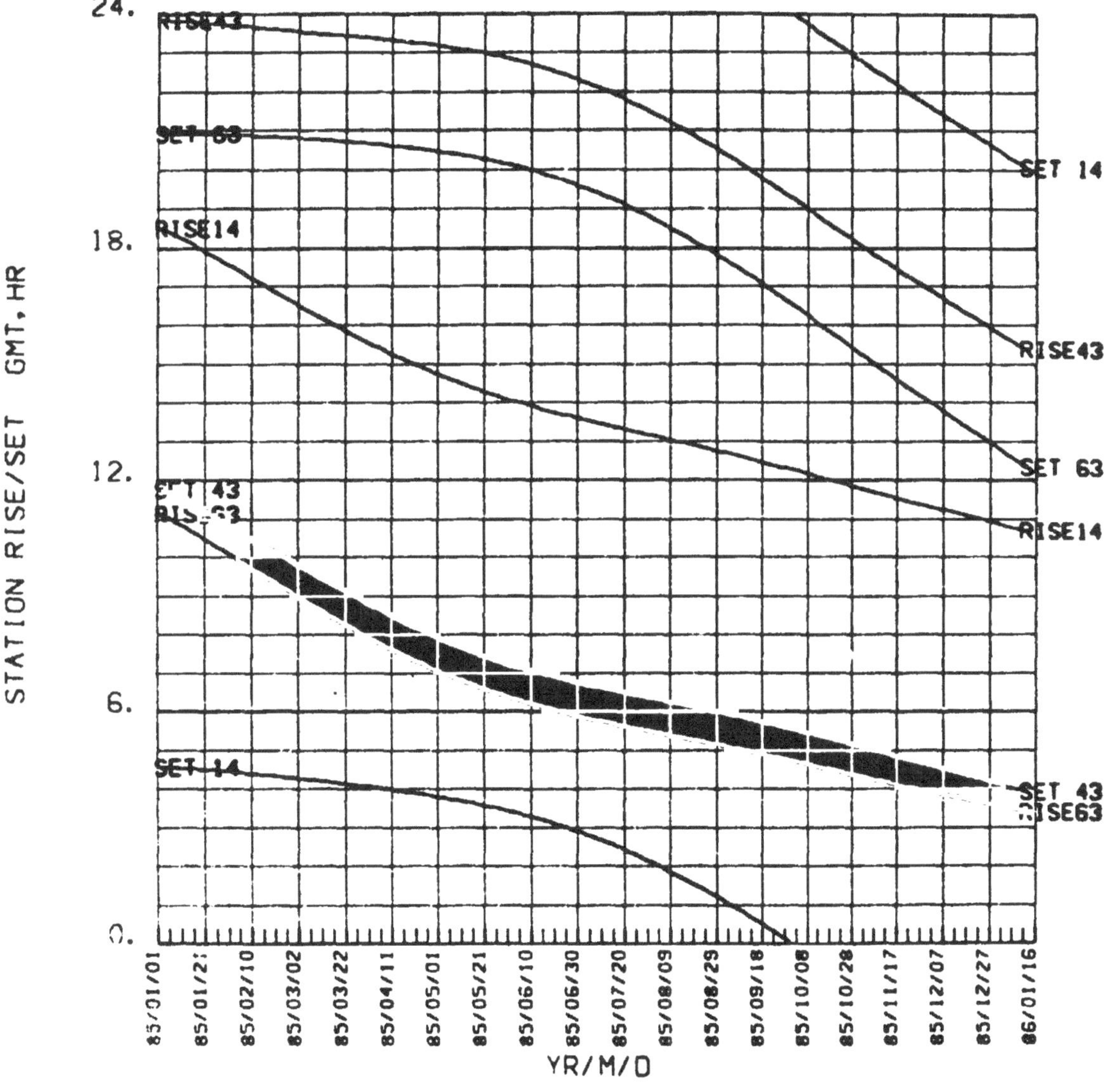
MARS 1985
STATION RISE/SET GMT, HR
YR/M/D
24.
18.
12.
6.
0.
RISE43
SET 63
RISE14
SET 43
RISE63
SET 14
SET 14
RISE43
SET 63
RISE14
SET 43
RISE63
85/01/01
85/02/10
85/03/02
85/03/22
85/04/11
85/05/01
85/05/21
85/06/10
85/06/30
85/07/20
85/08/09
85/08/29
85/09/18
85/10/08
85/10/28
85/11/17
85/12/07
85/12/27
86/01/16

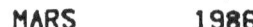
MARS 1986

40.
30.
20.
10.
0.
-10.
-20.
-30.
-40.
DECLIN
S
E
R
D
S
E
86/01/01
86/01/21
86/02/10
86/03/02
86/03/22
86/04/11
86/05/01
86/05/21
86/06/10
86/06/30
86/07/20
86/08/09
86/08/29
86/09/18
86/10/08
86/10/28
86/11/17
86/12/07
86/12/27
87/01/16
YR/M/D

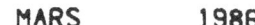

MARS 1986
360.
270.
180.
90.
0.
RT. ASC
S
E
P
E
E
P
86/01/01
86/01/21
86/02/10
86/03/02
86/03/22
86/04/11
86/05/01
86/05/21
86/06/10
86/06/30
86/07/20
86/08/09
86/08/29
86/09/18
86/10/08
86/10/28
86/11/17
86/12/07
86/12/27
87/01/16
YR/M/D

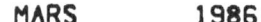
MARS 1986

360.
270.
180.
90.
0.
EC. LON
86/01/01
86/01/21
86/02/10
86/03/02
86/03/22
86/04/11
86/05/01
86/05/21
86/06/10
86/06/30
86/07/20
86/08/09
86/08/29
86/09/18
86/10/08
86/10/28
86/11/17
86/12/07
86/12/27
87/01/16
YR/M/D

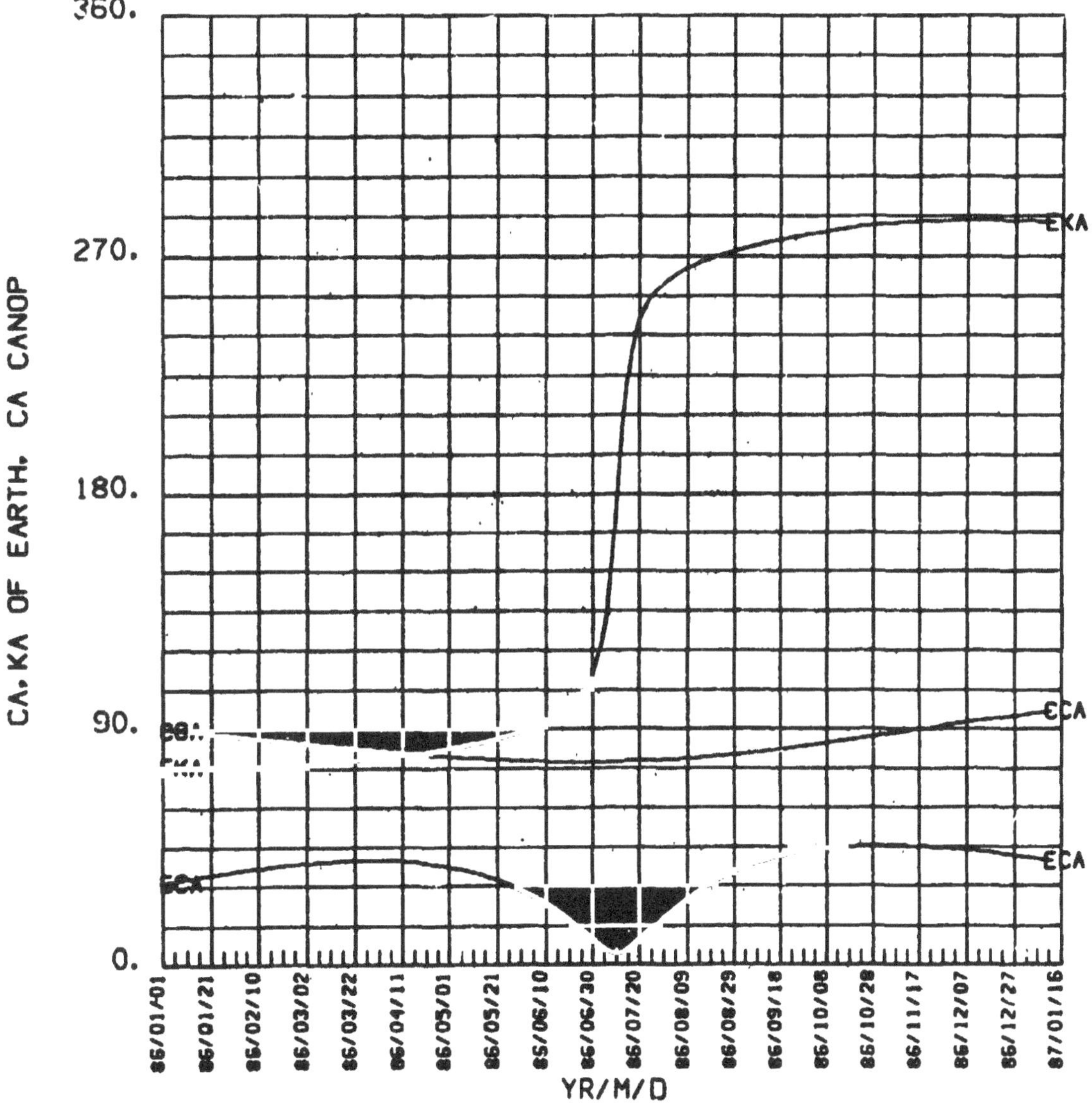

MARS 1986
CA, KA OF EARTH, CA CANOP
360.
270.
180.
90.
0.
EKA
ECA
ECA
ECA
86/01/01
86/01/21
86/02/10
86/03/02
86/03/22
86/04/11
86/05/01
86/05/21
86/06/10
86/06/30
86/07/20
86/08/09
86/08/29
86/09/18
86/10/08
86/10/28
86/11/17
86/12/07
86/12/27
87/01/16
YR/M/D

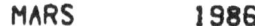
MARS 1986

300.
250.
200.
150.
100.
50.
DISTANCE, KM (*10** 6)
SP
SR
86/01/01
86/01/21
86/02/10
86/03/02
86/03/22
86/04/11
86/05/01
86/05/21
86/06/10
86/06/30
86/07/20
86/08/09
86/08/29
86/09/18
86/10/08
86/10/28
86/11/17
86/12/07
86/12/27
87/01/16
YR/M/D

SUN-EARTH-PLANET, DEG

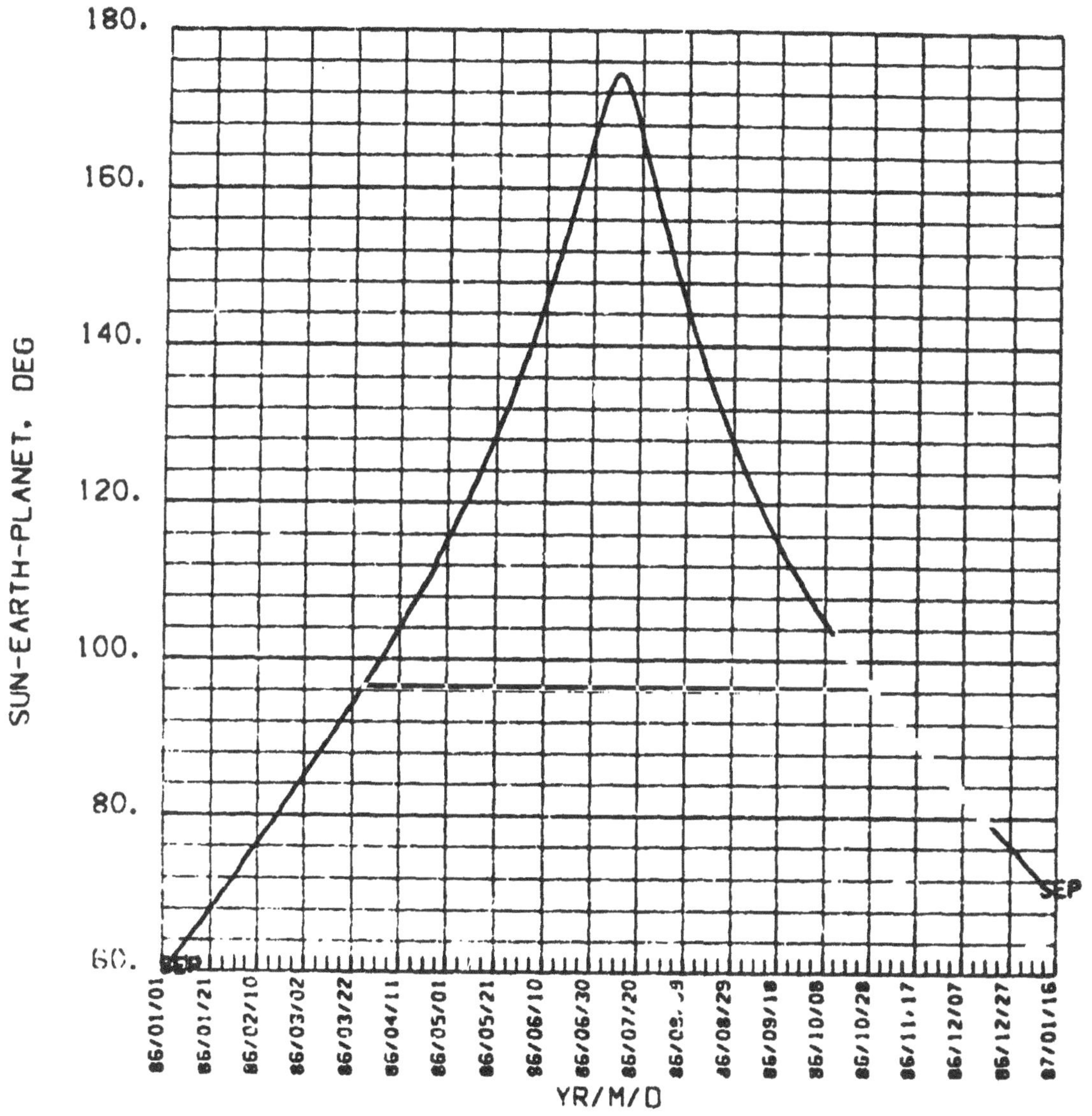

180.
160.
140.
120.
100.
80.
60.

SEP

SEP

86/01/01 86/01/21 86/02/10 86/03/02 86/03/22 86/04/11 86/05/01 86/05/21 86/06/10 86/06/30 86/07/20 86/08/[illegible] 86/08/29 86/09/18 86/10/08 86/10/28 86/11/17 86/12/07 86/12/27 87/01/16

YR/M/D

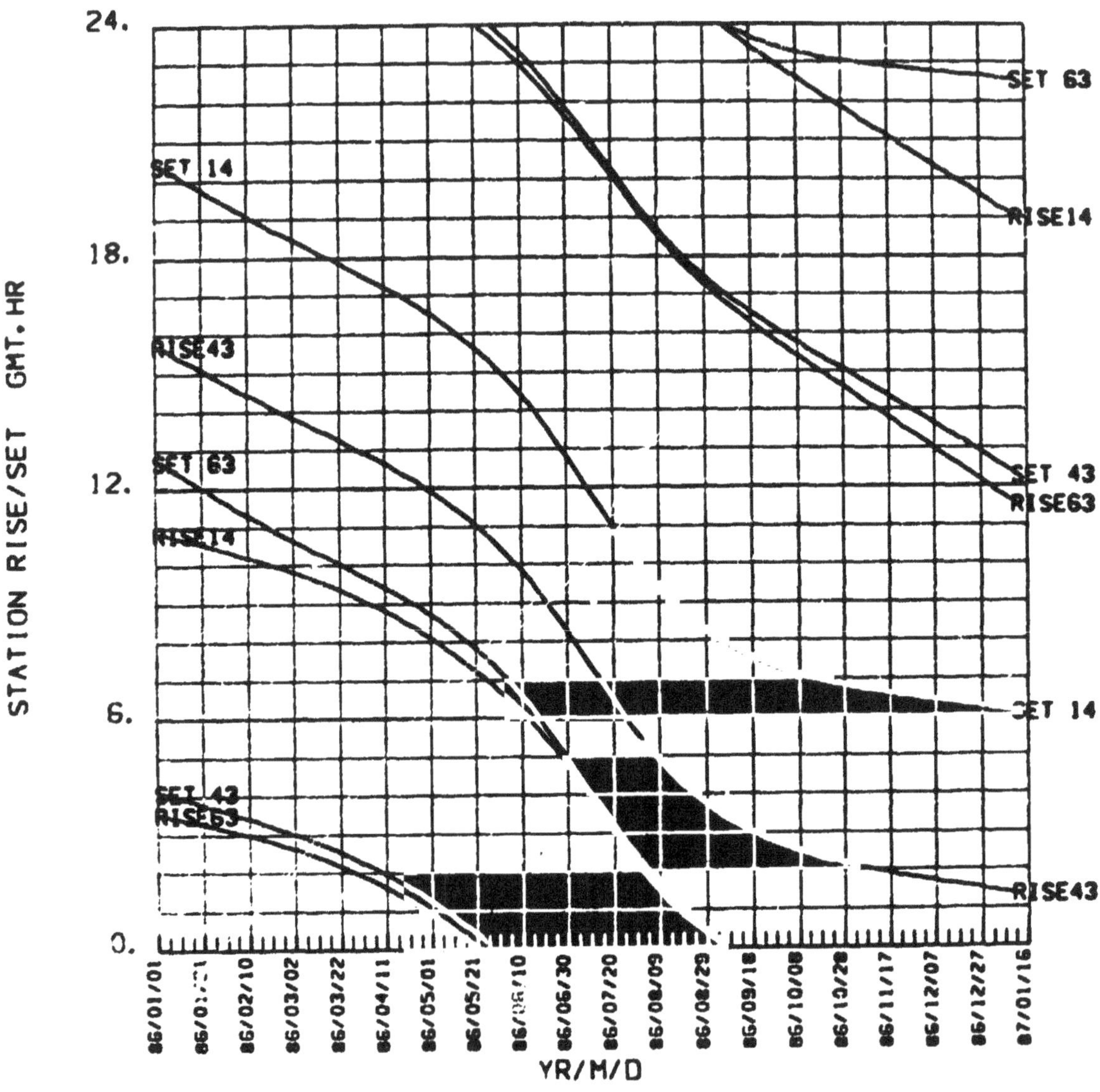
MARS 1986
STATION RISE/SET GMT.HR
24.
18.
12.
6.
0.
SET 14
RISE43
SET 63
RISE14
SET 43
RISE63
SET 63
RISE14
SET 43
RISE63
SET 14
RISE43
YR/M/D
86/01/01
86/02/10
86/03/02
86/03/22
86/04/11
86/05/01
86/05/21
86/06/30
86/07/20
86/08/09
86/08/29
86/09/18
86/10/08
86/10/28
86/11/17
86/12/07
86/12/27
87/01/16

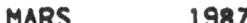

DECLIN

40.
30.
20.
10.
0.
-10.
-20.
-30.
-40.

87/01/0?
87/01/21
87/02/10
87/03/02
87/03/22
87/04/11
87/05/01
87/05/21
87/06/10
87/06/30
87/07 20
87/08/09
87/08/29
87/09/18
87/10/08
87/10/28
87/1?/17
87/12/07
87/?2/27
8?/01/16

YR/M/D

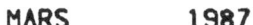
MARS 1987
RT. ASC
360.
270.
130.
90.
0.
87/01/0?
87/01/21
87/02/10
87/03/02
87/03/22
87/04/11
87/05/01
87/05/21
87/06/10
87/06/30
87/07/20
87/?9/09
87/08/29
87/09/18
87/10/08
87/10/28
87/11/17
87/12/07
87/12/27
88/01/16
YR/M/D
P
S
E

 1987

EC. LON

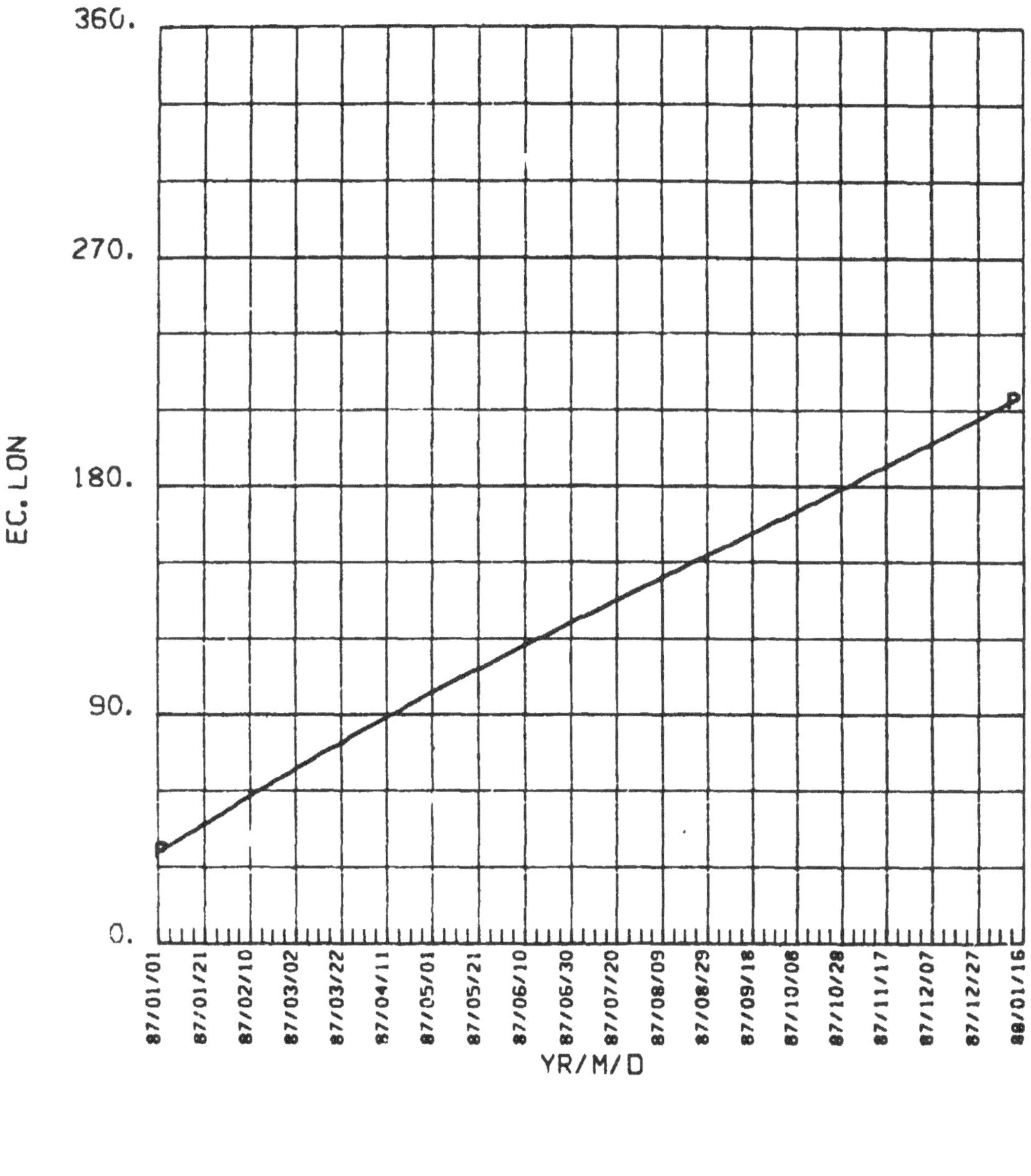

360.
270.
180.
90.
0.

87/01/01
87/01/21
87/02/10
87/03/02
87/03/22
87/04/11
87/05/01
87/05/21
87/06/10
87/06/30
87/07/20
87/08/09
87/08/29
87/09/18
87/10/08
87/10/28
87/11/17
87/12/07
87/12/27
88/01/16

YR/M/D

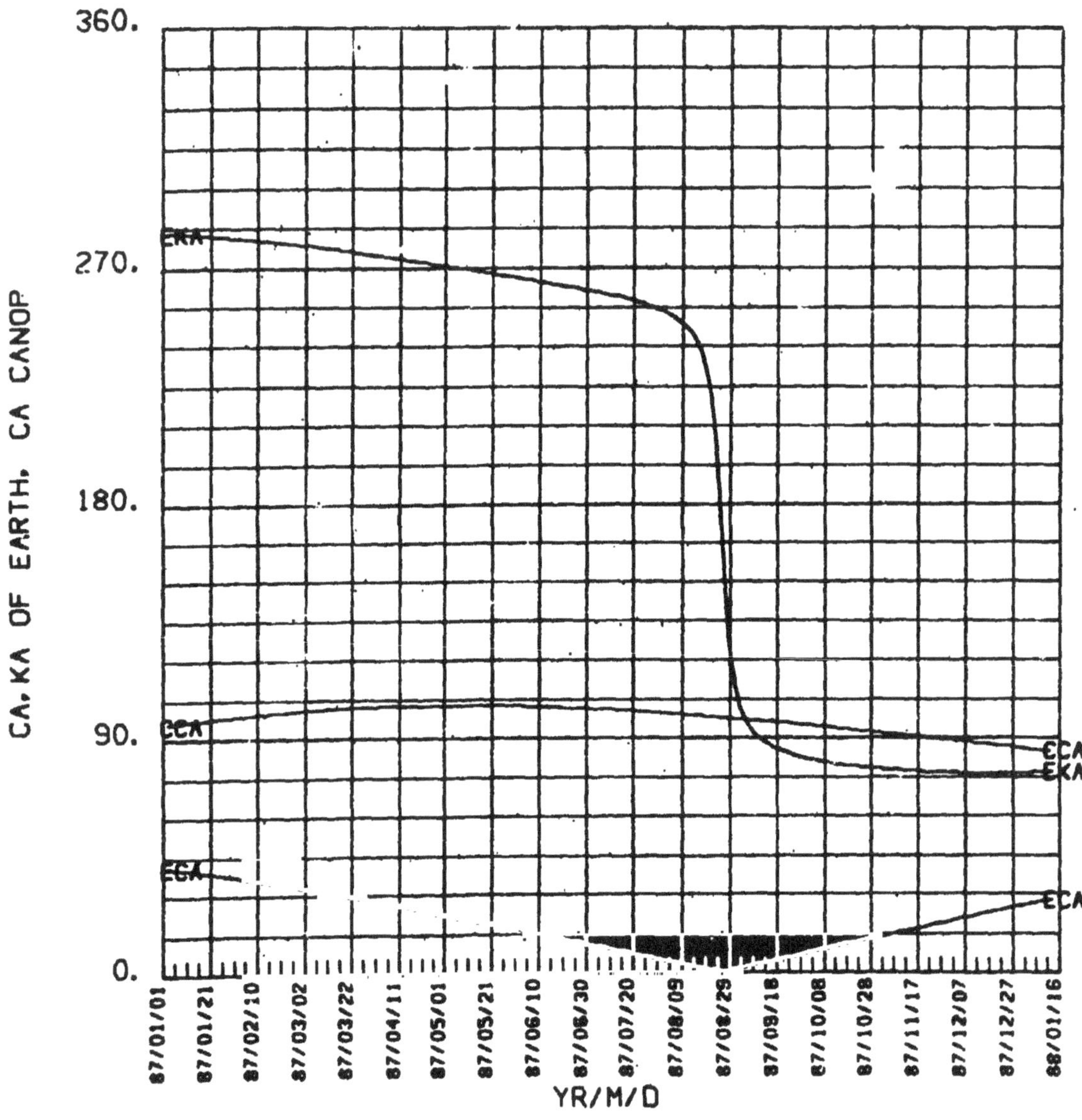

MARS 1987
360.
270.
180.
90.
0.
CA, KA OF EARTH, CA CANOP
EKA
CCA
ECA
CCA
EKA
ECA
87/01/01
87/01/21
87/02/10
87/03/02
87/03/22
87/04/11
87/05/01
87/05/21
87/06/10
87/06/30
87/07/20
87/08/09
87/08/29
87/09/18
87/10/08
87/10/28
87/11/17
87/12/07
87/12/27
88/01/16
YR/M/D

MARS 1987

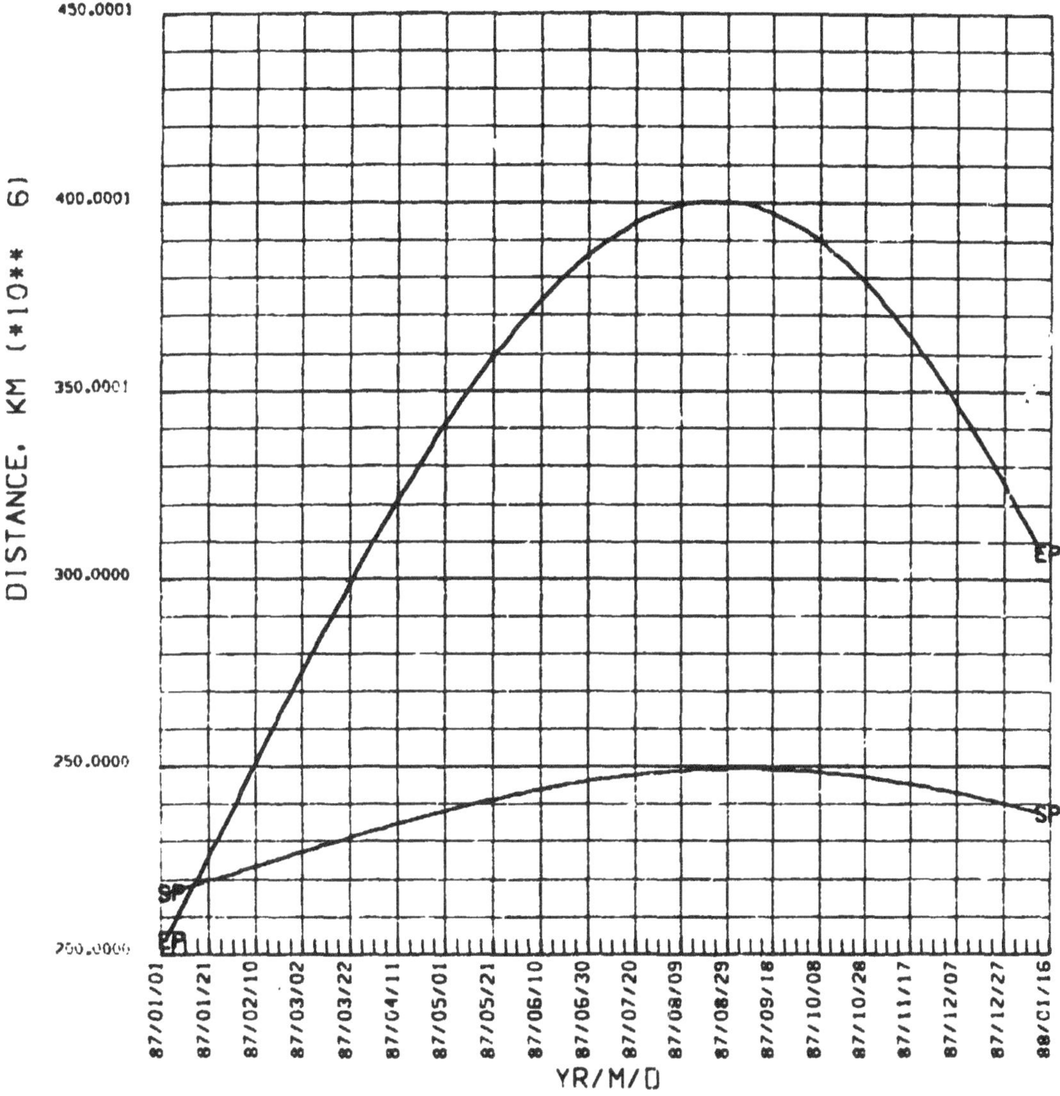

450.0001
400.0001
350.0001
300.0000
250.0000
200.0000
DISTANCE. KM (*10** 6)
EP
SP
87/01/01
87/01/21
87/02/10
87/03/02
87/03/22
87/04/11
87/05/01
87/05/21
87/06/10
87/06/30
87/07/20
87/08/09
87/08/29
87/09/18
87/10/08
87/10/28
87/11/17
87/12/07
87/12/27
88/01/16
YR/M/D

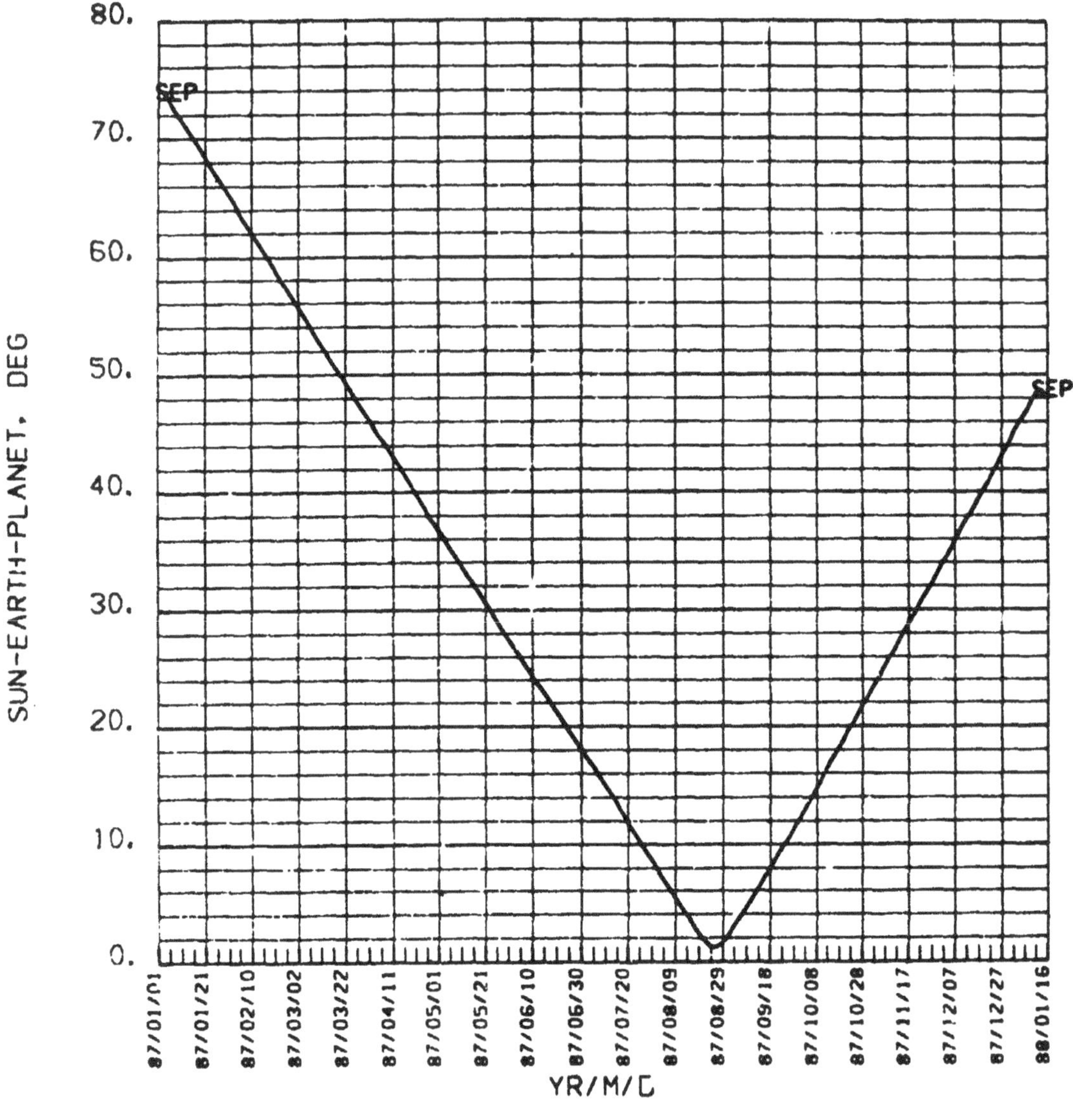

MARS 1987
SUN-EARTH-PLANET, DEG
80.
70.
60.
50.
40.
30.
20.
10.
0.
SEP
SEP
87/01/01
87/01/21
87/02/10
87/03/02
87/03/22
87/04/11
87/05/01
87/05/21
87/06/10
87/06/30
87/07/20
87/08/09
87/08/29
87/09/18
87/10/08
87/10/28
87/11/17
87/12/07
87/12/27
88/01/16
YR/M/D

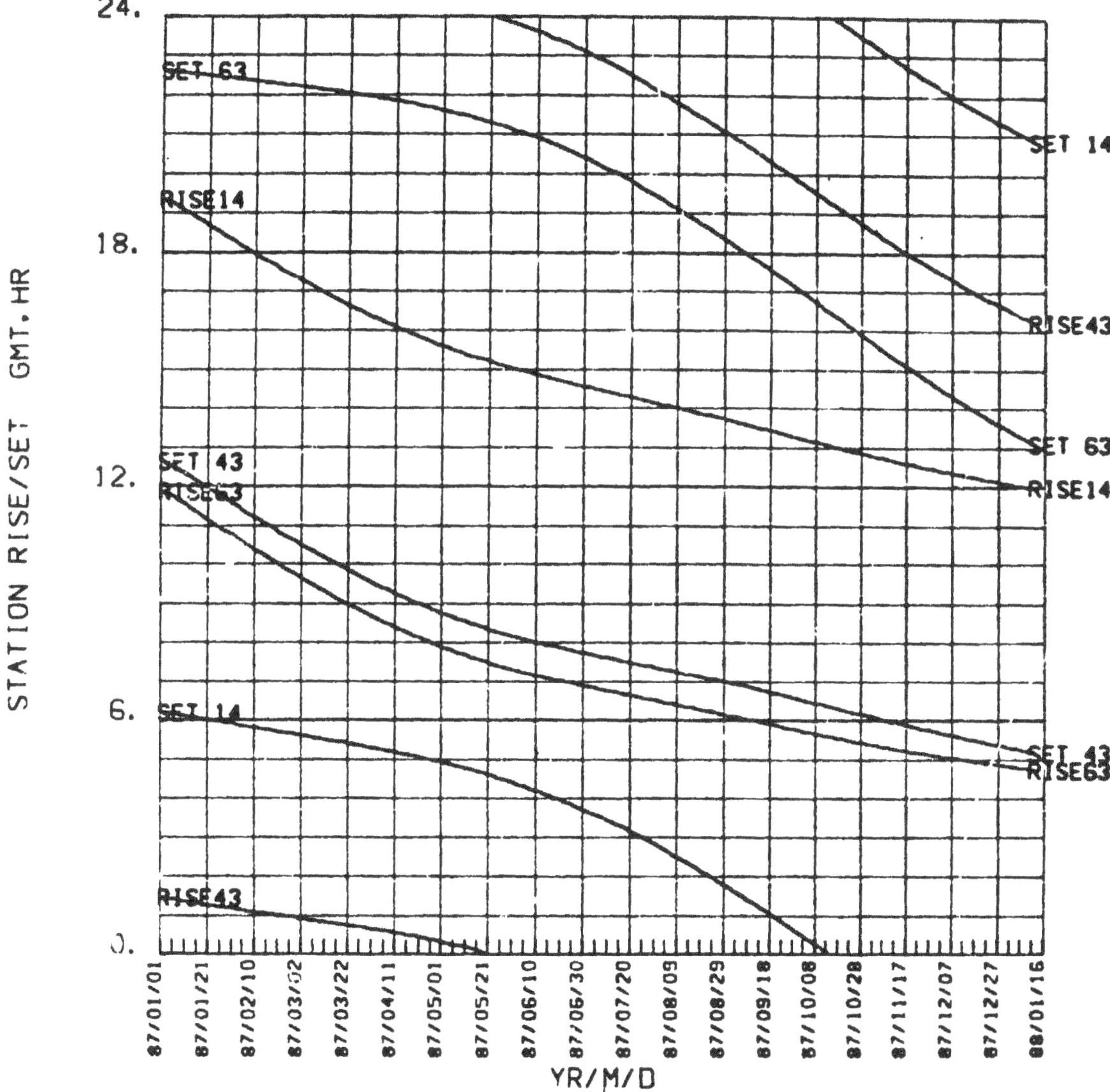
MARS 1987
STATION RISE/SET GMT.HR
24.
18.
12.
6.
0.
SET 63
RISE14
SET 43
RISE63
SET 14
RISE43
SET 14
RISE43
SET 63
RISE14
SET 43
RISE63
87/01/01
87/01/21
87/02/10
87/03/02
87/03/22
87/04/11
87/05/01
87/05/21
87/06/10
87/06/30
87/07/20
87/08/09
87/08/29
87/09/18
87/10/08
87/10/28
87/11/17
87/12/07
87/12/27
88/01/16
YR/M/D

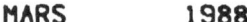

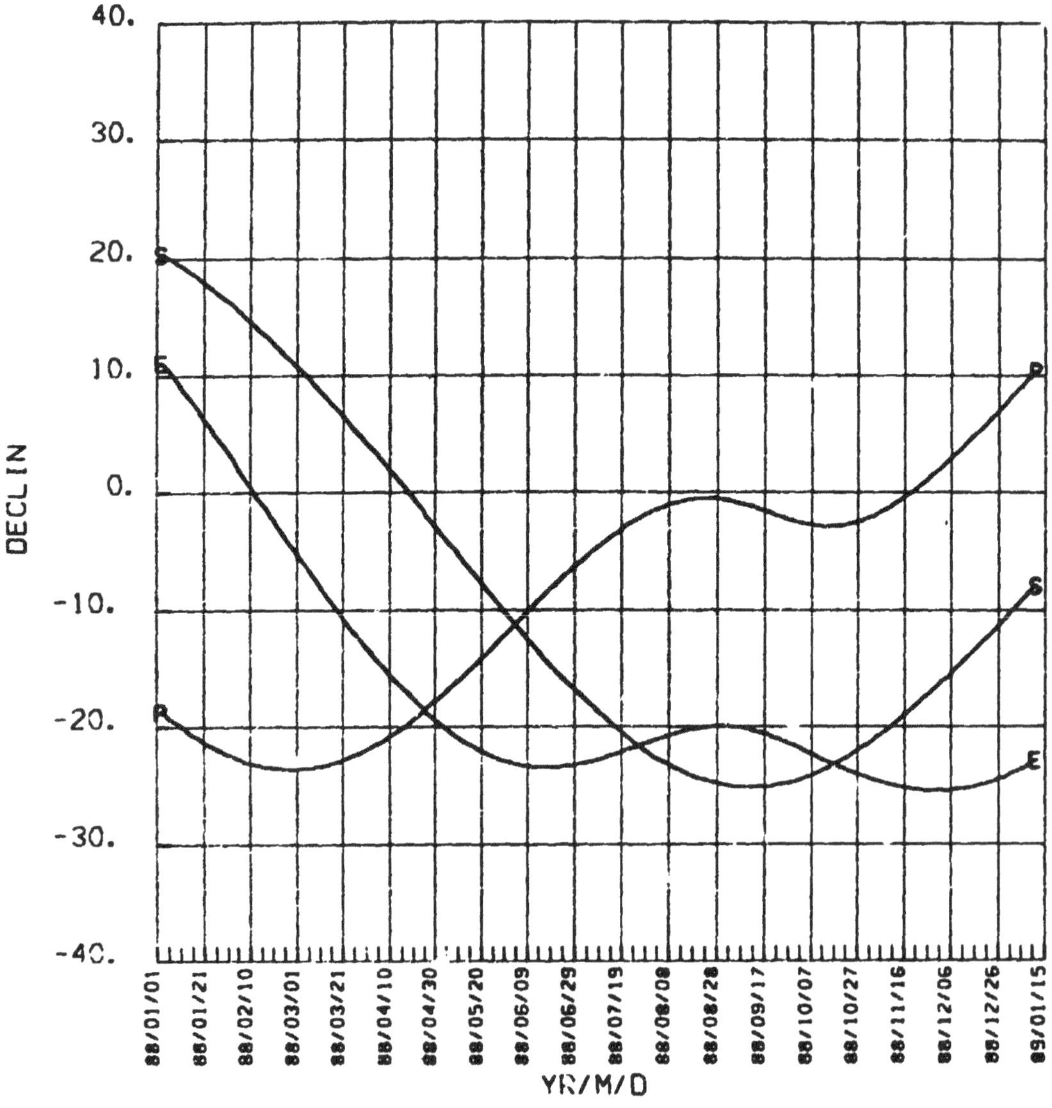

40.
30.
20.
10.
0.
-10.
-20.
-30.
-40.

DECLIN

S
E
R
D
S
E

88/01/01
88/01/21
88/02/10
88/03/01
88/03/21
88/04/10
88/04/30
88/05/20
88/06/09
88/06/29
88/07/19
88/08/08
88/08/28
88/09/17
88/10/07
88/10/27
88/11/16
88/12/06
88/12/26
89/01/15

YR/M/D

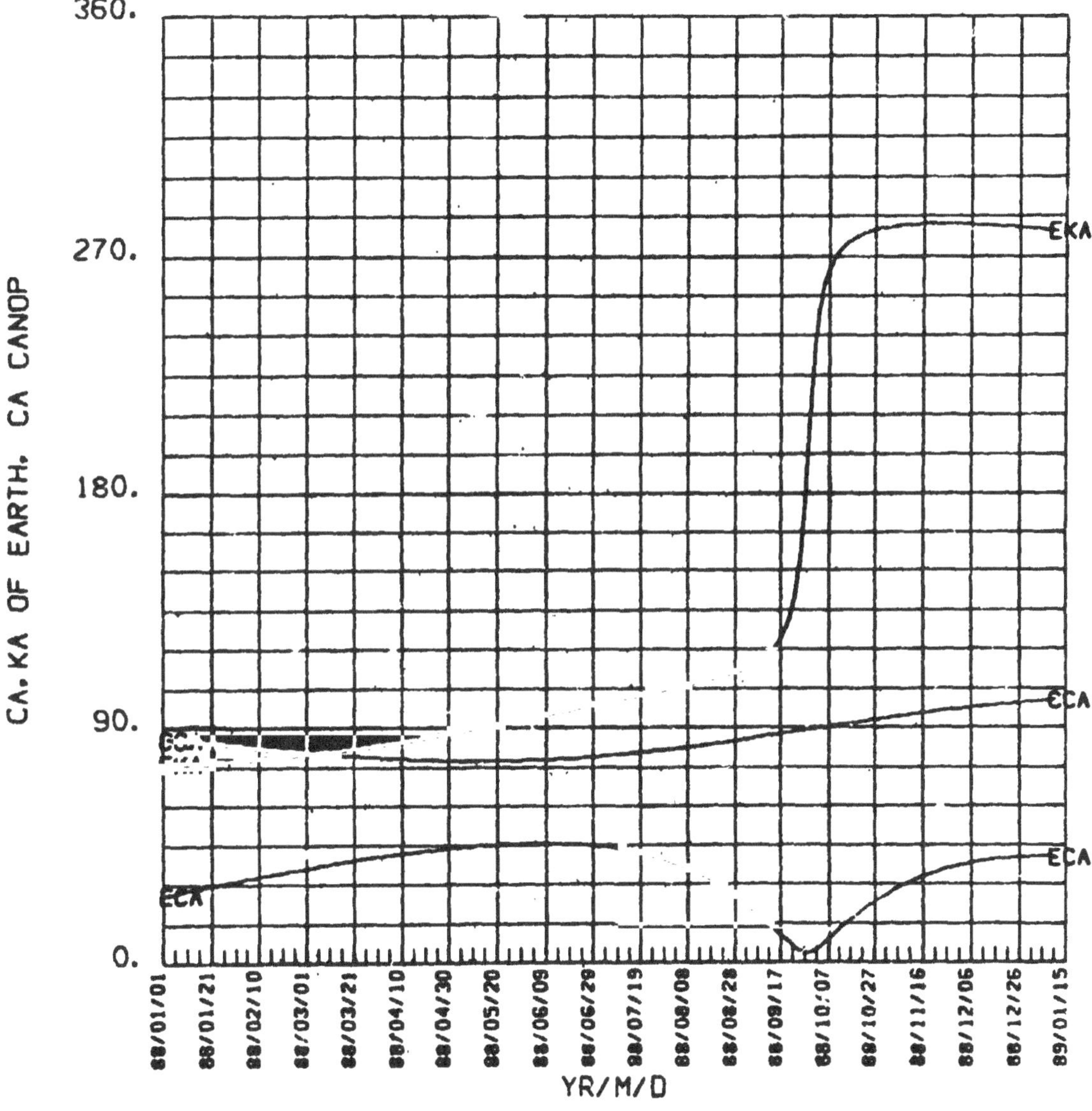

MARS 1988
360.
270.
180.
90.
0.
CA.KA OF EARTH, CA CANOP
EKA
ECA
ECA
ECA
88/01/01
88/01/21
88/02/10
88/03/01
88/03/21
88/04/10
88/04/30
88/05/20
88/06/09
88/06/29
88/07/19
88/08/08
88/08/28
88/09/17
88/10/07
88/10/27
88/11/16
88/12/06
88/12/26
89/01/15
YR/M/D

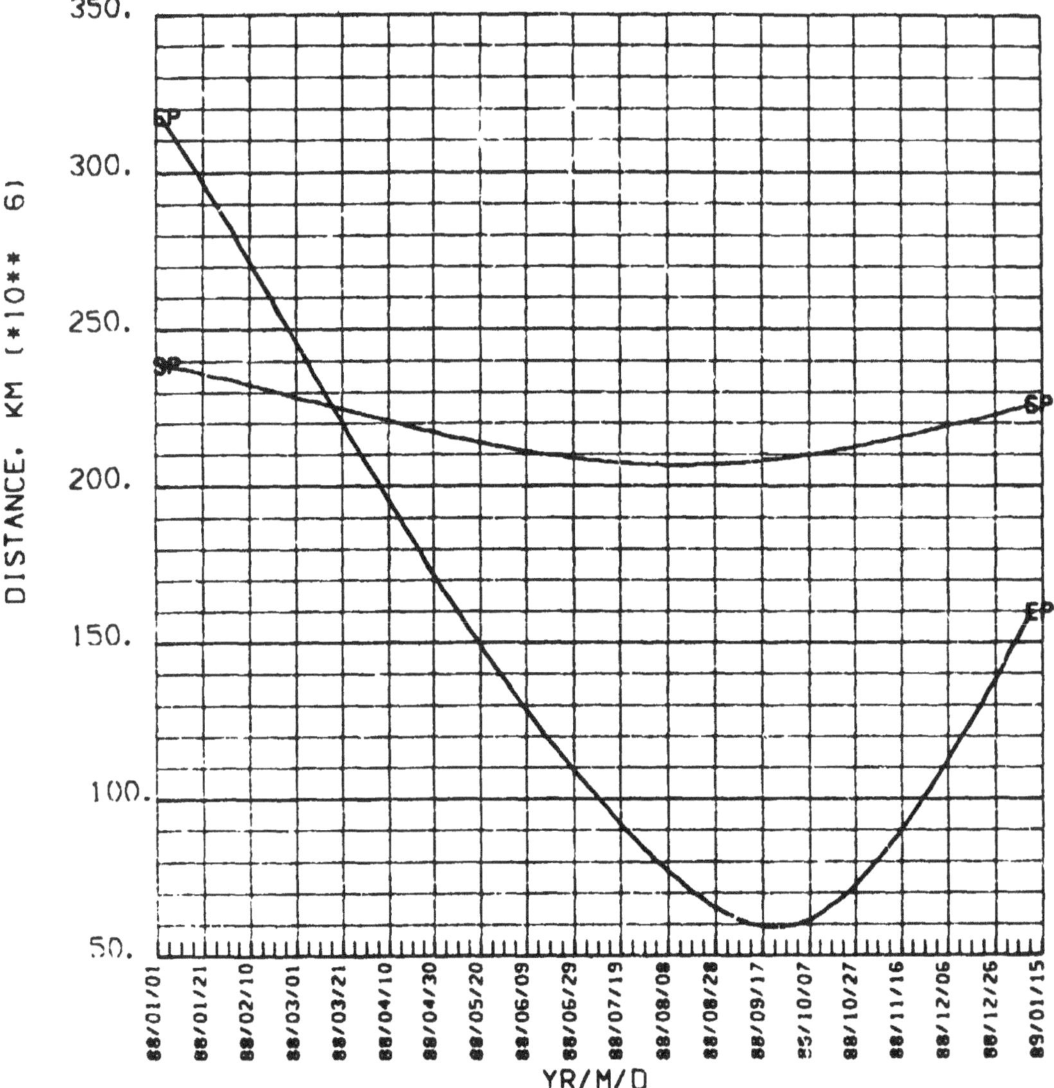
MARS 1988
350.
300.
250.
200.
150.
100.
50.
DISTANCE, KM (*10** 6)
SP
SP
SP
EP
88/01/01
88/01/21
88/02/10
88/03/01
88/03/21
88/04/10
88/04/30
88/05/20
88/06/09
88/06/29
88/07/19
88/08/08
88/08/28
88/09/17
88/10/07
88/10/27
88/11/16
88/12/06
88/12/26
89/01/15
YR/M/D

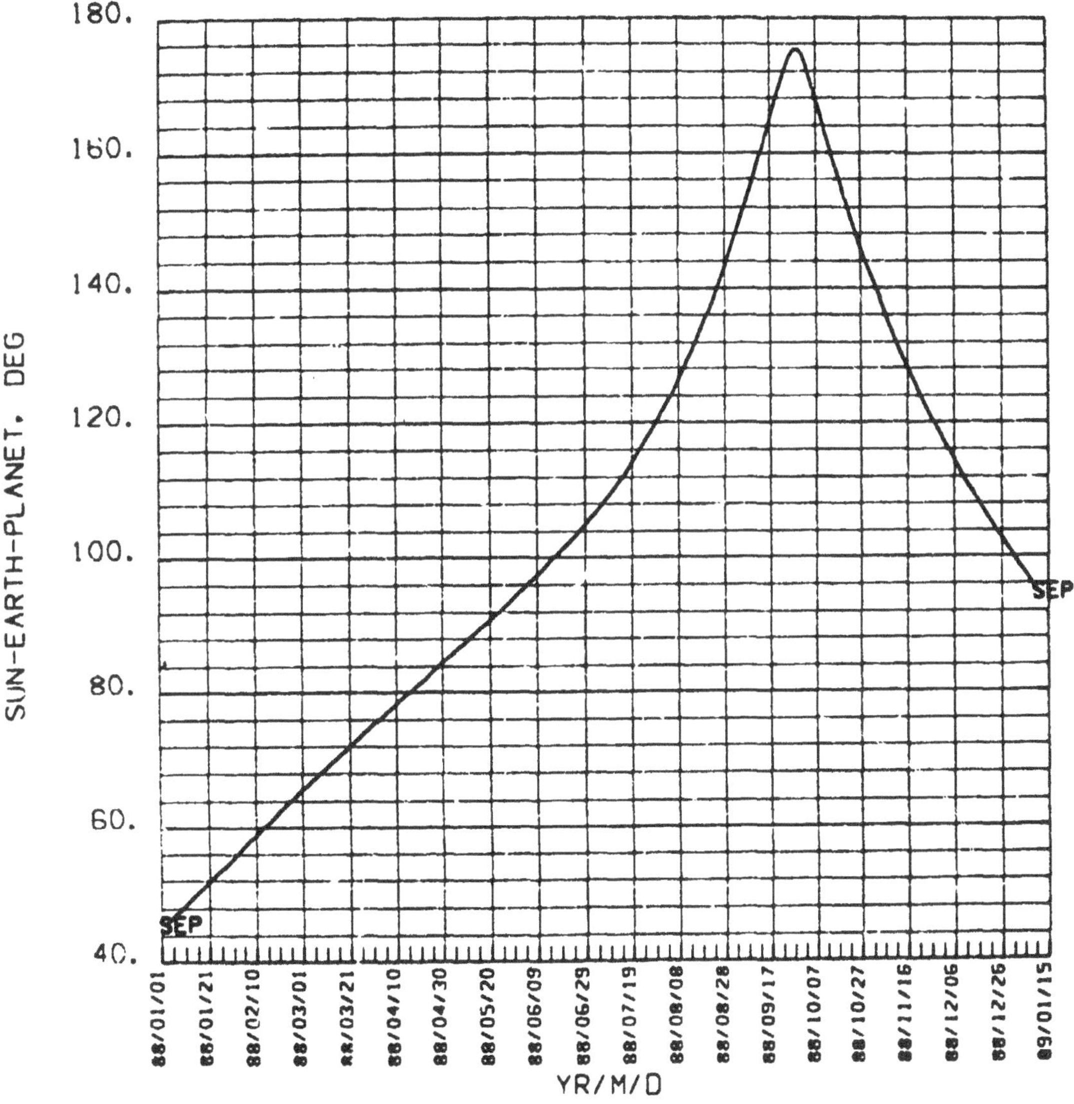
MARS 1988
180.
160.
140.
120.
100.
80.
60.
40.
SUN-EARTH-PLANET, DEG
SEP
SEP
88/01/01
88/01/21
88/02/10
88/03/01
88/03/21
88/04/10
88/04/30
88/05/20
88/06/09
88/06/29
88/07/19
88/08/08
88/08/28
88/09/17
88/10/07
88/10/27
88/11/16
88/12/06
88/12/26
89/01/15
YR/M/D

MARS 1988

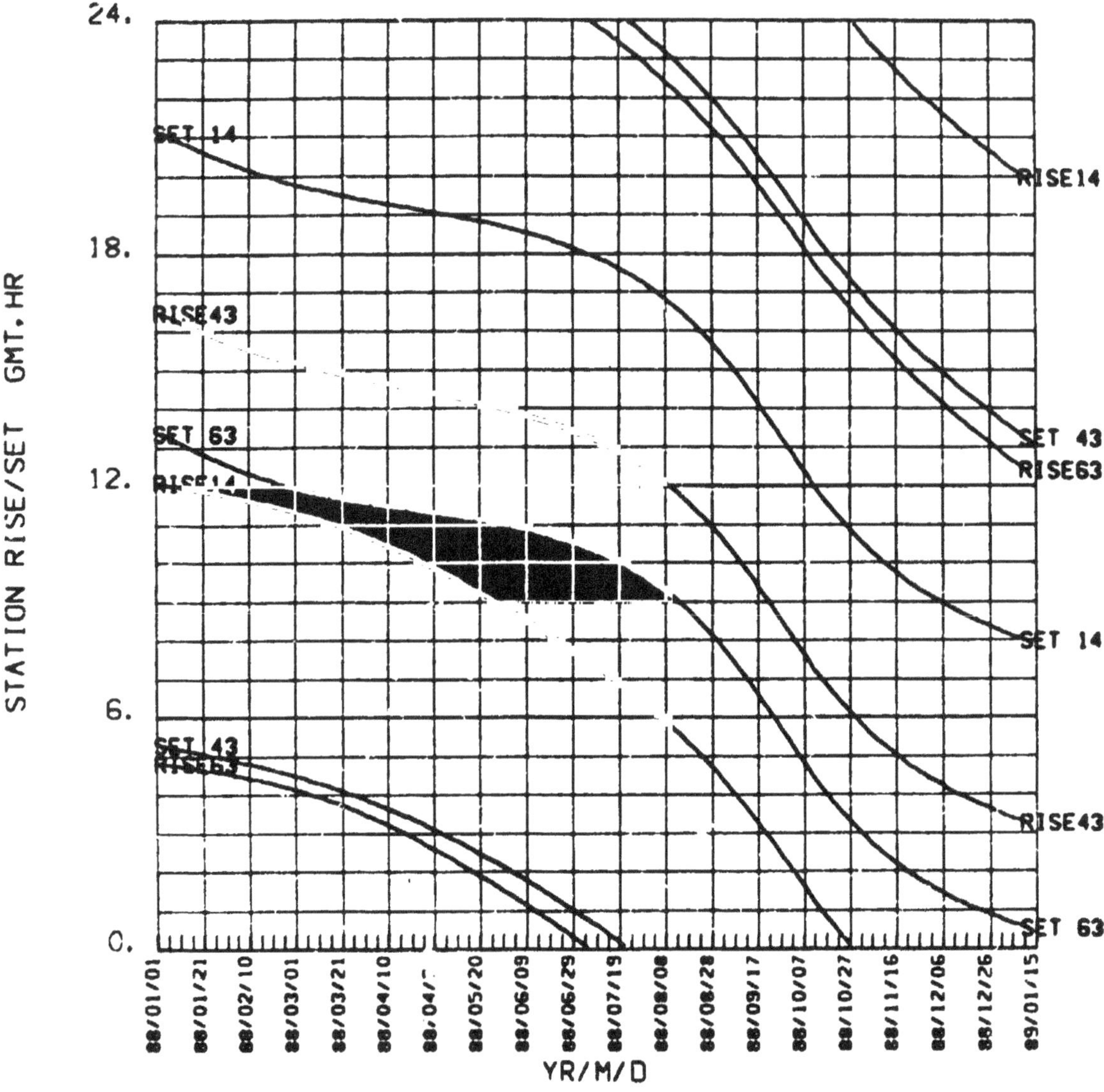

STATION RISE/SET GMT.HR
24.
18.
12.
6.
0.
SET 14
RISE43
SET 63
SET 43
RISE63
RISE14
SET 43
RISE63
SET 14
RISE43
SET 63
88/01/01
88/01/21
88/02/10
88/03/01
88/03/21
88/04/10
88/05/20
88/06/09
88/06/29
88/07/19
88/08/08
88/08/28
88/09/17
88/10/07
88/10/27
88/11/16
88/12/06
88/12/26
89/01/15
YR/M/D

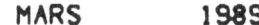

40.
30.
20.
10.
0.
-10.
-20.
-30.
-40.

DECLIN

89/01/01
89/01/21
89/02/10
89/03/02
89/03/22
89/04/11
89/05/01
89/05/21
89/06/10
89/06/30
89/07/20
89/08/09
89/08/29
89/09/18
89/10/08
89/10/28
89/11/17
89/12/07

YR/M/D

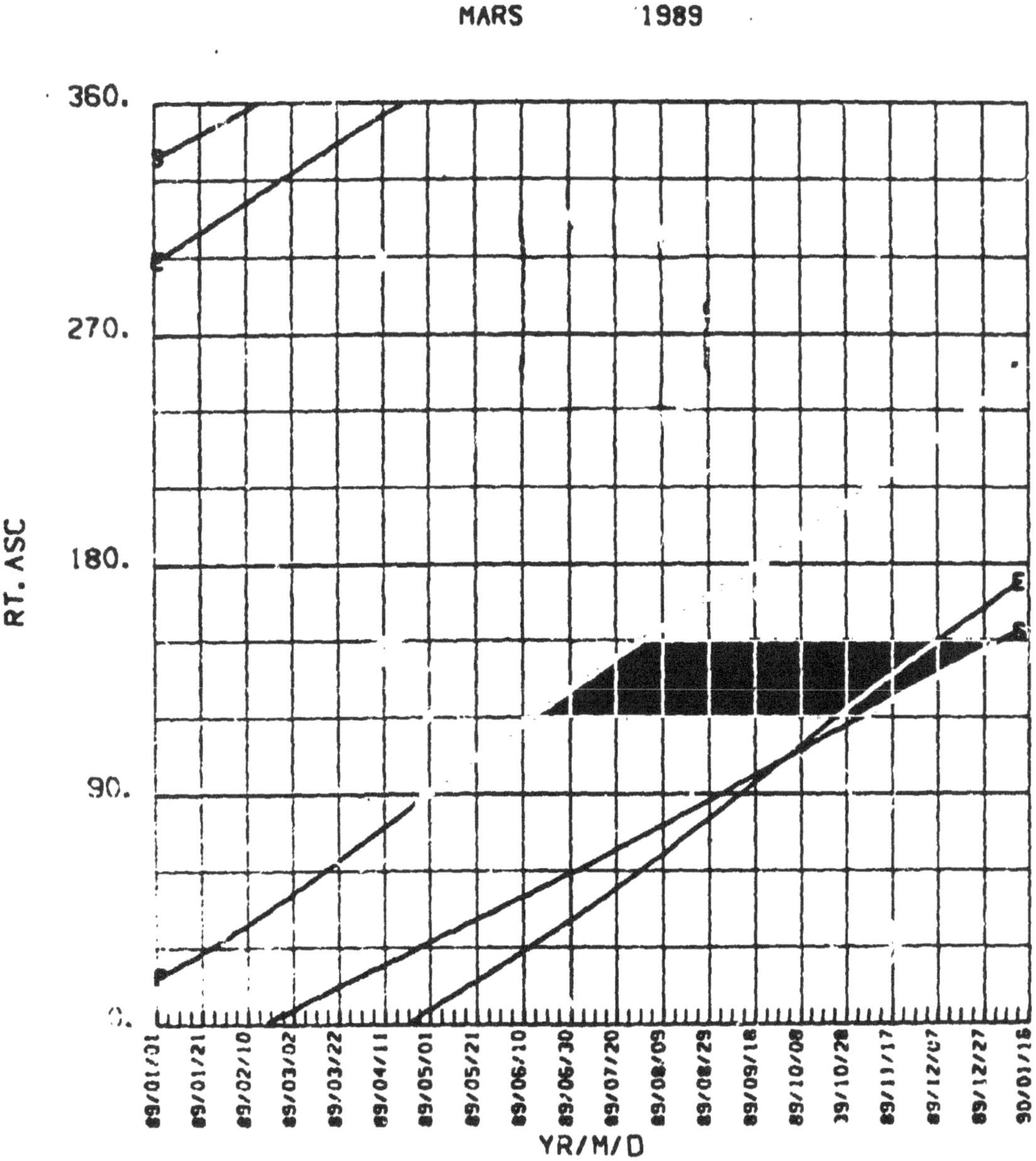

MARS 1989
360.
270.
180.
90.
0.
RT. ASC
89/01/01
89/01/21
89/02/10
89/03/02
89/03/22
89/04/11
89/05/01
89/05/21
89/06/10
89/06/30
89/07/20
89/08/09
89/08/29
89/09/18
89/10/08
89/10/28
89/11/17
89/12/07
89/12/27
90/01/16
YR/M/D

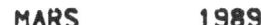
MARS 1989
EC. LON
360.
270.
180.
90.
0.
YR/M/D
89/01/01
89/01/21
89/02/10
89/03/02
89/03/22
89/04/11
89/05/01
89/06/30
89/07/20
89/08/09
89/08/29
89/09/18
89/10/08
89/10/28
89/11/17
89/12/27
90/01/16

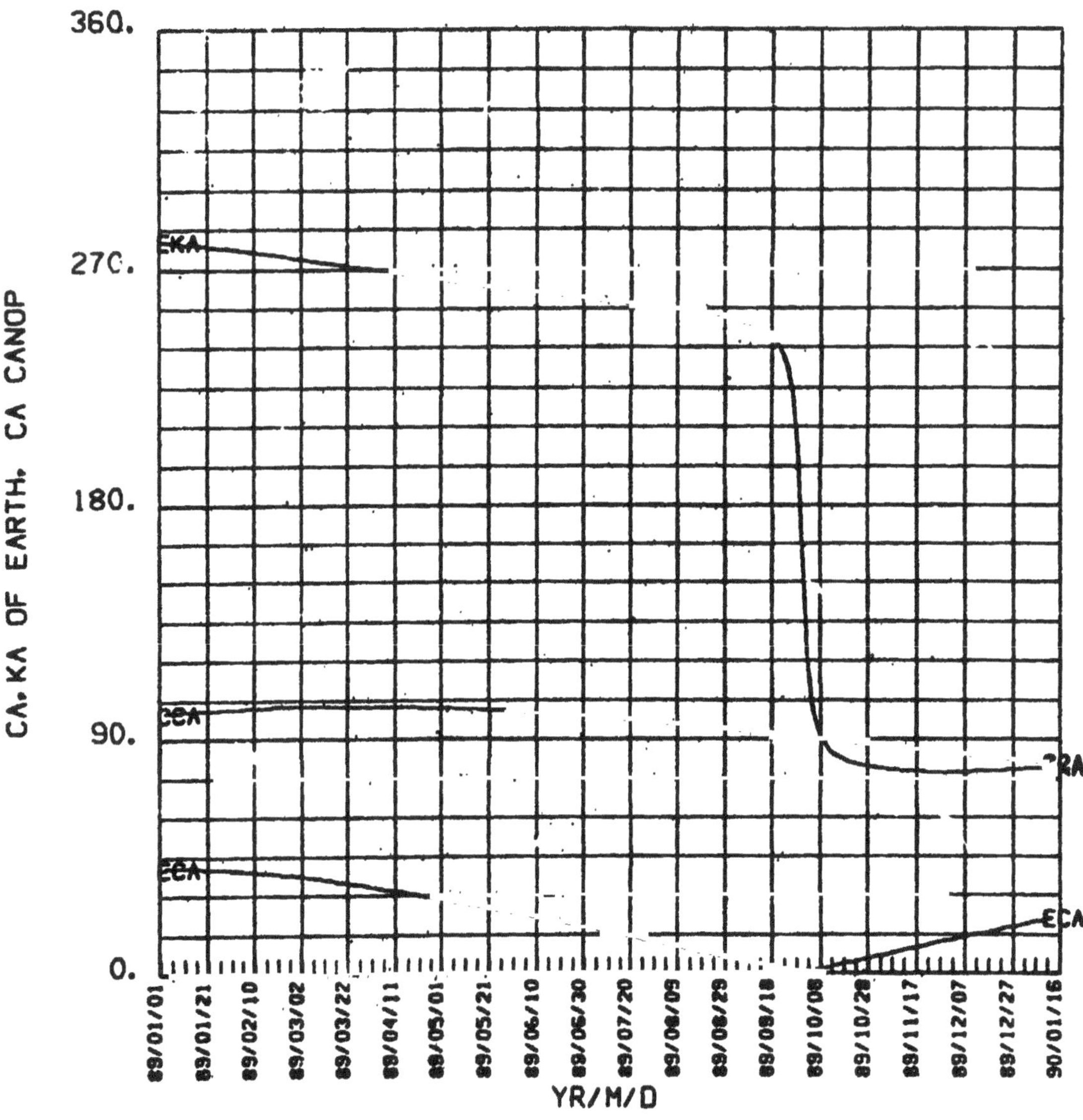

MARS 1989
360.
270.
180.
90.
0.
CA. KA OF EARTH, CA CANOP
EKA
ECA
ECA
ECA
89/01/01
89/01/21
89/02/10
89/03/02
89/03/22
89/04/11
89/05/01
89/05/21
89/06/10
89/06/30
89/07/20
89/08/09
89/08/29
89/09/18
89/10/08
89/10/28
89/11/17
89/12/07
89/12/27
90/01/16
YR/M/D

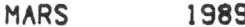

DISTANCE, KM (*10** 6)

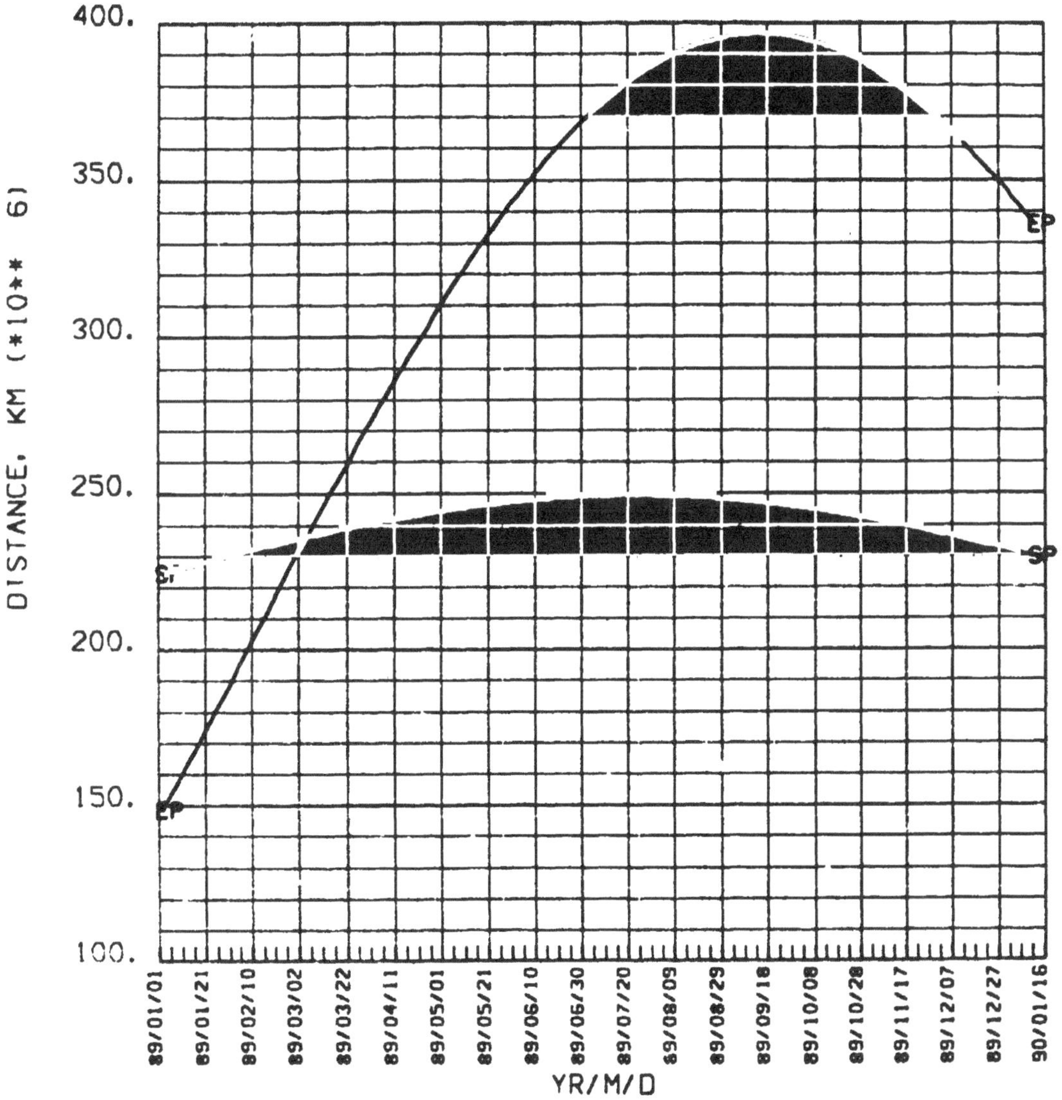

400.
350.
300.
250.
200.
150.
100.

EP
SP
EP
SP

89/01/01
89/01/21
89/02/10
89/03/02
89/03/22
89/04/11
89/05/01
89/05/21
89/06/10
89/06/30
89/07/20
89/08/09
89/08/29
89/09/18
89/10/08
89/10/28
89/11/17
89/12/07
89/12/27
90/01/16

YR/M/D

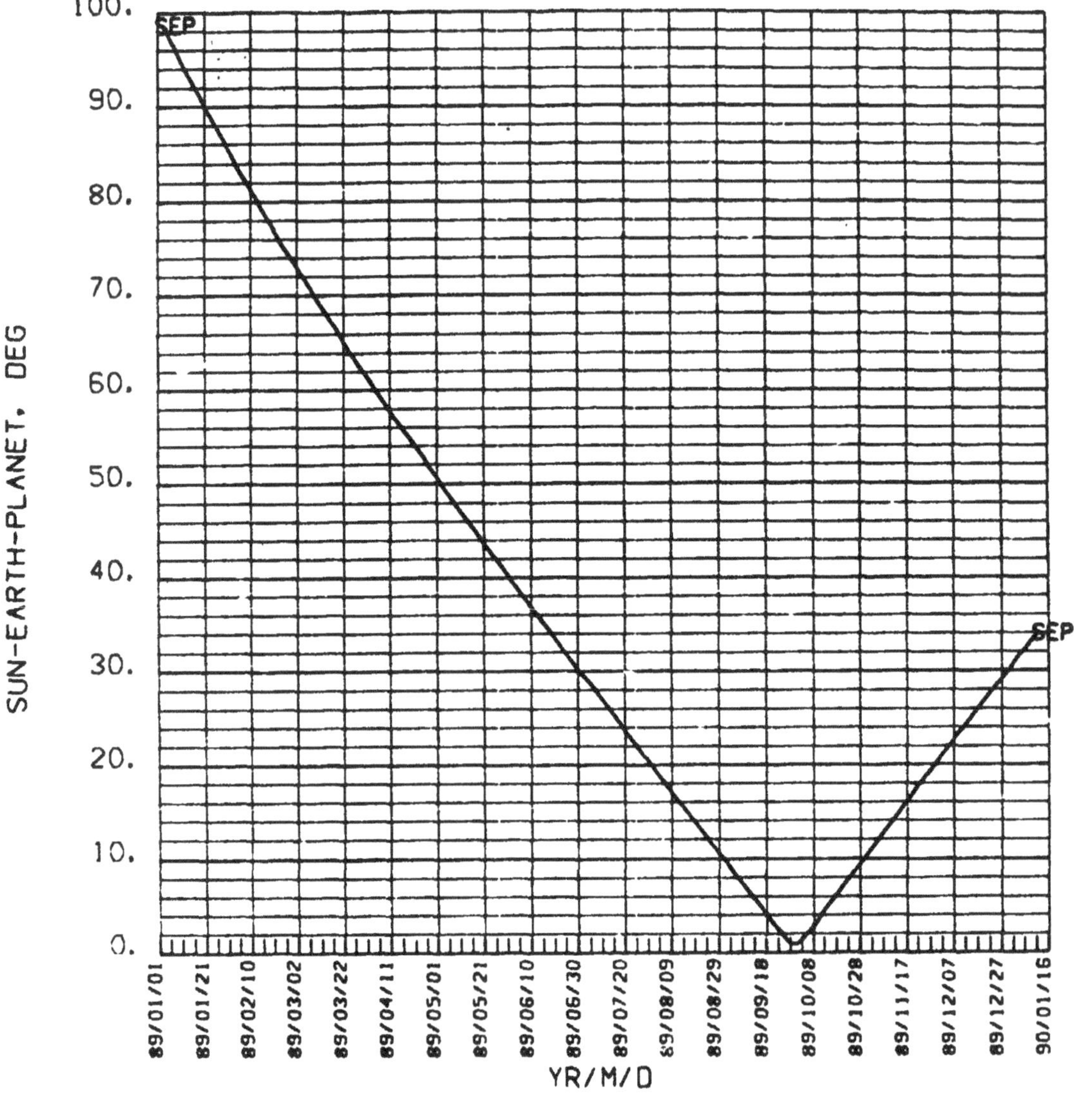
MARS 1989
SUN-EARTH-PLANET. DEG
100.
90.
80.
70.
60.
50.
40.
30.
20.
10.
0.
SEP
SEP
89/01/01
89/01/21
89/02/10
89/03/02
89/03/22
89/04/11
89/05/01
89/05/21
89/06/10
89/06/30
89/07/20
89/08/09
89/08/29
89/09/18
89/10/08
89/10/28
89/11/17
89/12/07
89/12/27
90/01/16
YR/M/D

MARS 1989
STATION RISE/SET GMT.HR
YR/M/D
24.
18.
12.
6.
0.
RISE14
SET 43
RISE63
SET 14
RISE43
SET 63
RISE14
SET 43
RISE63
89/01/01
89/01/21
89/02/10
89/03/02
89/03/22
89/04/11
89/05/01
89/05/21
89/06/10
89/06/30
89/07/20
89/08/09
89/08/29
89/09/18
89/10/08
89/10/28
89/11/17
89/12/07
89/12/27
90/01/16

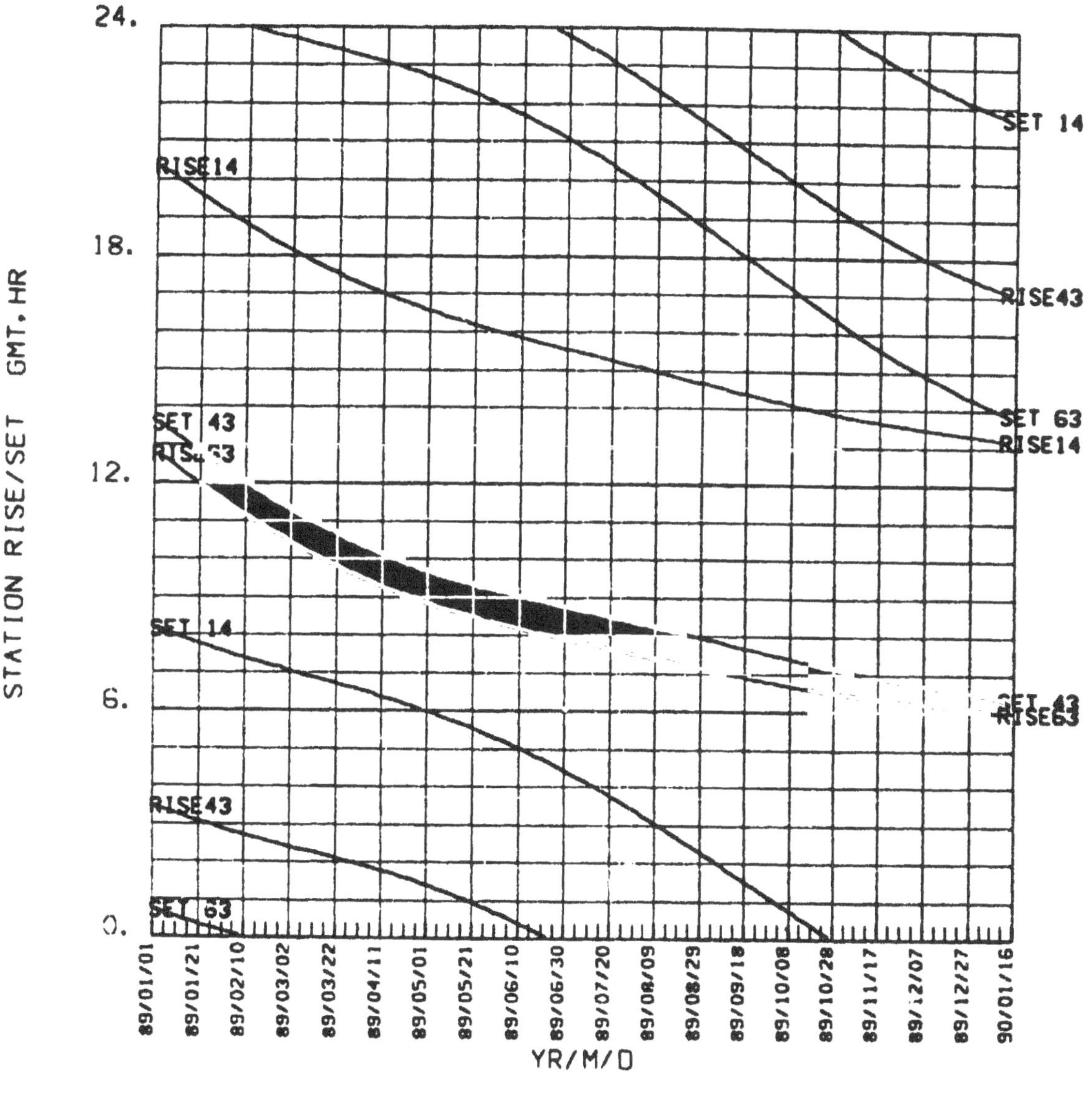

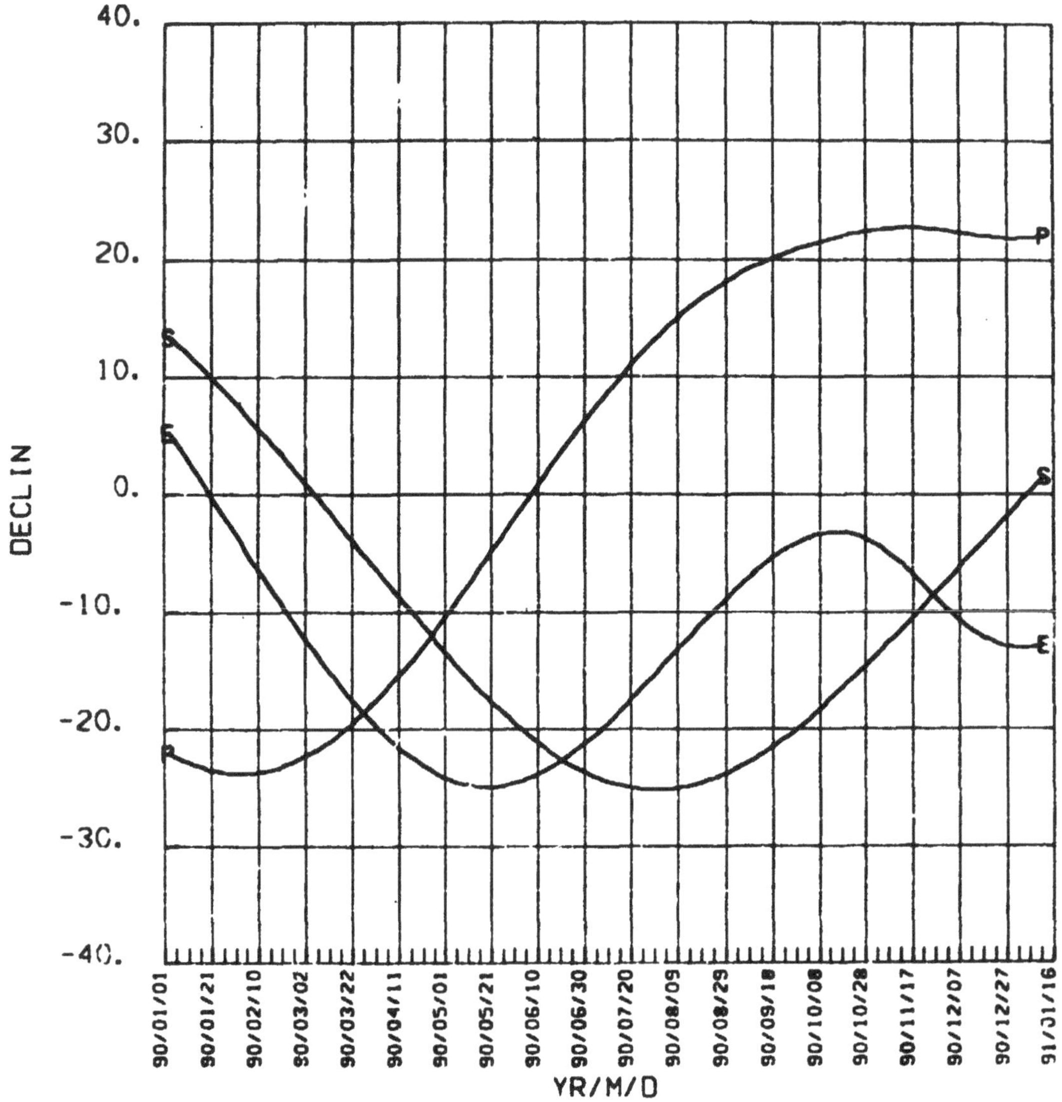

40.
30.
20.
10.
0.
-10.
-20.
-30.
-40.

DECLIN

P
S
E
S
E
P

90/01/01
90/01/21
90/02/10
90/03/02
90/03/22
90/04/11
90/05/01
90/05/21
90/06/10
90/06/30
90/07/20
90/08/09
90/08/29
90/09/18
90/10/08
90/10/28
90/11/17
90/12/07
90/12/27
91/01/16

YR/M/D

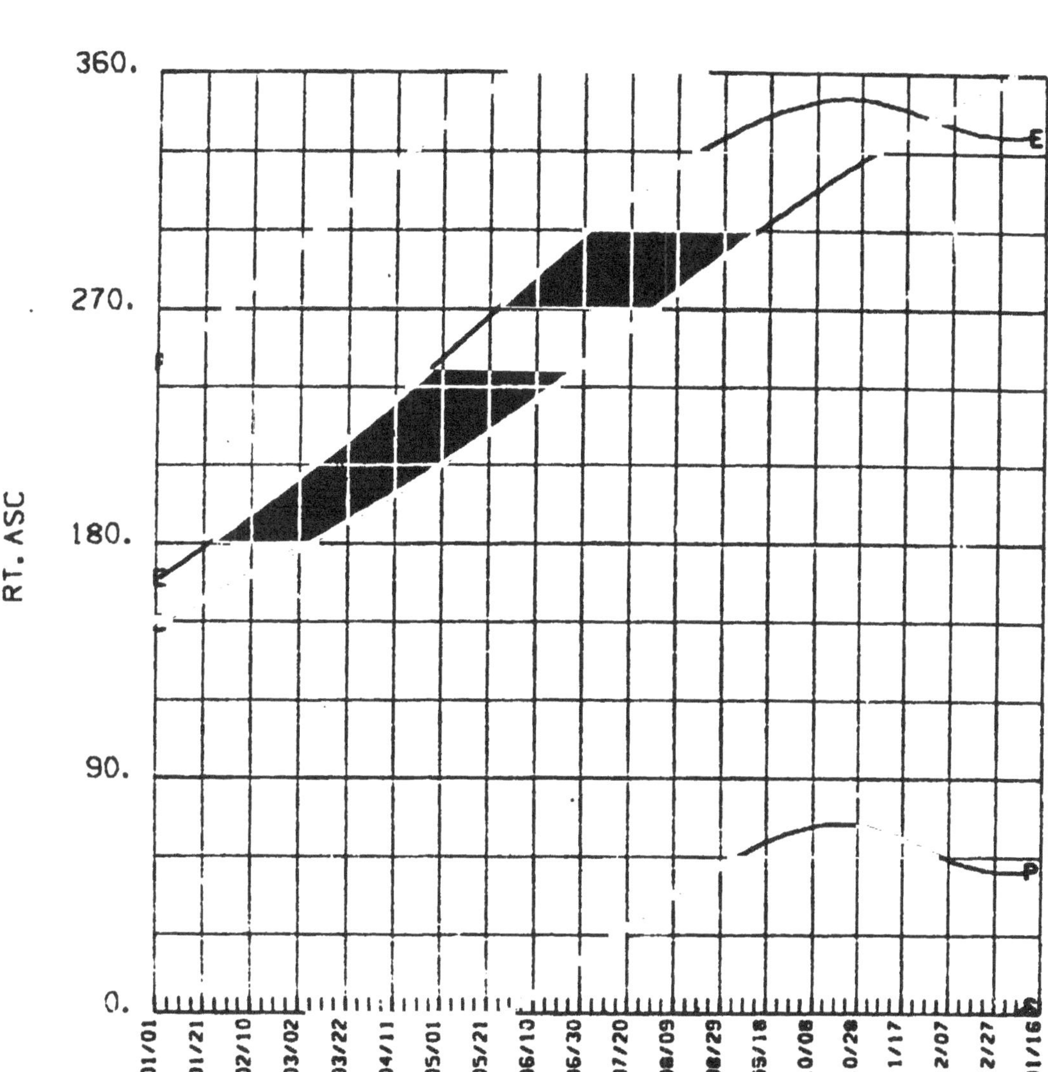

MARS 1990
360.
270.
180.
90.
0.
RT. ASC
E
P
90/01/01
90/01/21
90/02/10
90/03/02
90/03/22
90/04/11
90/05/01
90/05/21
90/06/10
90/06/30
90/07/20
90/08/09
90/08/29
90/10/08
90/10/28
90/11/17
90/12/07
90/12/27
91/01/16
YR/M/D

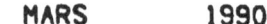

MARS 1990

360.
270.
180.
90.
0.
EC. LON
90/01/01
90/01/21
90/02/10
90/03/02
90/03/22
90/04/11
90/05/01
90/05/21
90/06/10
90/06/30
90/07/20
90/08/09
90/08/29
90/09/18
90/10/08
90/10/28
90/11/17
90/12/07
90/12/27
91/01/16
YR/M/D

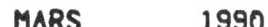

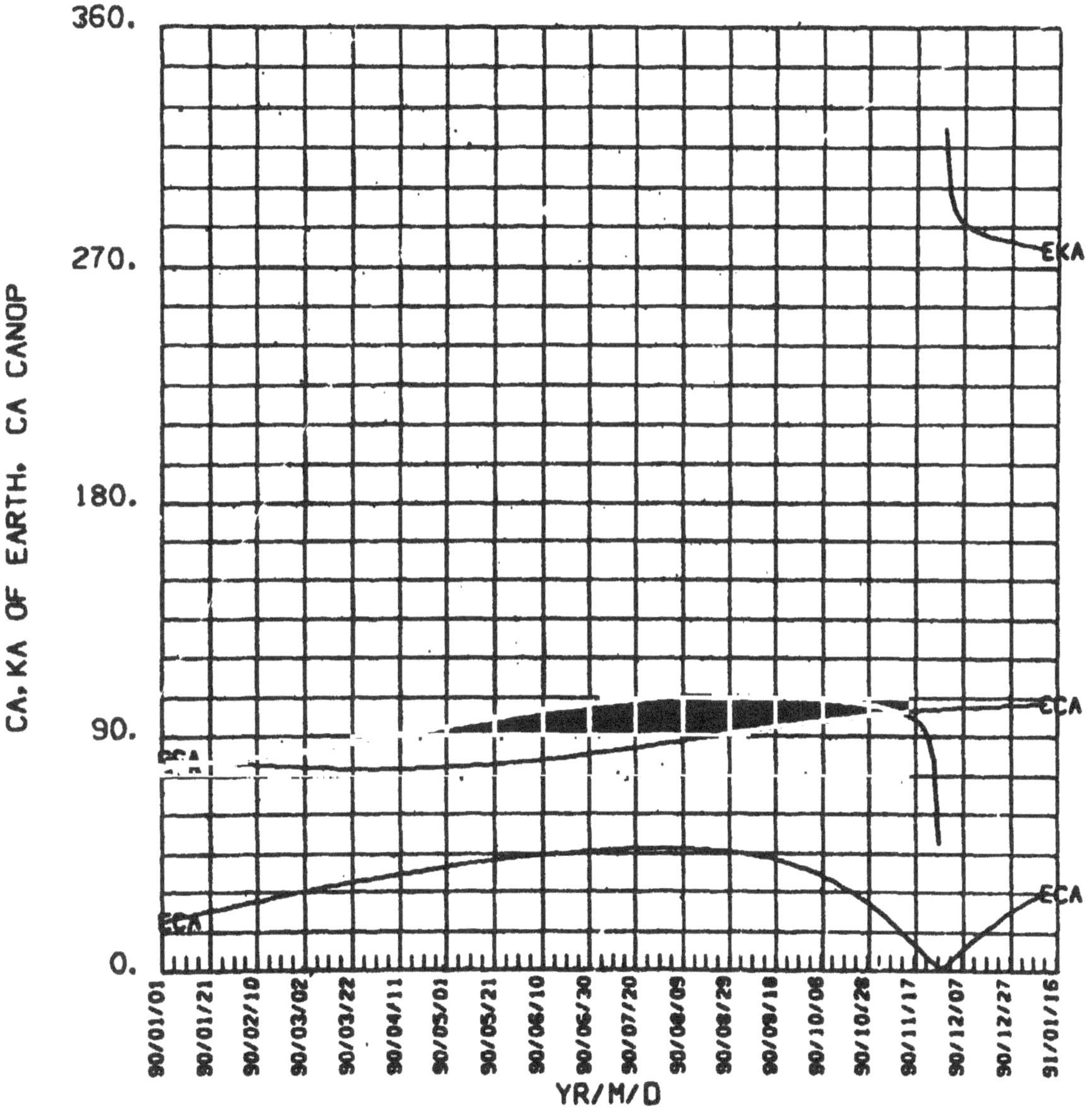

360.

270.

180.

90.

0.

CA, KA OF EARTH, CA CANOP

EKA

ECA

ECA

ECA

ECA

90/01/01 90/01/21 90/02/10 90/03/02 90/03/22 90/04/11 90/05/01 90/05/21 90/06/10 90/06/30 90/07/20 90/08/09 90/08/29 90/09/18 90/10/08 90/10/28 90/11/17 90/12/07 90/12/27 91/01/16

YR/M/D

MARS 1990

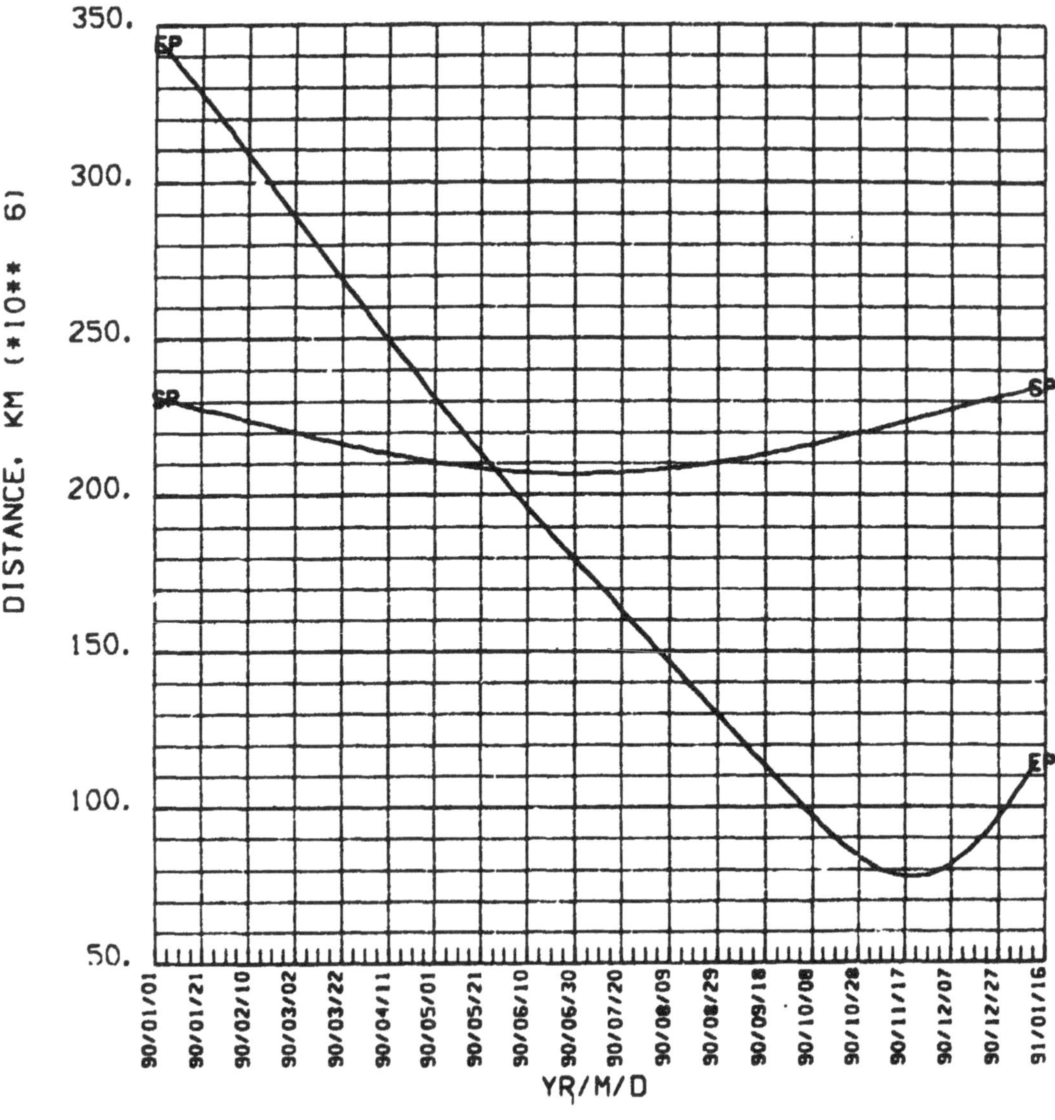

350.
300.
250.
200.
150.
100.
50.
DISTANCE, KM (*10** 6)
SP
SP
SP
EP
90/01/01
90/01/21
90/02/10
90/03/02
90/03/22
90/04/11
90/05/01
90/05/21
90/06/10
90/06/30
90/07/20
90/08/09
90/08/29
90/09/18
90/10/08
90/10/28
90/11/17
90/12/07
90/12/27
91/01/16
YR/M/D

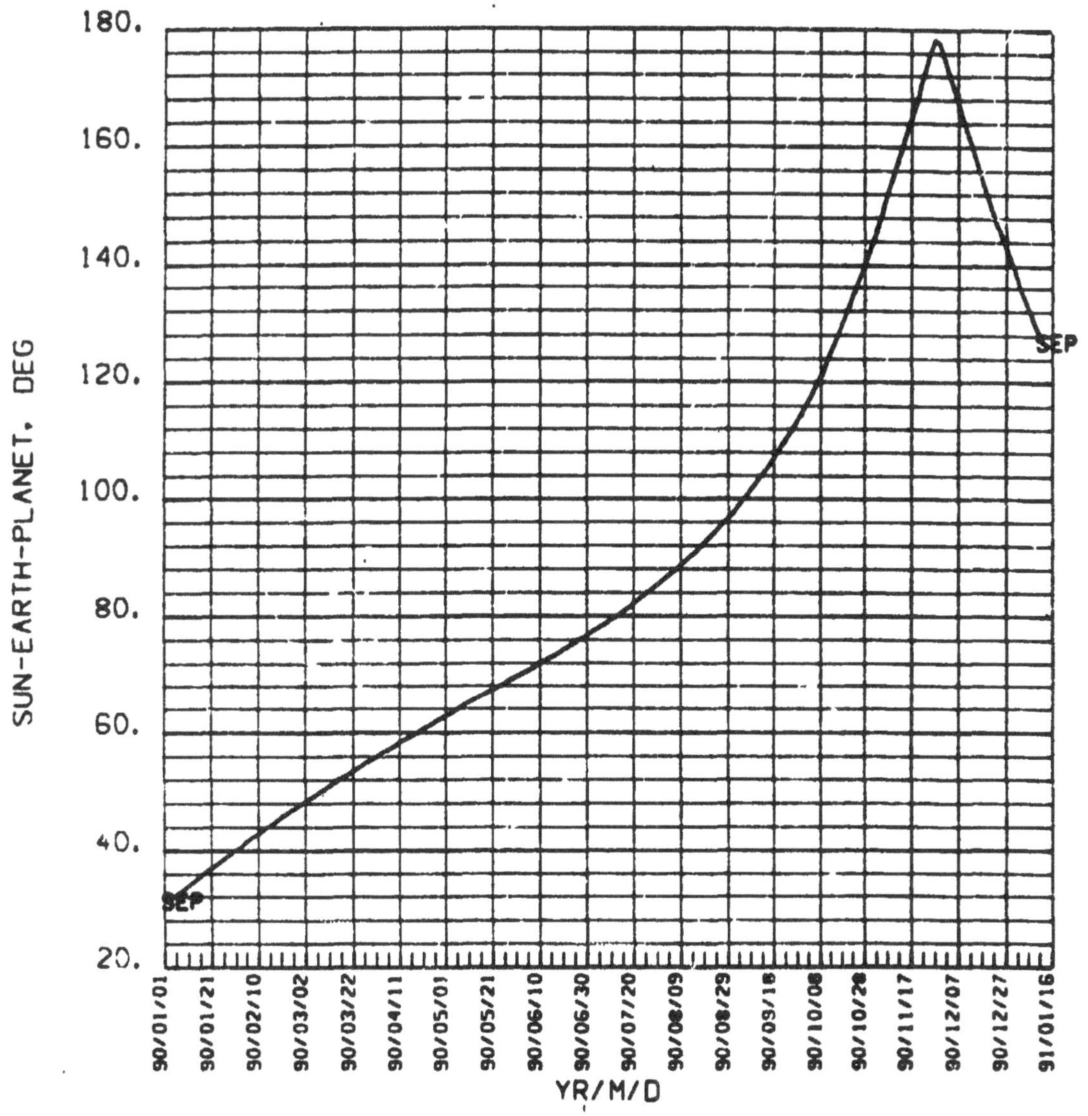

180.
160.
140.
120.
100.
80.
60.
40.
20.

SUN-EARTH-PLANET, DEG

SEP
SEP

90/01/01
90/01/21
90/02/10
90/03/02
90/03/22
90/04/11
90/05/01
90/05/21
90/06/10
90/06/30
90/07/20
90/08/09
90/08/29
90/09/18
90/10/08
90/10/28
90/11/17
90/12/07
90/12/27
91/01/16

YR/M/D

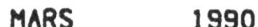

STATION RISE/SET GMT.HR

24.
18.
12.
6.
0.

SET 14
RISE43
SET 63
RISE14
SET 43
RISE63

RISE14
SET 43
RISE63
SET 14
RISE43
SET 63

90/01/01
90/01/21
90/02/10
90/03/02
90/03/22
90/04/11
90/05/01
90/05/21
90/06/10
90/06/30
90/07/20
90/08/09
90/08/29
90/09/18
90/10/08
90/10/28
90/11/17
90/12/07
90/12/27
91/01/16

YR/M/D

MARS 1991

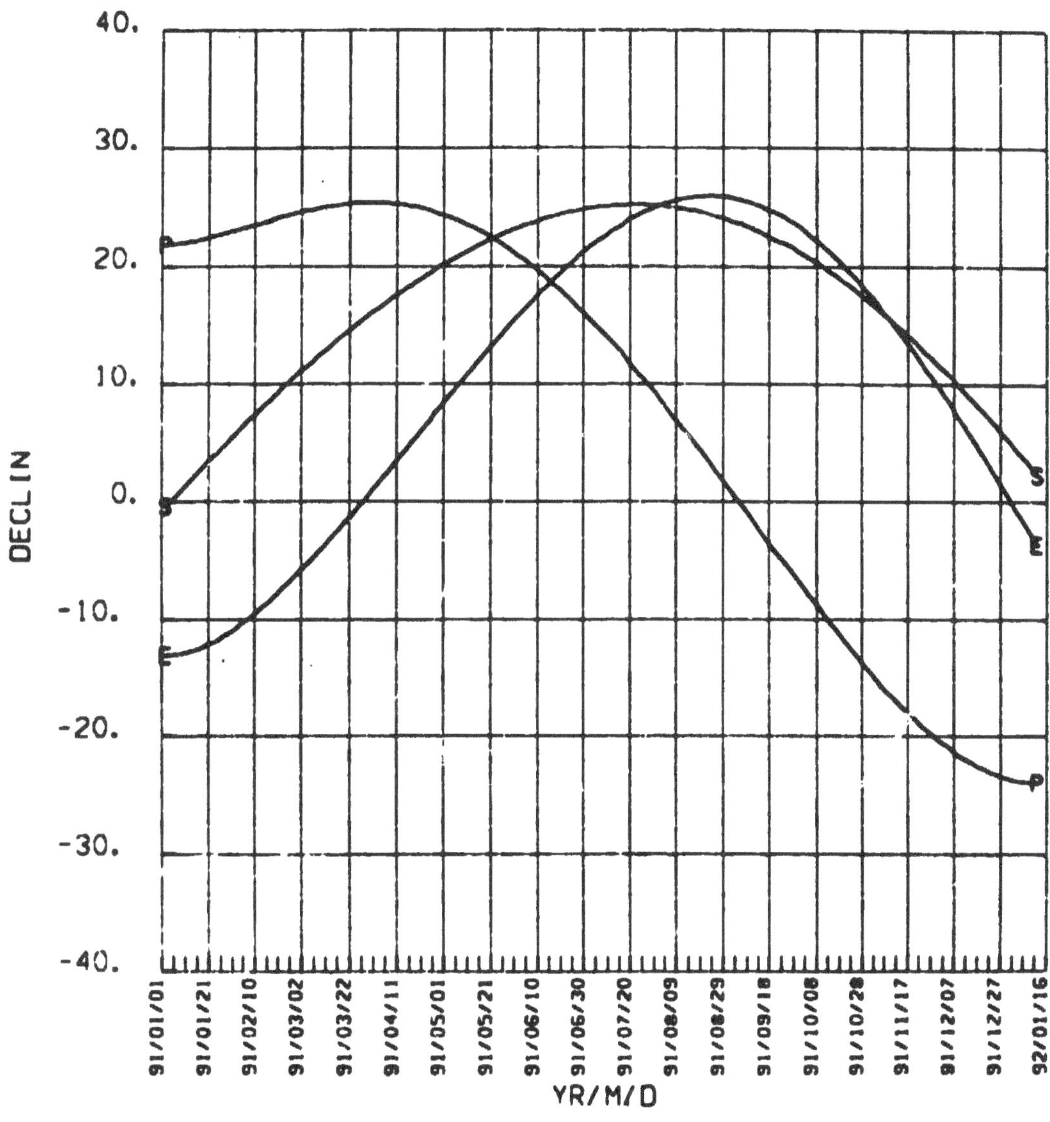

40.
30.
20.
10.
0.
-10.
-20.
-30.
-40.
DECLIN
91/01/01
91/01/21
91/02/10
91/03/02
91/03/22
91/04/11
91/05/01
91/05/21
91/06/10
91/06/30
91/07/20
91/08/09
91/08/29
91/09/18
91/10/08
91/10/28
91/11/17
91/12/07
91/12/27
92/01/16
YR/M/D
P
S
E
S
E
P

MARS 1991 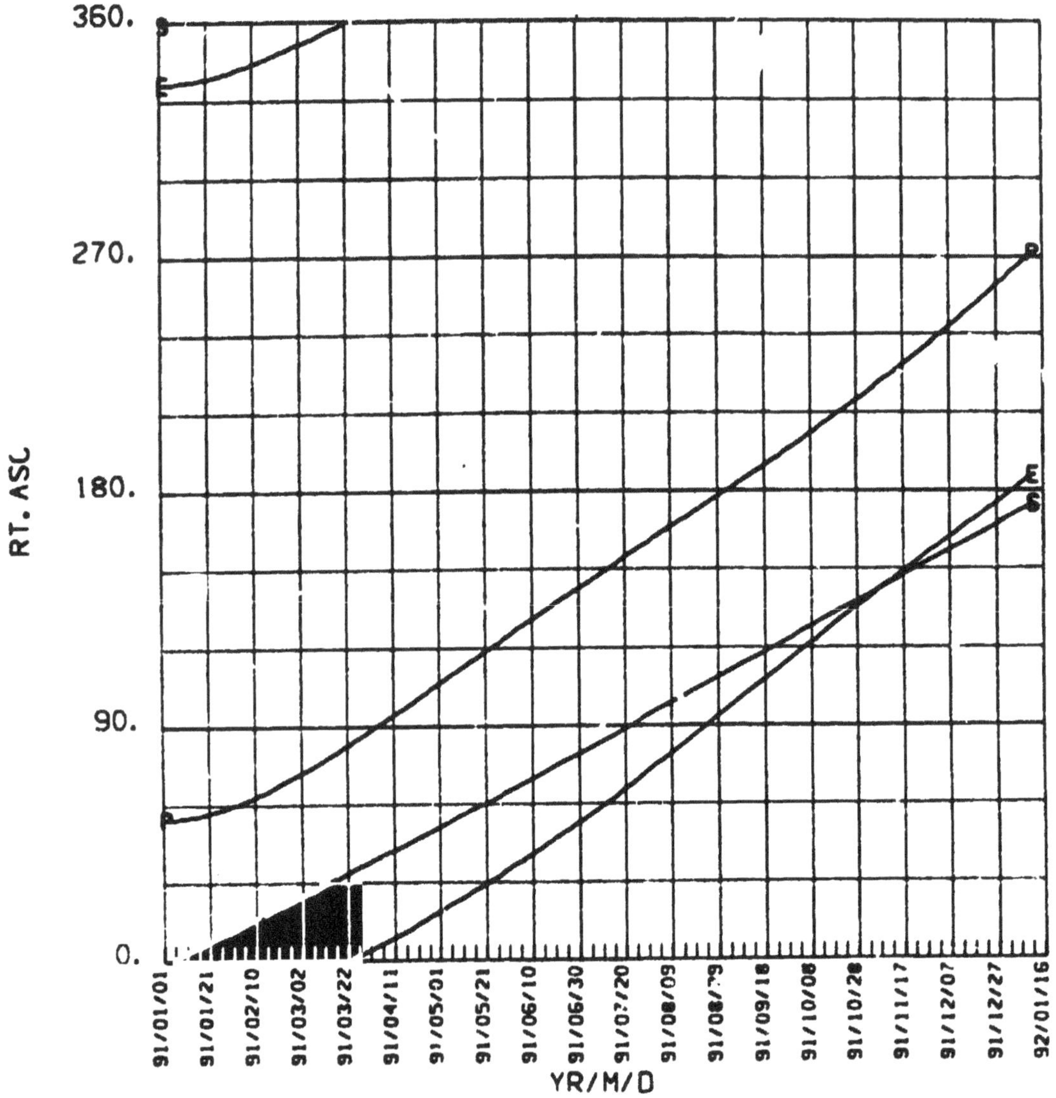
RT. ASC
360.
270.
180.
90.
0.
91/01/01
91/01/21
91/02/10
91/03/02
91/03/22
91/04/11
91/05/01
91/05/21
91/06/10
91/06/30
91/07/20
91/08/09
91/09/18
91/10/08
91/10/28
91/11/17
91/12/07
91/12/27
92/01/16
YR/M/D

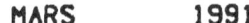

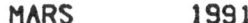
MARS 1991
360.
270.
180.
90.
0.
EC. LON
91/01/01
91/01/21
91/02/10
91/03/02
91/03/22
91/04/11
91/05/01
91/05/21
91/06/10
91/06/30
91/07/20
91/08/09
91/08/29
91/09/18
91/10/08
91/10/28
91/11/17
91/12/07
91/12/27
92/01/16
YR/M/D
P
P

MARS 1991

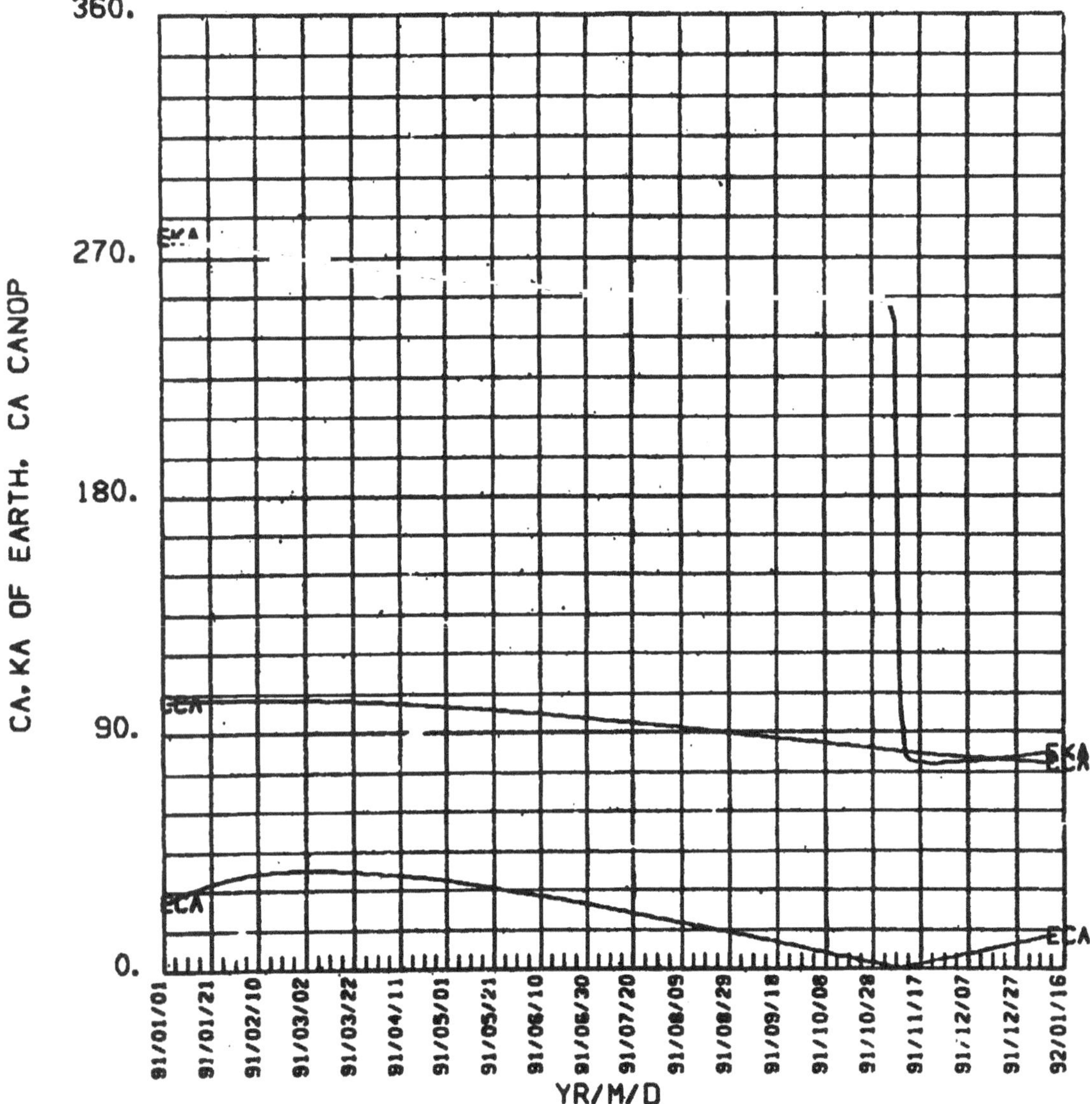

360.
270.
180.
90.
0.
CA, KA OF EARTH, CA CANOP
EKA
ECA
91/01/01
91/01/21
91/02/10
91/03/02
91/03/22
91/04/11
91/05/01
91/05/21
91/06/10
91/06/30
91/07/20
91/08/09
91/08/29
91/09/18
91/10/08
91/10/28
91/11/17
91/12/07
91/12/27
92/01/16
YR/M/D

MARS 1991

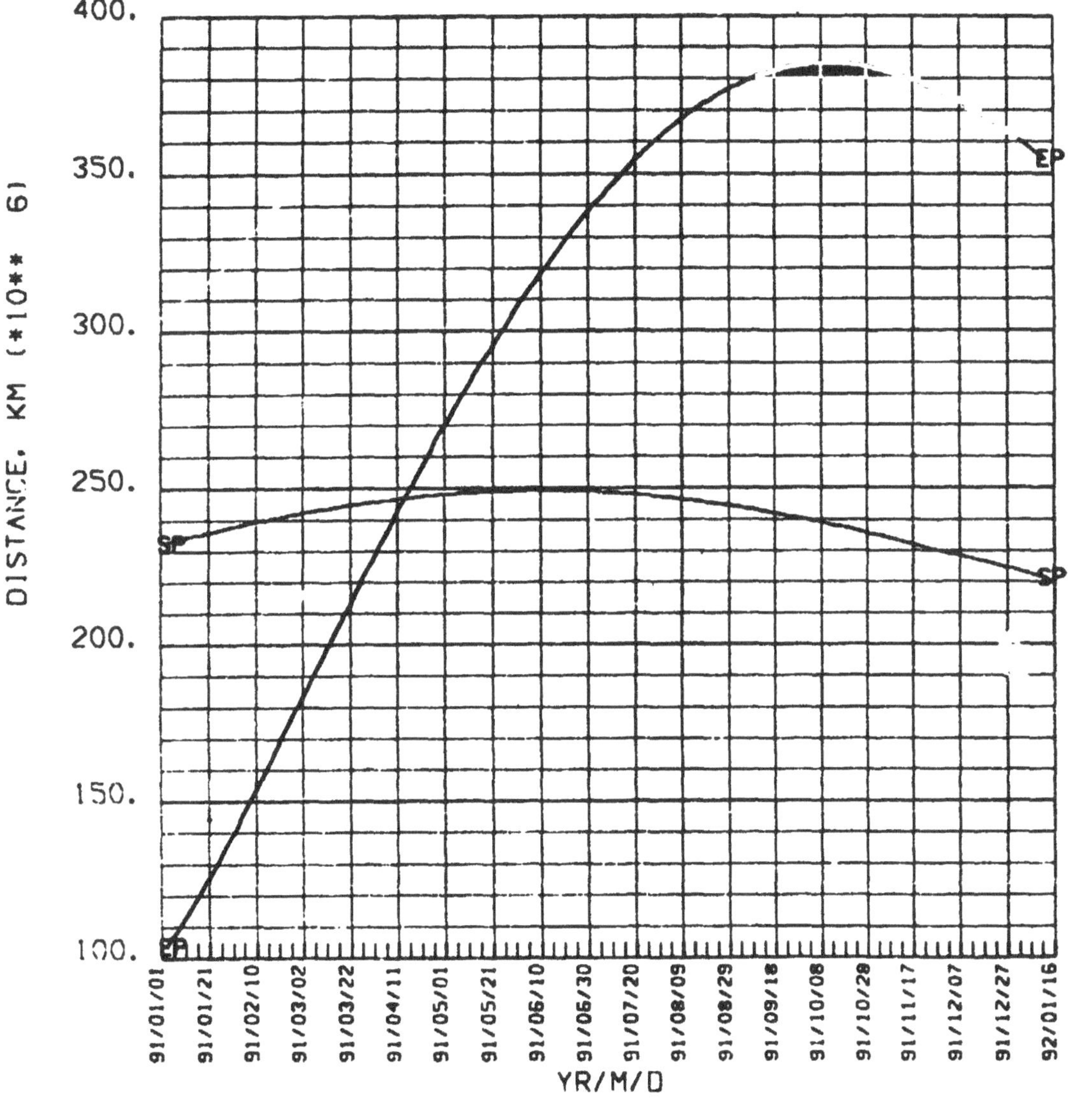

400.
350.
300.
250.
200.
150.
100.
DISTANCE, KM (*10** 6)
EP
SP
SP
EP
91/01/01
91/01/21
91/02/10
91/03/02
91/03/22
91/04/11
91/05/01
91/05/21
91/06/10
91/06/30
91/07/20
91/08/09
91/08/29
91/09/18
91/10/08
91/10/28
91/11/17
91/12/07
91/12/27
92/01/16
YR/M/D

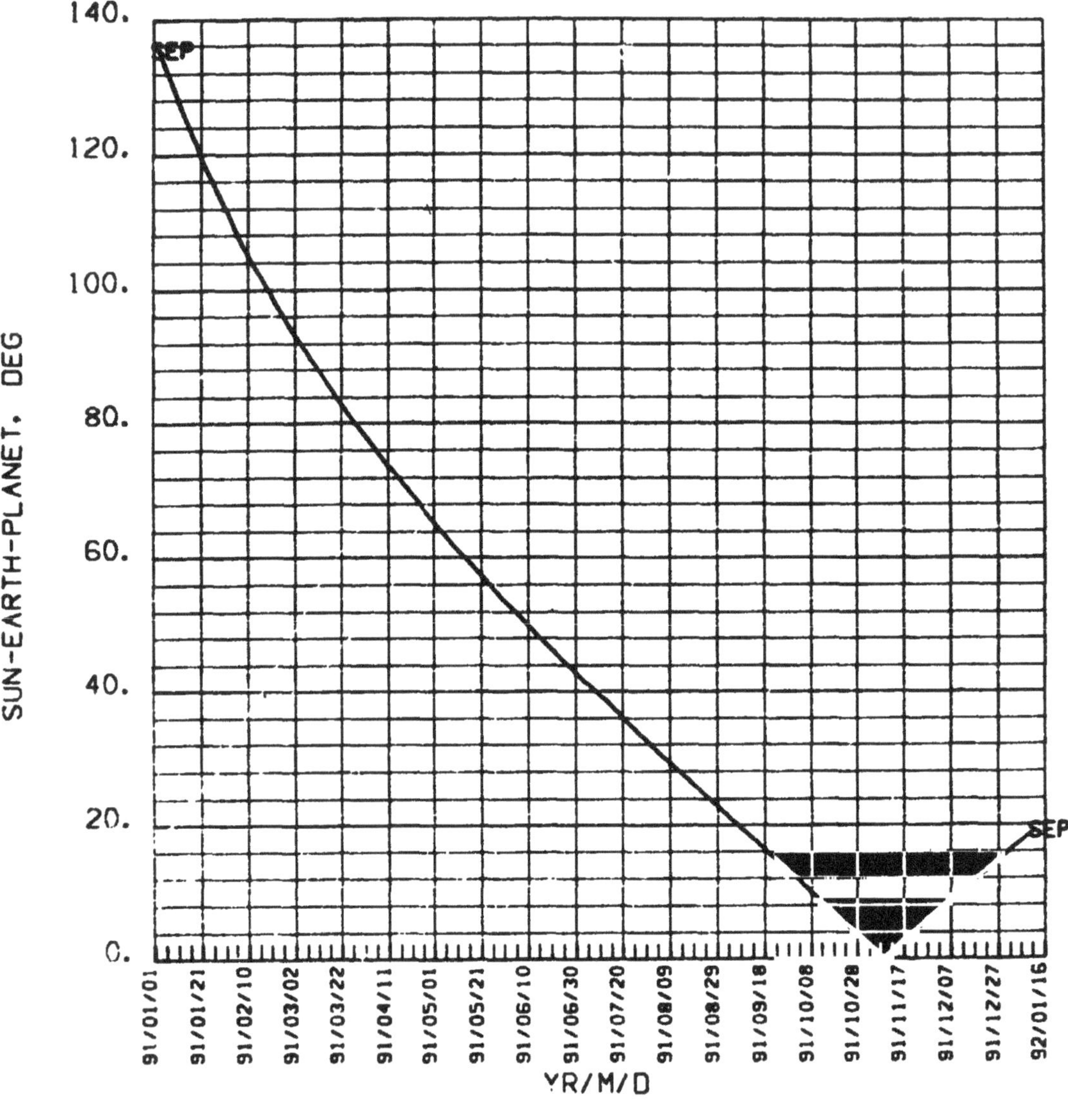
MARS 1991
140.
120.
100.
80.
60.
40.
20.
0.
SEP
SEP
SUN-EARTH-PLANET, DEG
91/01/01
91/01/21
91/02/10
91/03/02
91/03/22
91/04/11
91/05/01
91/05/21
91/06/10
91/06/30
91/07/20
91/08/09
91/08/29
91/09/18
91/10/08
91/10/28
91/11/17
91/12/07
91/12/27
92/01/16
YR/M/D

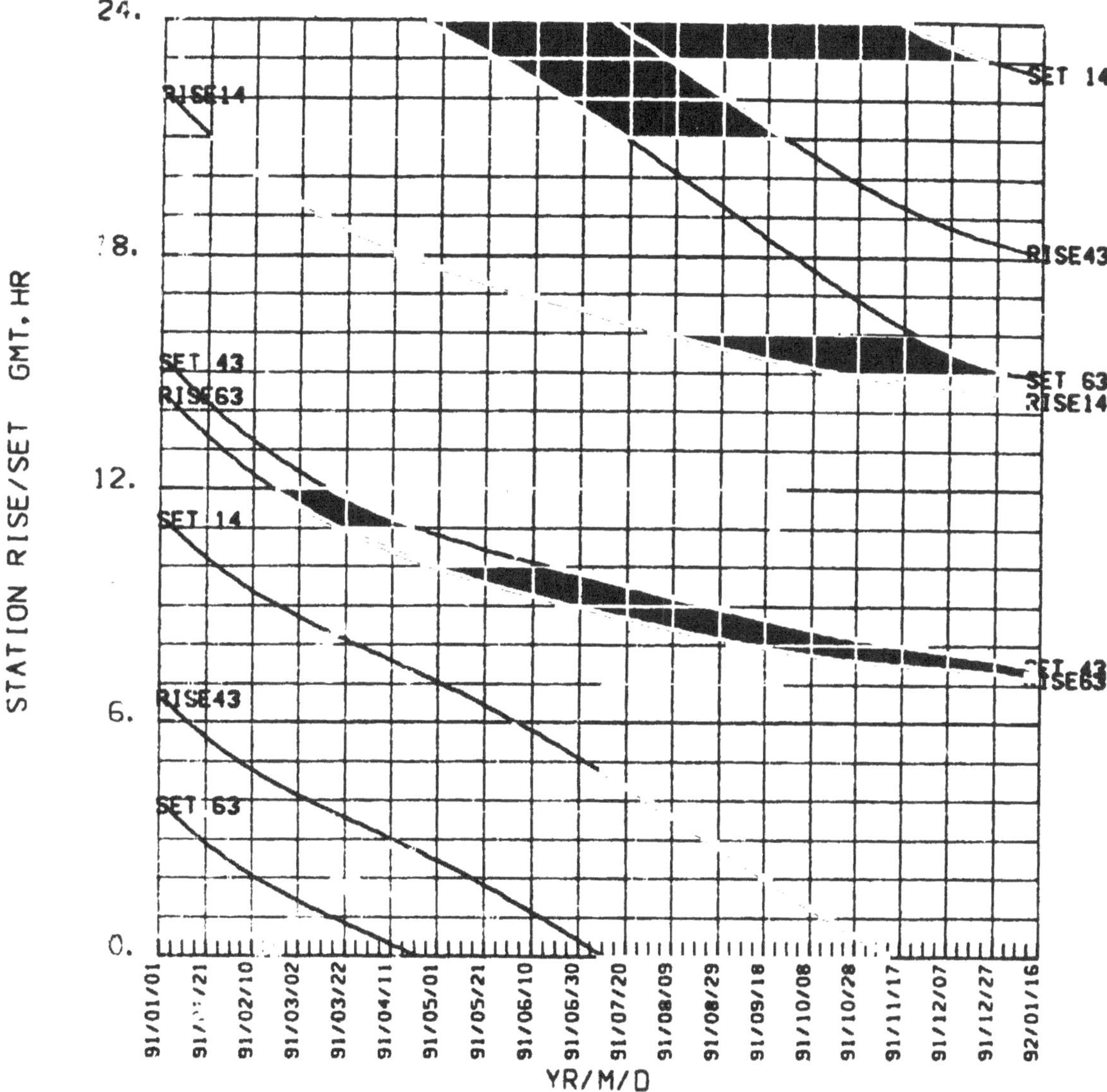
MARS 1991
STATION RISE/SET GMT, HR
YR/M/D
24.
18.
12.
6.
0.
RISE14
SET 14
RISE43
SET 63
RISE14
SET 43
RISE63
SET 14
RISE43
SET 63
SET 43
RISE63
91/01/01
91/02/10
91/03/02
91/03/22
91/04/11
91/05/01
91/05/21
91/06/10
91/06/30
91/07/20
91/08/09
91/08/29
91/09/18
91/10/08
91/10/28
91/11/17
91/12/07
91/12/27
92/01/16

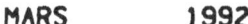

DECLIN

40.
30.
20.
10.
0.
-10.
-20.
-30.
-40.

92/01/01 92/01/21 92/02/10 92/03/01 92/03/21 92/04/10 92/04/30 92/05/20 92/06/09 92/06/29 92/07/19 92/08/08 92/08/28 92/09/17 92/10/07 92/10/27 92/11/16 92/12/06 92/12/26 93/01/15

YR/M/D

P
S
E

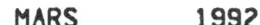
MARS 1992
RT. ASC
360.
270.
180.
90.
0.
92/01/01
92/01/21
92/02/10
92/03/01
92/03/21
92/04/10
92/04/30
92/05/20
92/06/09
92/06/29
92/07/19
92/08/08
92/08/28
92/09/17
92/10/07
92/10/27
92/11/16
92/12/06
92/12/26
93/01/15
YR/M/D

MARS 1992

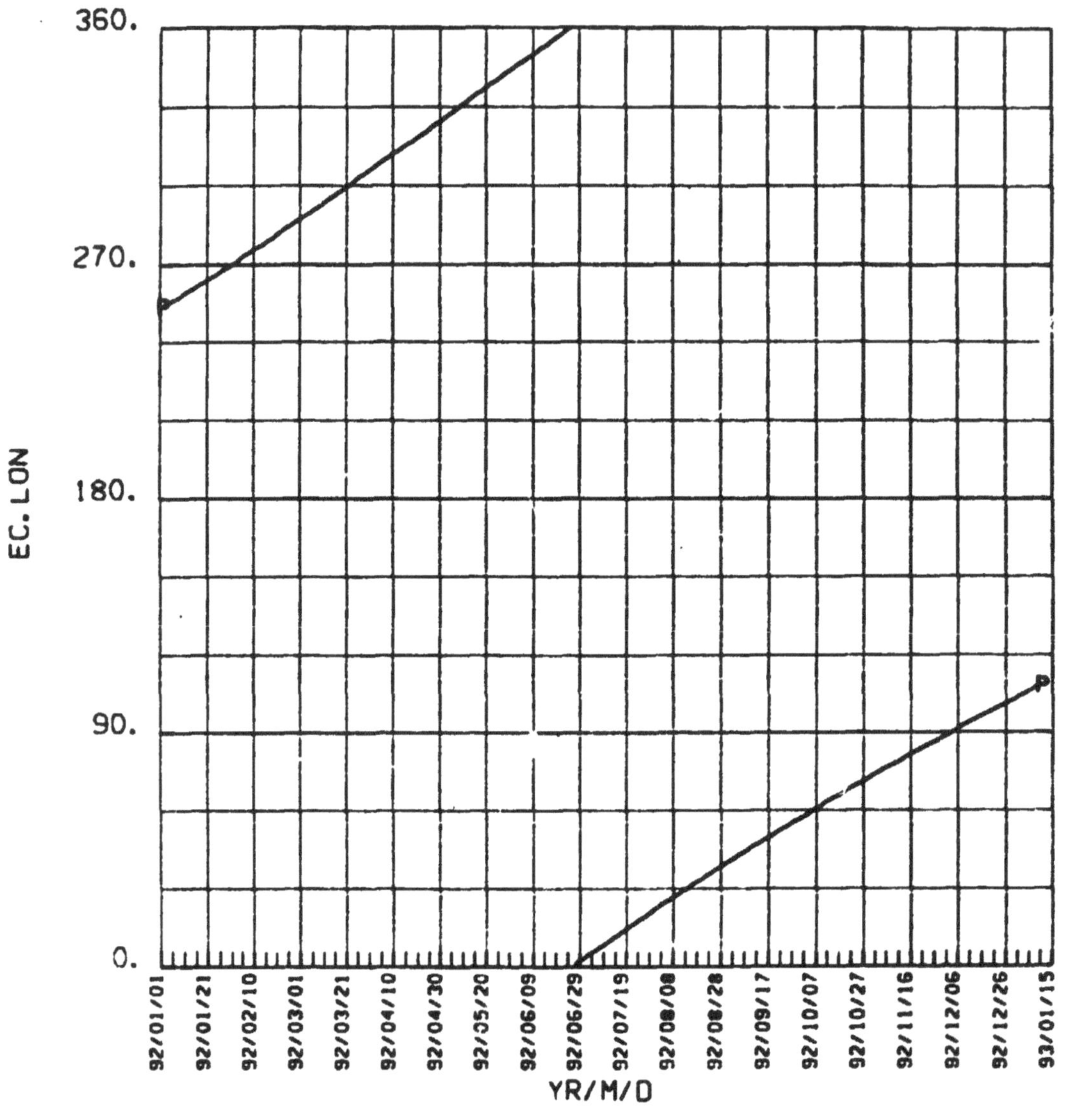

360.
270.
180.
90.
0.
EC. LON
92/01/01
92/01/21
92/02/10
92/03/01
92/03/21
92/04/10
92/04/30
92/05/20
92/06/09
92/06/29
92/07/19
92/08/08
92/08/28
92/09/17
92/10/07
92/10/27
92/11/16
92/12/06
92/12/26
93/01/15
YR/M/D

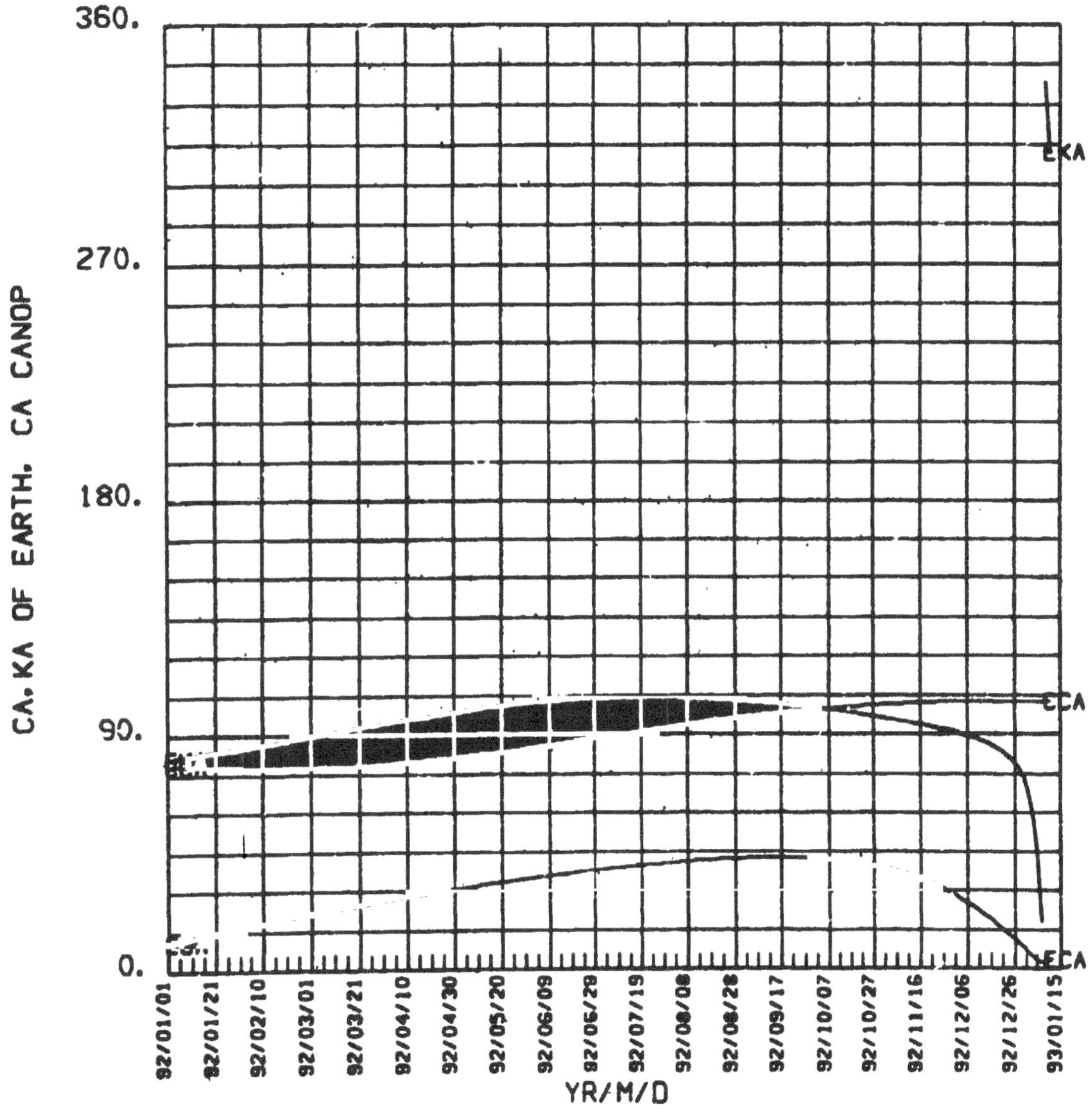

MARS 1992
360.
270.
180.
90.
0.
CA.KA OF EARTH. CA CANOP
EKA
ECA
ECA
92/01/01
92/01/21
92/02/10
92/03/01
92/03/21
92/04/10
92/04/30
92/05/20
92/06/09
92/06/29
92/07/19
92/08/08
92/08/28
92/09/17
92/10/07
92/10/27
92/11/16
92/12/06
92/12/26
93/01/15
YR/M/D

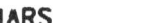

DISTANCE, KM (*10** 6)

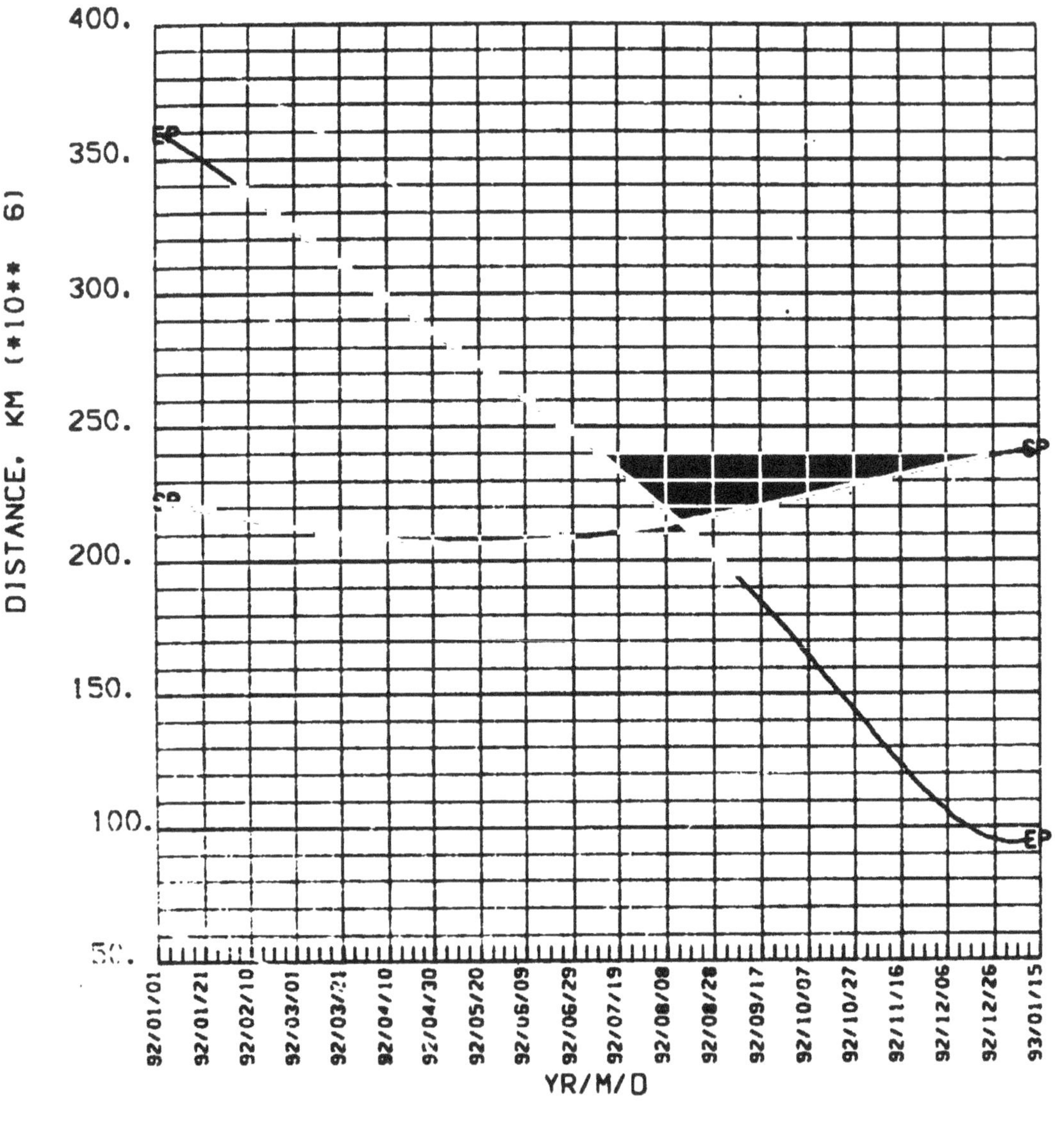

400.
350.
300.
250.
200.
150.
100.
50.

EP

YR/M/D

92/01/01 92/01/21 92/02/10 92/03/01 92/03/21 92/04/10 92/04/30 92/05/20 92/06/09 92/06/29 92/07/19 92/08/08 92/08/28 92/09/17 92/10/07 92/10/27 92/11/16 92/12/06 92/12/26 93/01/15

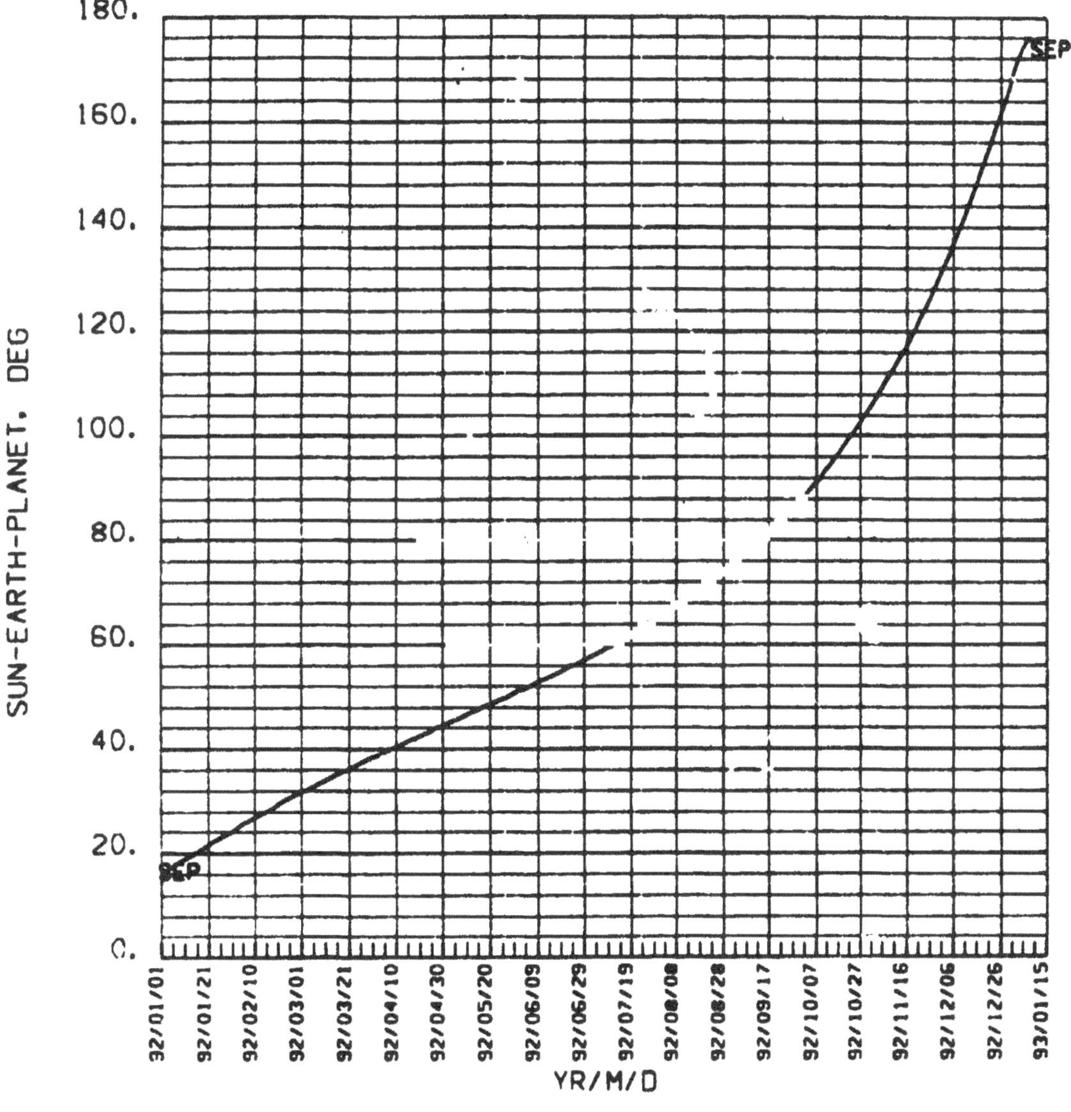

MARS 1992
180.
160.
140.
120.
100.
80.
60.
40.
20.
0.
SUN-EARTH-PLANET, DEG
SEP
SEP
92/01/01
92/01/21
92/02/10
92/03/01
92/03/21
92/04/10
92/04/30
92/05/20
92/06/09
92/06/29
92/07/19
92/08/08
92/08/28
92/09/17
92/10/07
92/10/27
92/11/16
92/12/06
92/12/26
93/01/15
YR/M/D

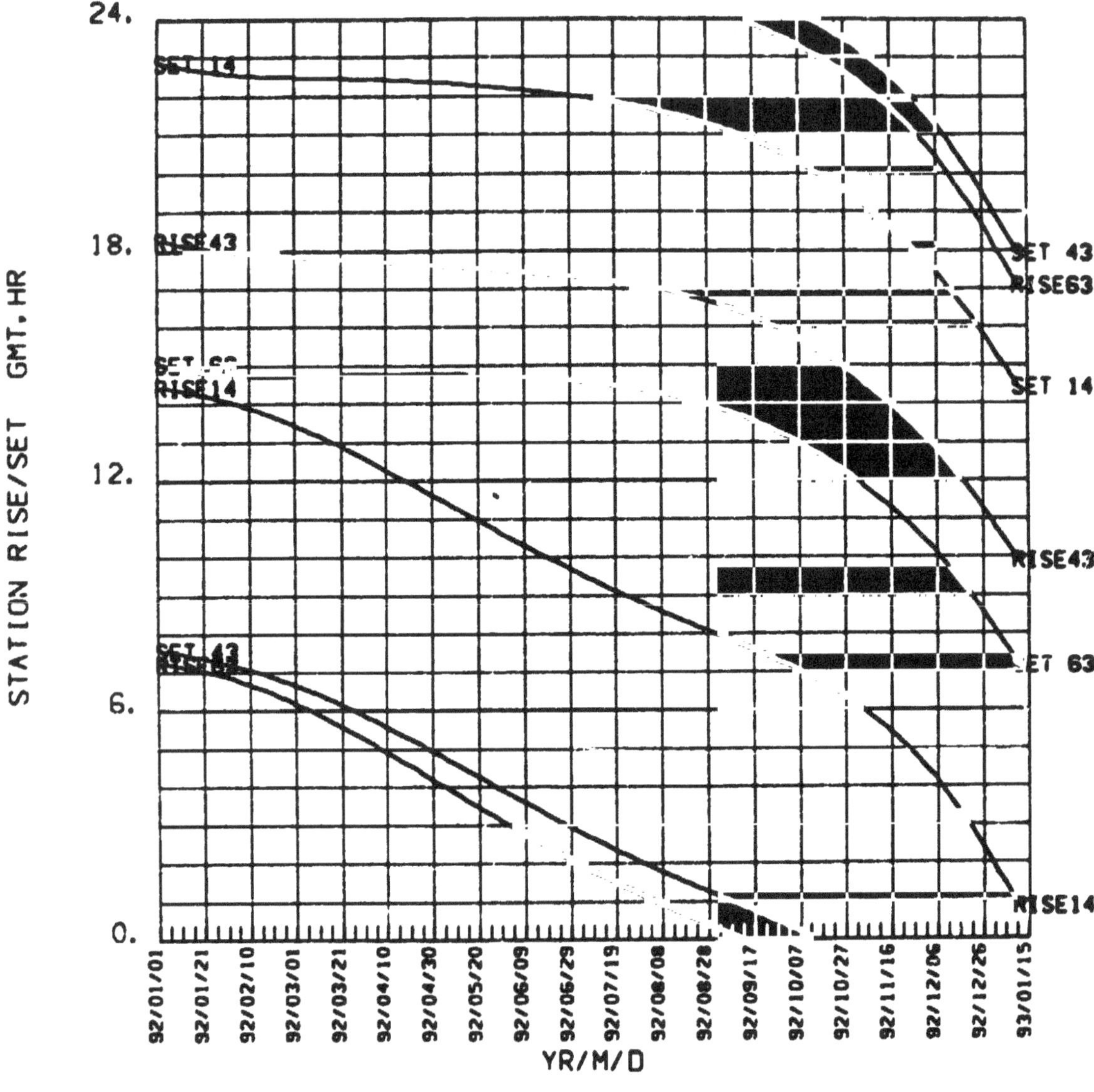

MARS 1992
STATION RISE/SET GMT.HR
YR/M/D
24.
18.
12.
6.
0.
SET 14
RISE43
RISE14
SET 43
SET 43
RISE63
SET 14
RISE43
SET 63
RISE14
92/01/01
92/01/21
92/02/10
92/03/01
92/03/21
92/04/10
92/04/30
92/05/20
92/06/09
92/06/29
92/07/19
92/08/08
92/08/28
92/09/17
92/10/07
92/10/27
92/11/16
92/12/06
92/12/26
93/01/15

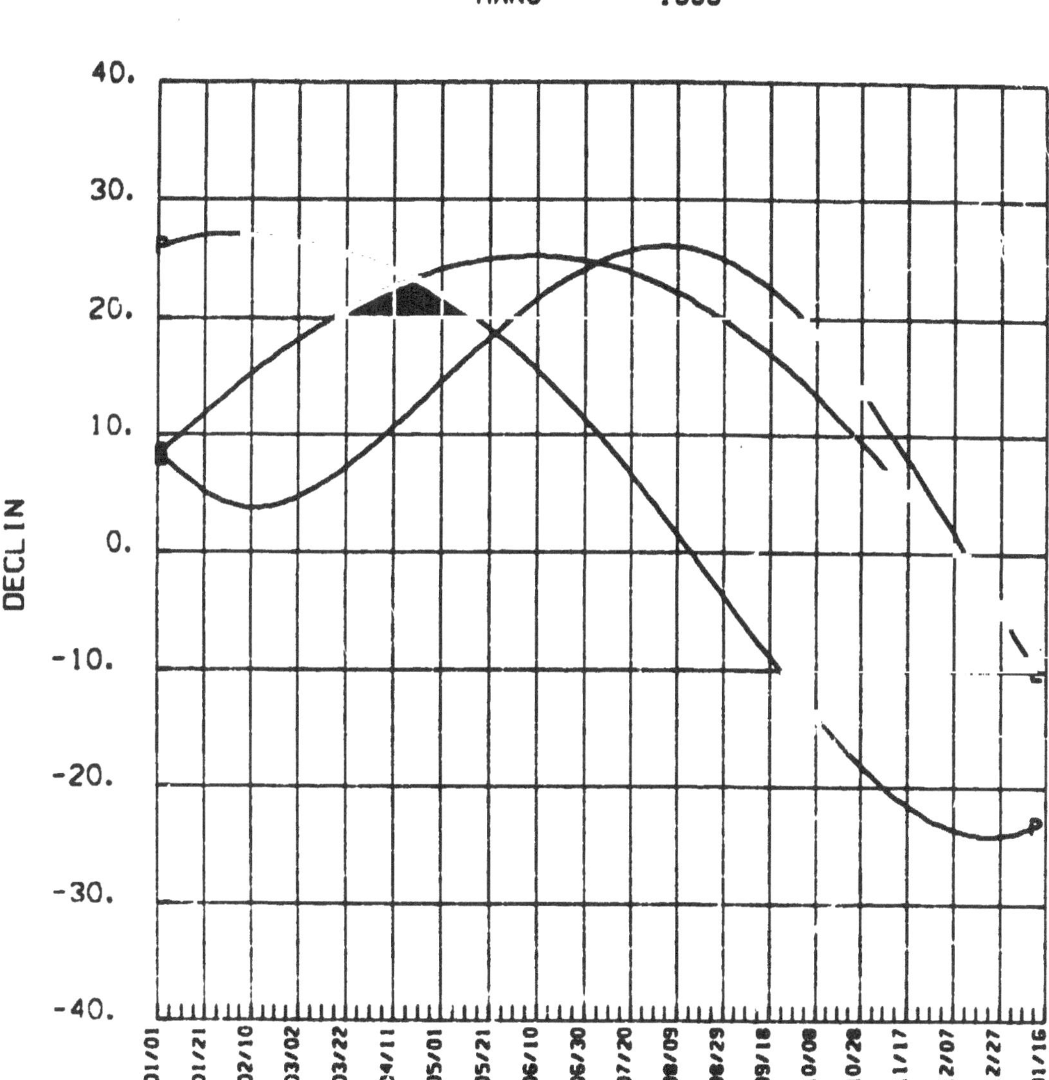

MARS 1993
40.
30.
20.
10.
0.
-10.
-20.
-30.
-40.
DECLIN
93/01/01
93/01/21
93/02/10
93/03/02
93/03/22
93/04/11
93/05/01
93/05/21
93/06/10
93/06/30
93/07/20
93/08/09
93/08/29
93/09/18
93/10/08
93/10/28
93/11/17
93/12/07
93/12/27
94/01/16
YR/M/D

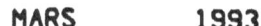
MARS 1993
360.
270.
180.
90.
0.
RT. ASC
93/01/01
93/01/21
93/02/10
93/03/02
93/03/22
93/04/11
93/05/01
93/05/21
93/06/10
93/06/30
93/07/20
93/08/09
93/08/29
93/09/18
93/10/08
93/10/28
93/11/17
93/12/07
93/12/27
94/01/16
YR/M/D

MARS 1993 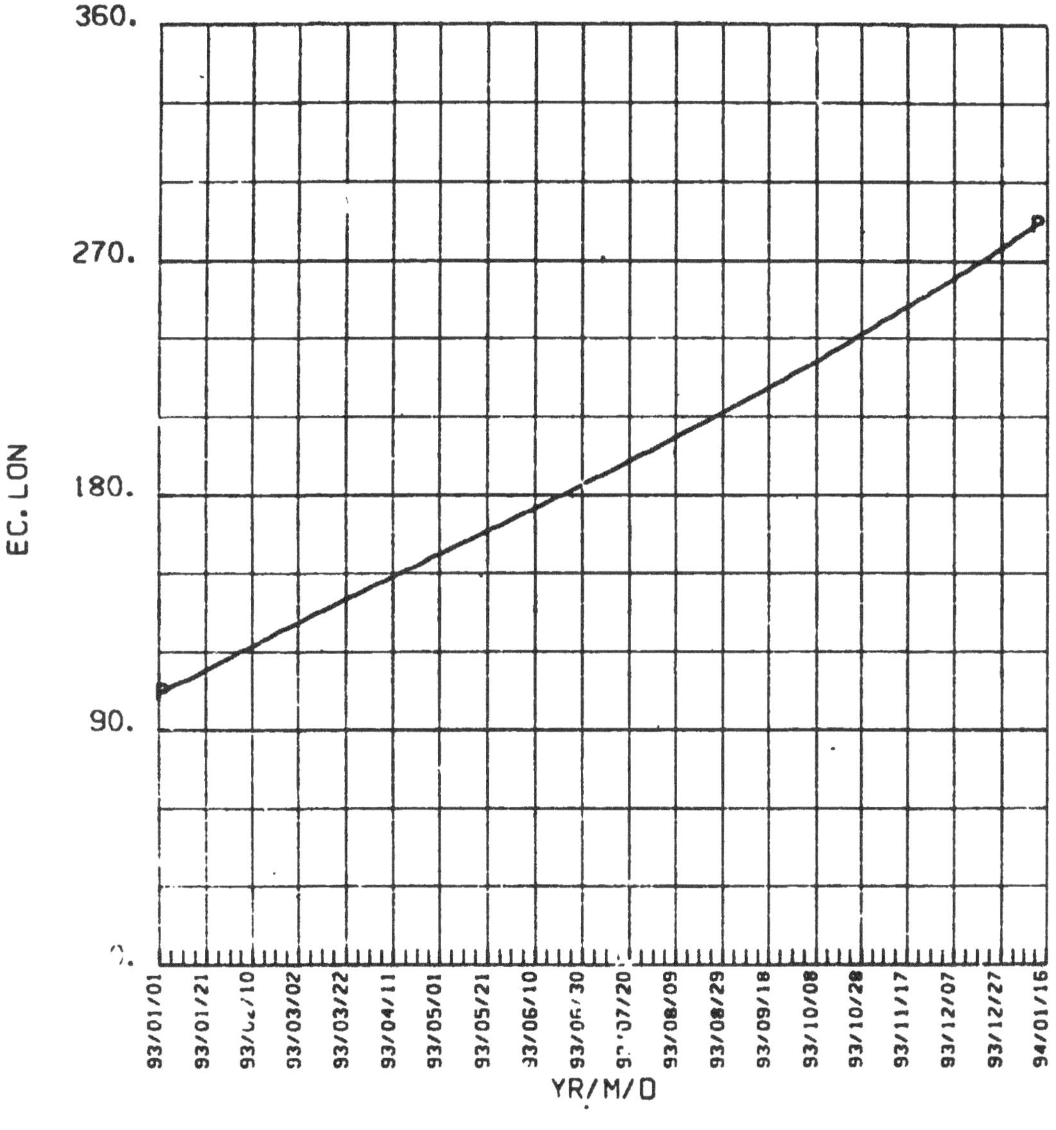
360.
270.
180.
90.
0.
EC. LON
93/01/01
93/01/21
93/02/10
93/03/02
93/03/22
93/04/11
93/05/01
93/05/21
93/06/10
93/06/30
93/07/20
93/08/09
93/08/29
93/09/18
93/10/08
93/10/28
93/11/17
93/12/07
93/12/27
94/01/16
YR/M/D

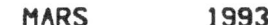

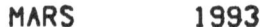
MARS 1993

360.
270.
180.
90.
0.
CA. KA OF EARTH. CA CANOP
93/01/01
93/01/21
93/02/10
93/03/02
93/03/22
93/04/11
93/05/01
93/05/21
93/06/10
93/06/30
93/07/20
93/08/09
93/08/29
93/09/18
93/10/08
93/10/28
93/11/17
93/12/07
93/12/27
94/01/16
YR/M/D

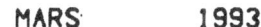

MARS 1993

DISTANCE, KM (*10** 6)

400.
350.
300.
250.
200.
150.
100.
50.

EP
SP
SP
EP

93/01/01 93/01/21 93/02/10 93/03/02 93/03/22 93/04/11 93/05/01 93/05/21 93/06/10 93/06/30 93/07/20 93/08/09 93/08/29 93/09/18 93/10/08 93/10/28 93/11/17 93/12/07 93/12/27 94/01/16

YR/M/D

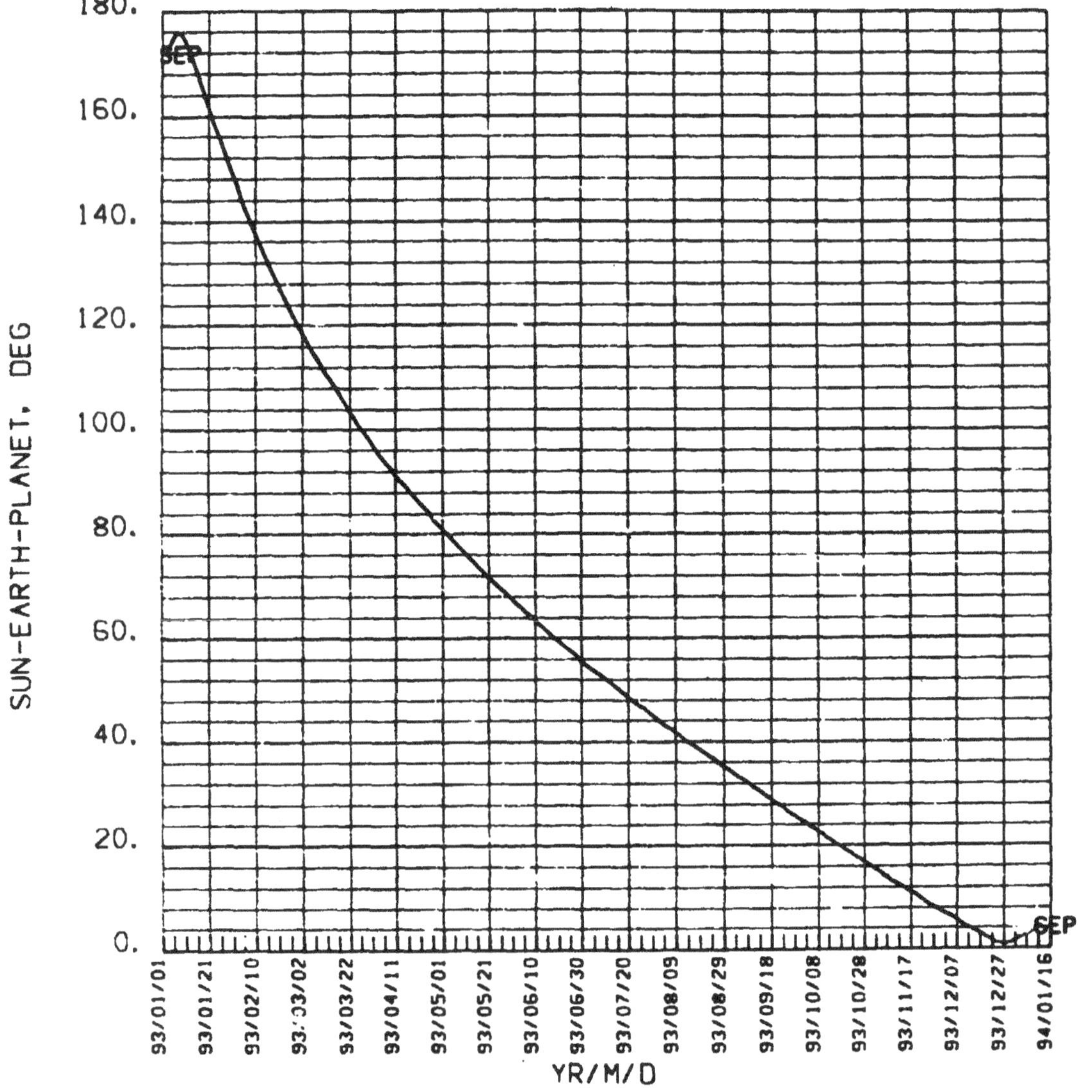
MARS 1993
180.
160.
140.
120.
100.
80.
60.
40.
20.
0.
SUN-EARTH-PLANET, DEG
SEP
SEP
93/01/01
93/01/21
93/02/10
93/03/02
93/03/22
93/04/11
93/05/01
93/05/21
93/06/10
93/06/30
93/07/20
93/08/09
93/08/29
93/09/18
93/10/08
93/10/28
93/11/17
93/12/07
93/12/27
94/01/16
YR/M/D

MARS 1993

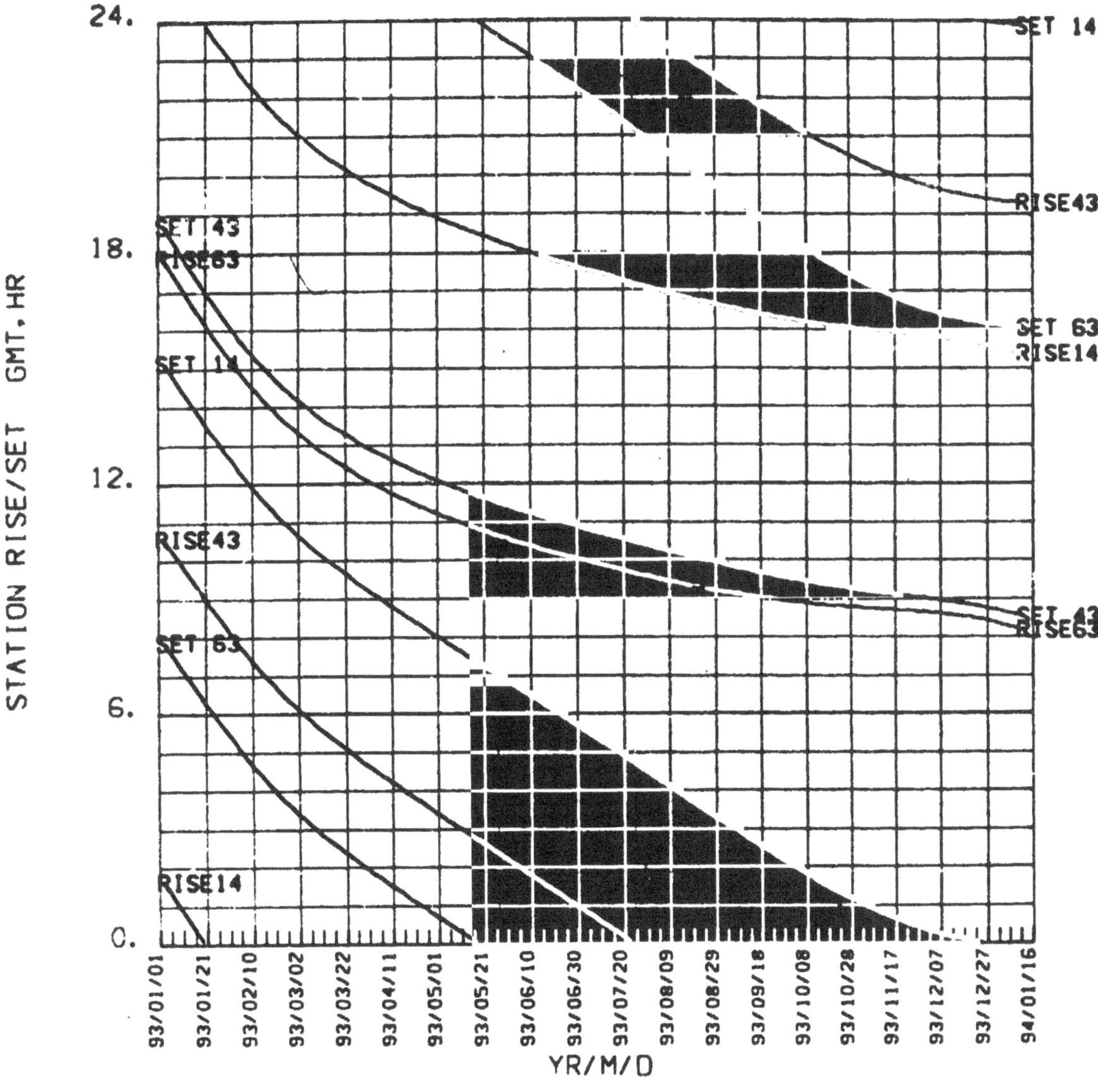

24.
18.
12.
6.
0.
STATION RISE/SET GMT, HR
SET 14
RISE43
SET 43
RISE63
SET 63
RISE14
SET 14
RISE43
SET 63
RISE14
SET 43
RISE63
93/01/01
93/01/21
93/02/10
93/03/02
93/03/22
93/04/11
93/05/01
93/05/21
93/06/10
93/06/30
93/07/20
93/08/09
93/08/29
93/09/18
93/10/08
93/10/28
93/11/17
93/12/07
93/12/27
94/01/16
YR/M/D

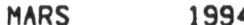
MARS 1994

40.
30.
20.
10.
0.
-10.
-20.
-30.
-40.
DECLIN
E
S
P
94/01/01
94/01/21
94/02/10
94/03/02
94/03/22
94/04/11
94/05/01
94/05/21
94/06/10
94/06/30
94/07/20
94/08/09
94/08/29
94/09/18
94/10/08
94/10/28
94/11/17
94/12/07
94/12/27
95/01/16
YR/M/D

MARS 1994 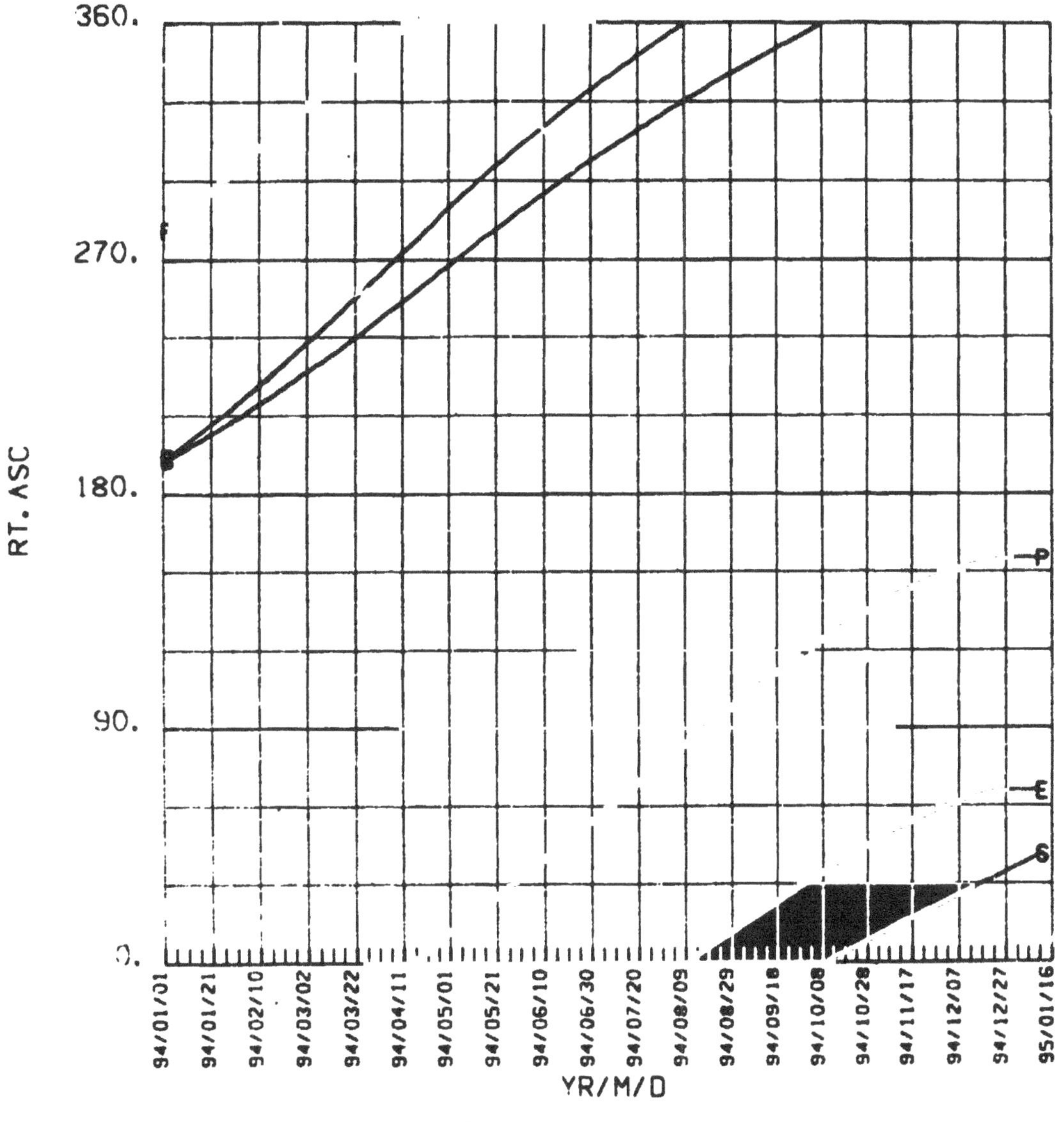
RT. ASC
360.
270.
180.
90.
0.
P
E
S
94/01/01
94/01/21
94/02/10
94/03/02
94/03/22
94/04/11
94/05/01
94/05/21
94/06/10
94/06/30
94/07/20
94/08/09
94/08/29
94/09/18
94/10/08
94/10/28
94/11/17
94/12/07
94/12/27
95/01/16
YR/M/D

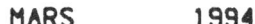

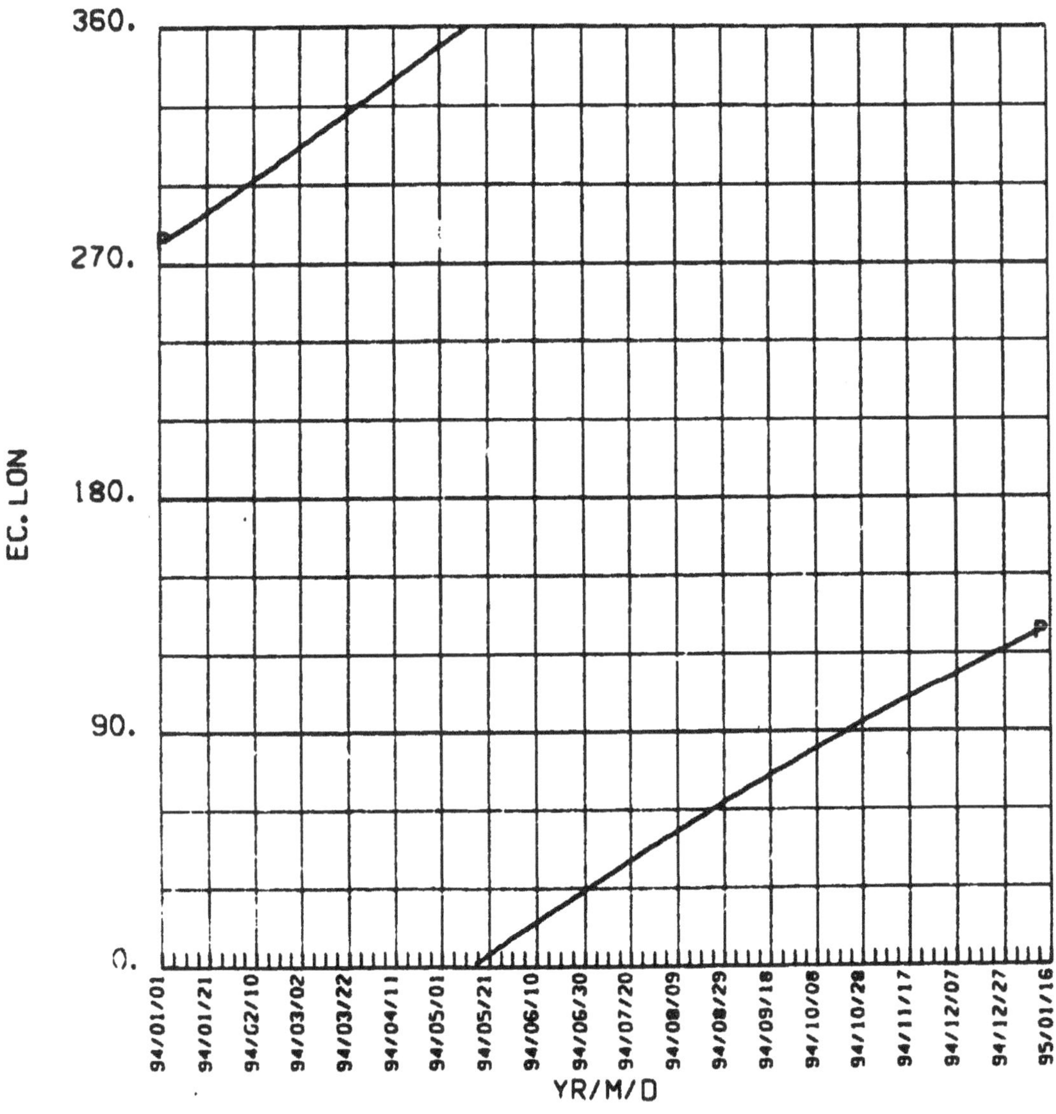
MARS 1994
360.
270.
180.
90.
0.
EC.LON
94/01/01
94/01/21
94/G2/10
94/03/02
94/03/22
94/04/11
94/05/01
94/05/21
94/06/10
94/06/30
94/07/20
94/08/09
94/08/29
94/09/18
94/10/08
94/10/28
94/11/17
94/12/07
94/12/27
95/01/16
YR/M/D

MARS 1994

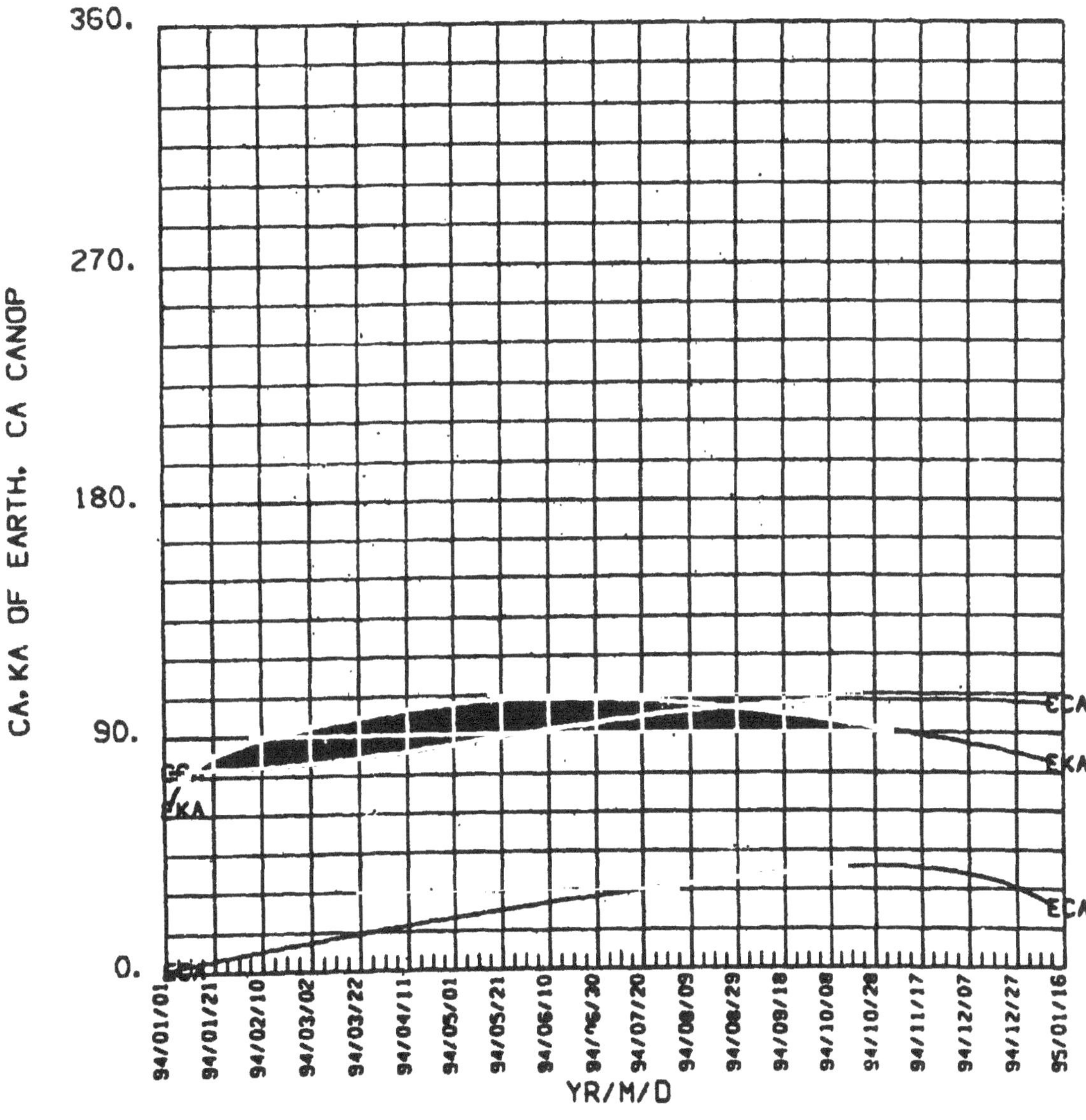

360.
270.
180.
90.
0.
CA.KA OF EARTH. CA CANOP
ECA
EKA
ECA
94/01/01
94/01/21
94/02/10
94/03/02
94/03/22
94/04/11
94/05/01
94/05/21
94/06/10
94/07/20
94/08/09
94/08/29
94/09/18
94/10/08
94/10/28
94/11/17
94/12/07
94/12/27
95/01/16
YR/M/D

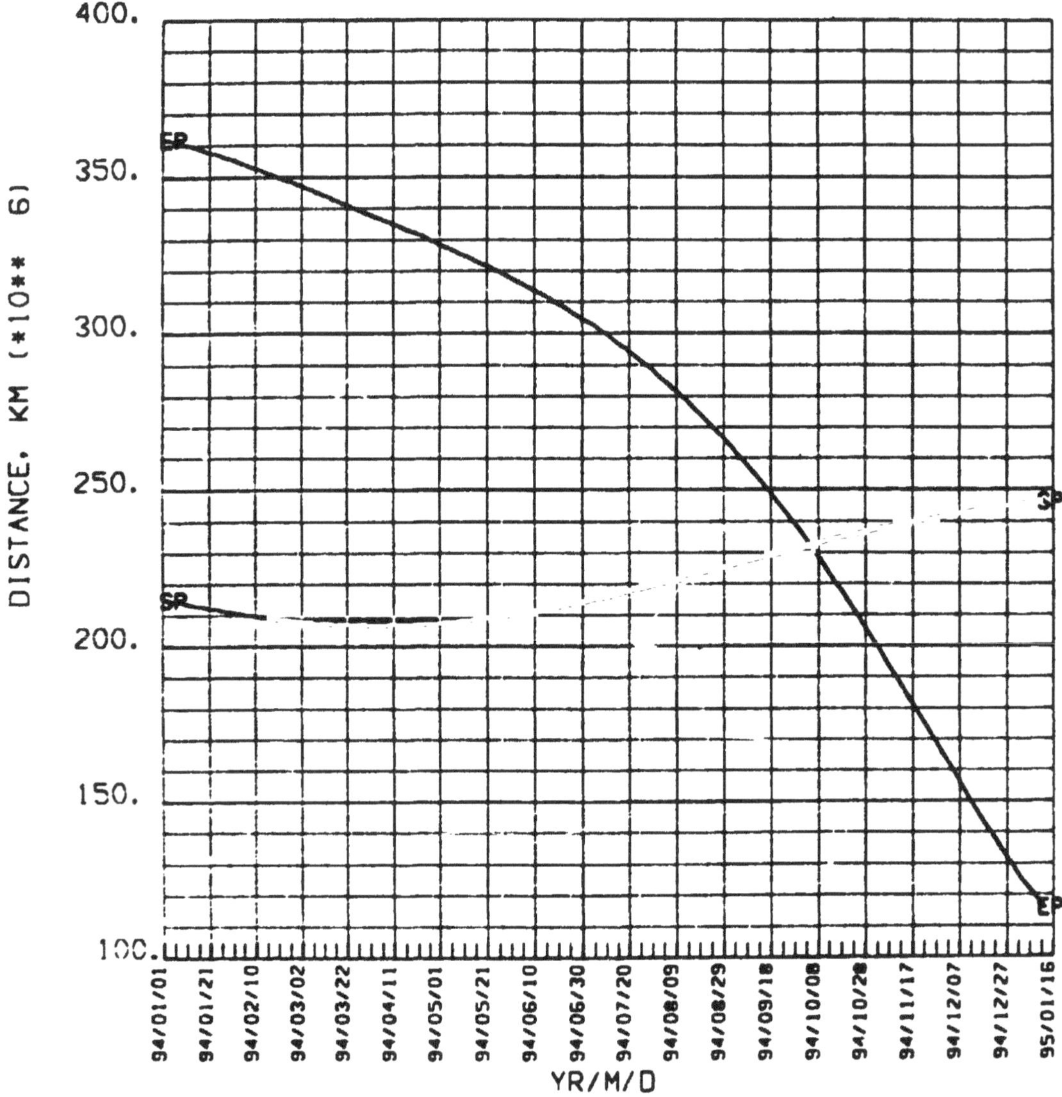

MARS 1994
400.
350.
300.
250.
200.
150.
100.
DISTANCE, KM (*10** 6)
SP
EP
94/01/01
94/01/21
94/02/10
94/03/02
94/03/22
94/04/11
94/05/01
94/05/21
94/06/10
94/06/30
94/07/20
94/08/09
94/08/29
94/09/18
94/10/08
94/10/28
94/11/17
94/12/07
94/12/27
95/01/16
YR/M/D

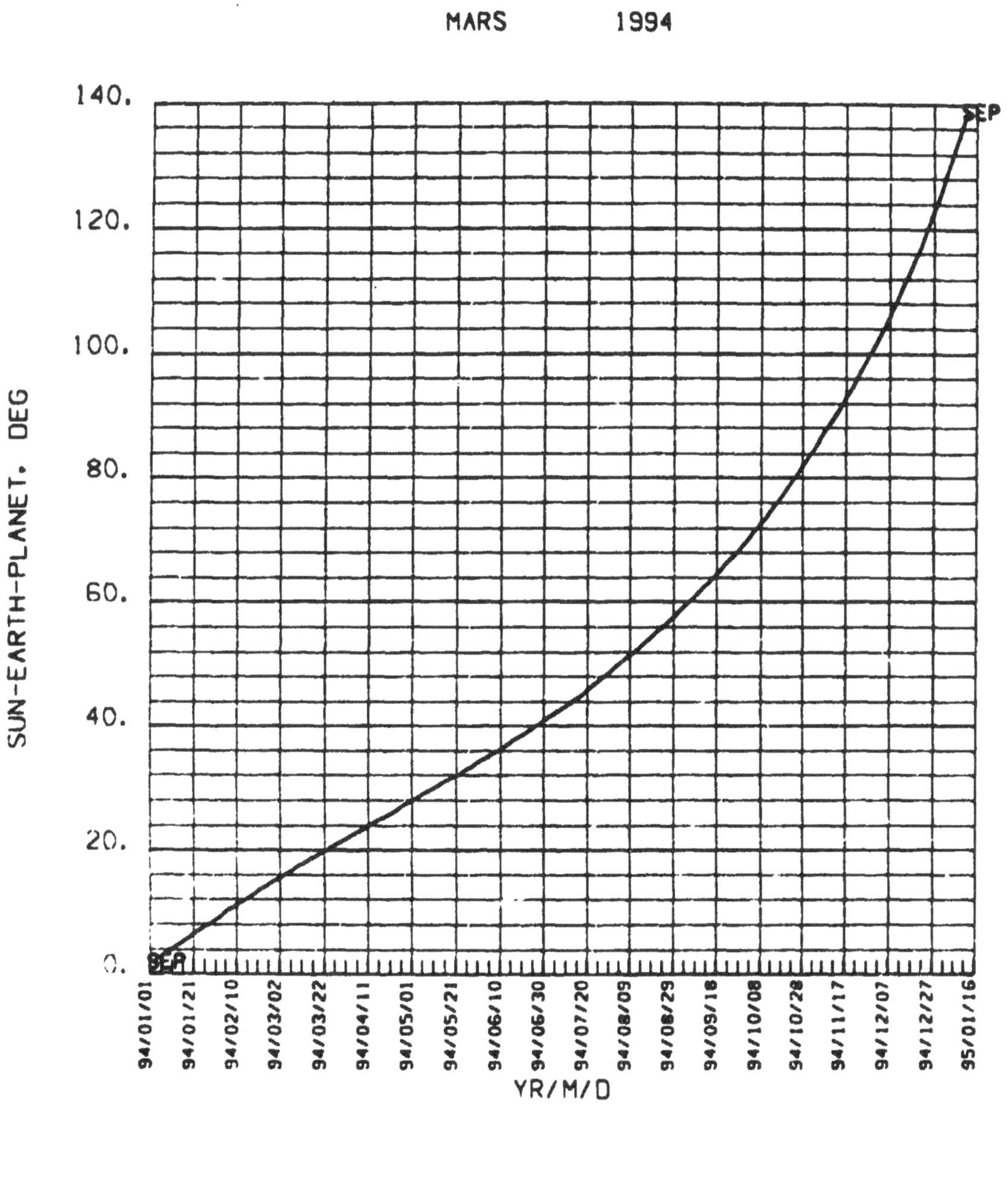

MARS 1994

SUN-EARTH-PLANET, DEG

140.
120.
100.
80.
60.
40.
20.
0.

SEP

SEP

94/01/01 94/01/21 94/02/10 94/03/02 94/03/22 94/04/11 94/05/01 94/05/21 94/06/10 94/06/30 94/07/20 94/08/09 94/08/29 94/09/18 94/10/08 94/10/28 94/11/17 94/12/07 94/12/27 95/01/16

YR/M/D

MARS 1994

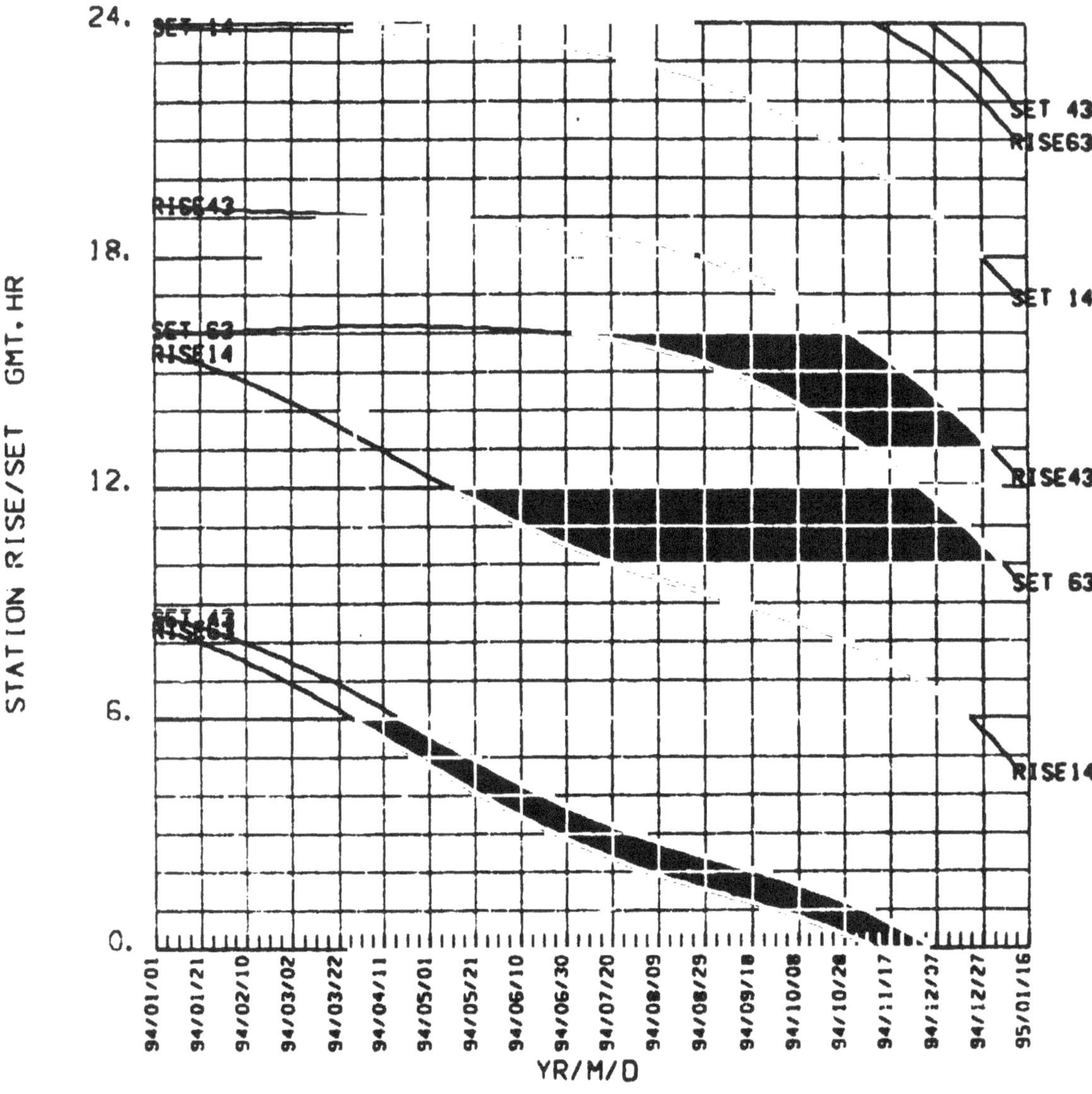

24.
18.
12.
6.
0.
STATION RISE/SET GMT.HR
SET 14
RISE43
SET 63
RISE14
SET 43
RISE63
SET 43
RISE63
SET 14
RISE43
SET 63
RISE14
94/01/01
94/01/21
94/02/10
94/03/02
94/03/22
94/04/11
94/05/01
94/05/21
94/06/10
94/06/30
94/07/20
94/08/09
94/08/29
94/09/18
94/10/08
94/10/28
94/11/17
94/12/07
94/12/27
95/01/16
YR/M/D

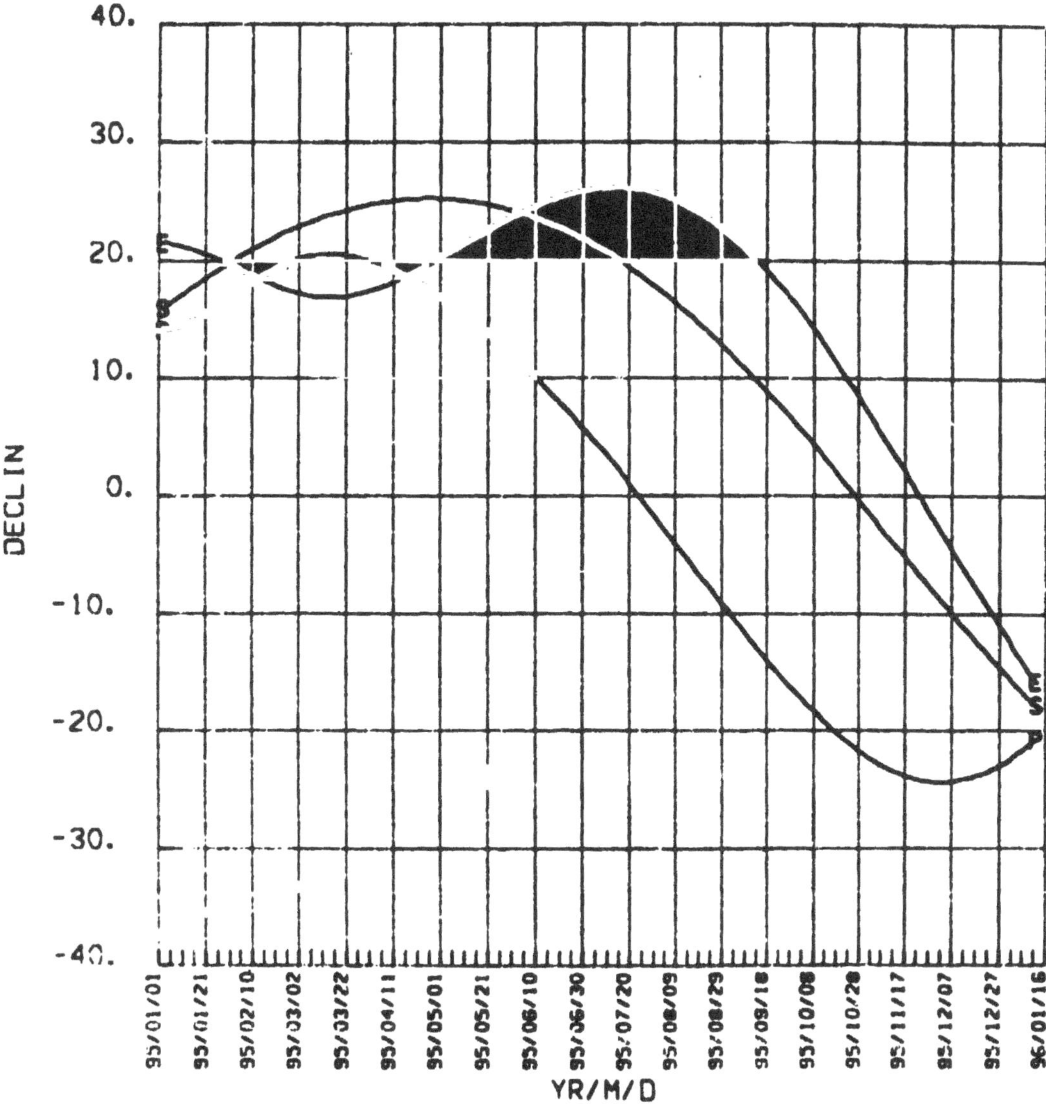

MARS 1995
DECLIN
YR/M/D
40.
30.
20.
10.
0.
-10.
-20.
-30.
-40.
95/01/01
95/01/21
95/02/10
95/03/02
95/03/22
95/04/11
95/05/01
95/05/21
95/06/10
95/06/30
95/07/20
95/08/09
95/08/29
95/09/18
95/10/08
95/10/28
95/11/17
95/12/07
95/12/27
96/01/16

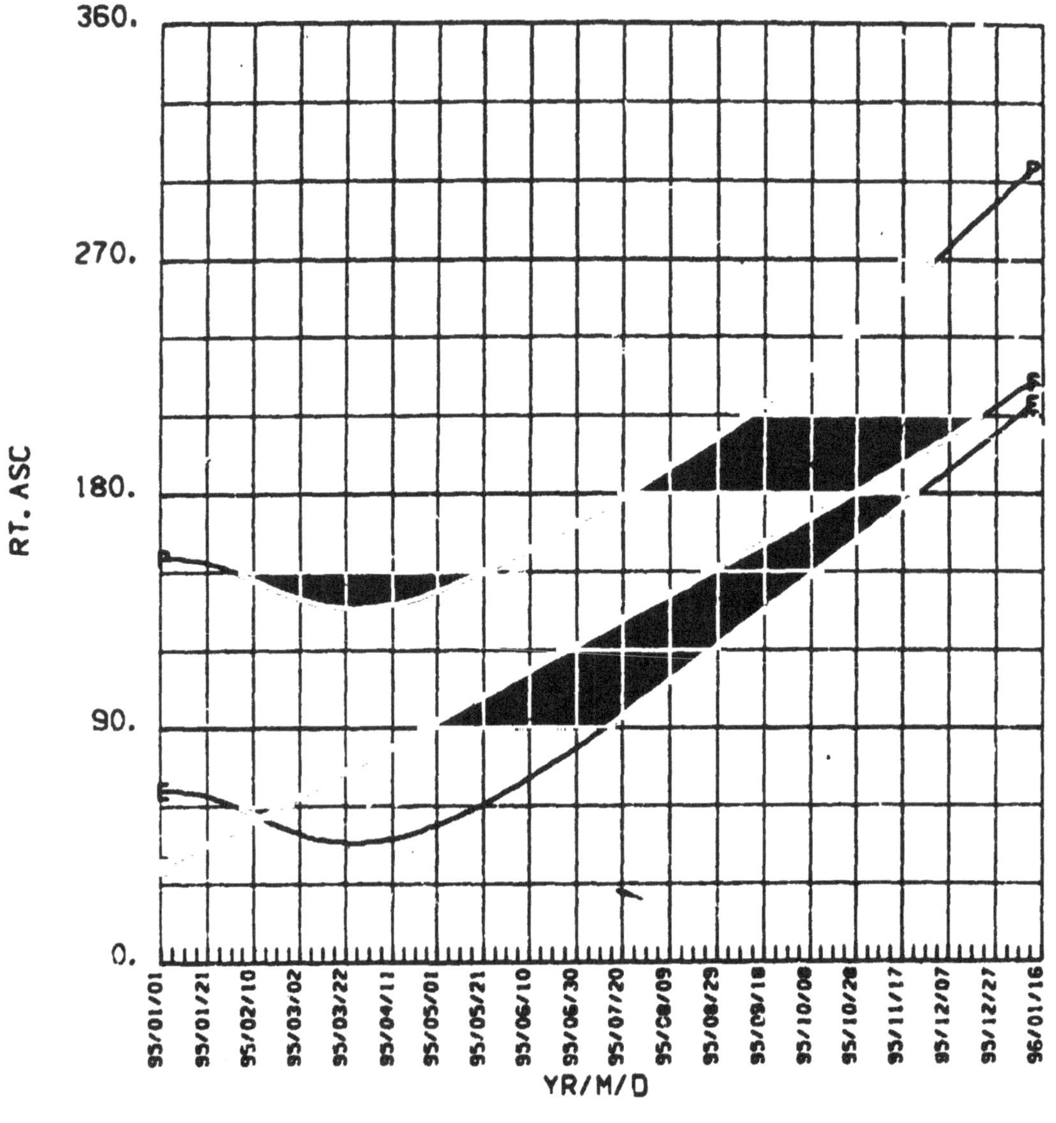

360.
270.
180.
90.
0.

RT. ASC

95/01/01 95/01/21 95/02/10 95/03/02 95/03/22 95/04/11 95/05/01 95/05/21 95/06/10 95/06/30 95/07/20 95/08/09 95/08/29 95/09/18 95/10/08 95/10/28 95/11/17 95/12/07 95/12/27 96/01/16

YR/M/D

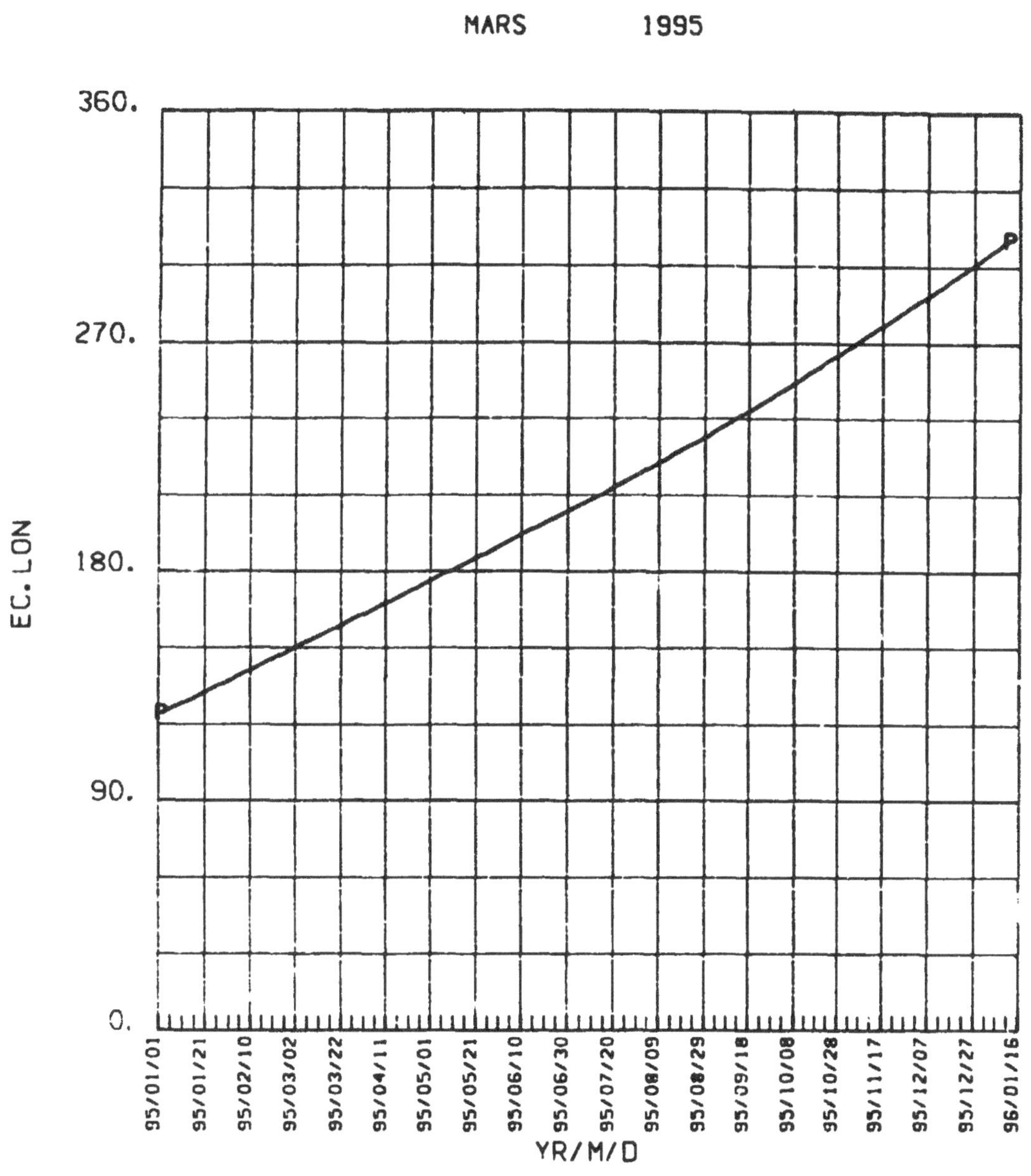

360.
270.
180.
90.
0.
EC. LON

95/01/01
95/01/21
95/02/10
95/03/02
95/03/22
95/04/11
95/05/01
95/05/21
95/06/10
95/06/30
95/07/20
95/08/09
95/08/29
95/09/18
95/10/08
95/10/28
95/11/17
95/12/07
95/12/27
96/01/16
YR/M/D

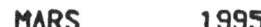

360.
270.
180.
90.
0.

CA, KA OF EARTH, CA CANOP

EKA
CCA
EKA
ECA
CCA
ECA

95/01/01 95/01/21 95/02/10 95/03/02 95/03/22 95/04/11 95/05/01 95/05/21 95/06/10 95/06/30 95/07/20 95/08/09 95/08/29 95/09/18 95/10/08 95/10/28 95/11/17 95/12/07 95/12/27 96/01/16

YR/M/D

MARS 1995

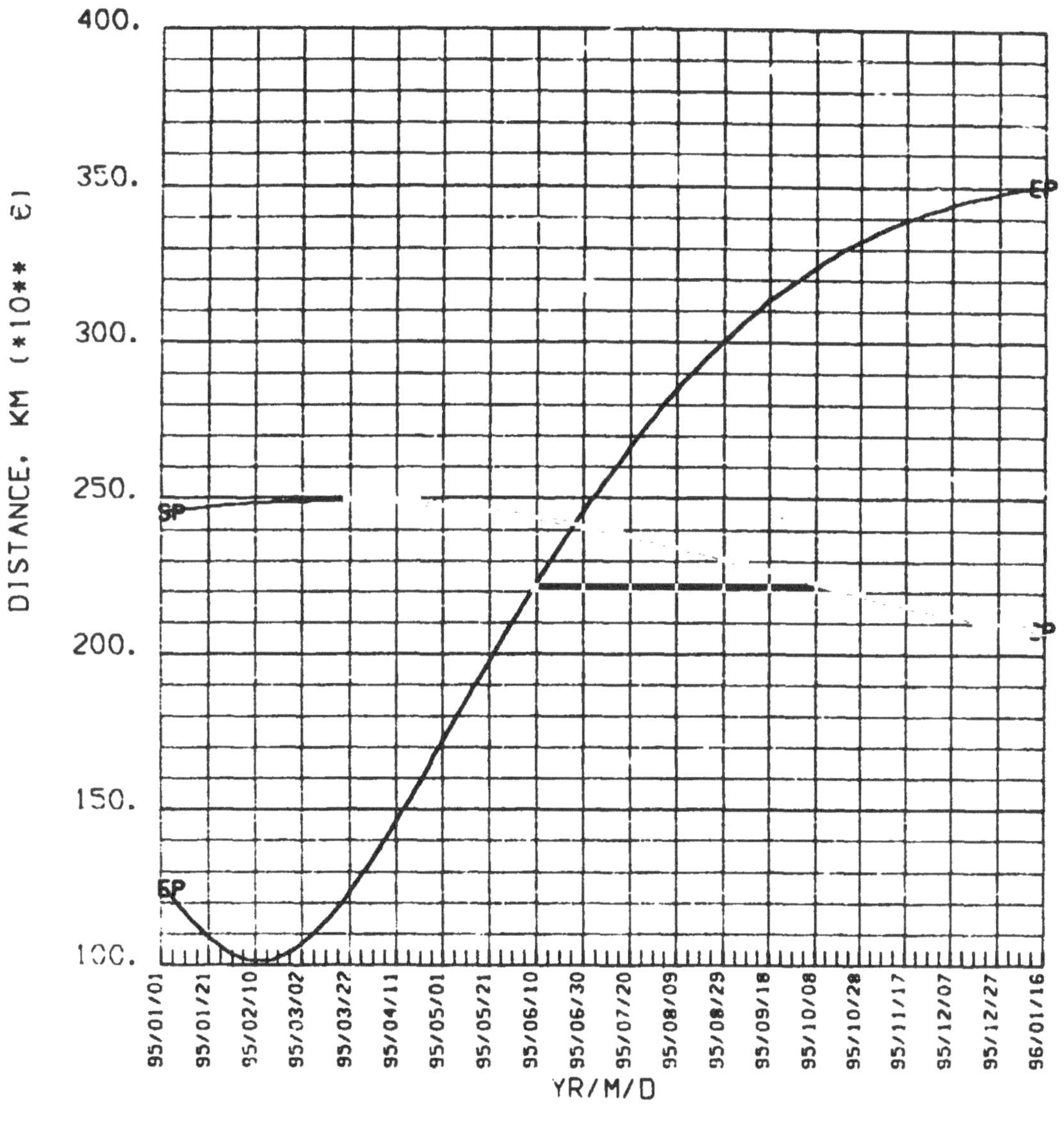

DISTANCE, KM (*10** 6)

400.
350.
300.
250.
200.
150.
100.

SP
EP
EP

95/01/01
95/01/21
95/02/10
95/03/02
95/03/22
95/04/11
95/05/01
95/05/21
95/06/10
95/06/30
95/07/20
95/08/09
95/08/29
95/09/18
95/10/08
95/10/28
95/11/17
95/12/07
95/12/27
96/01/16

YR/M/D

MARS 1995

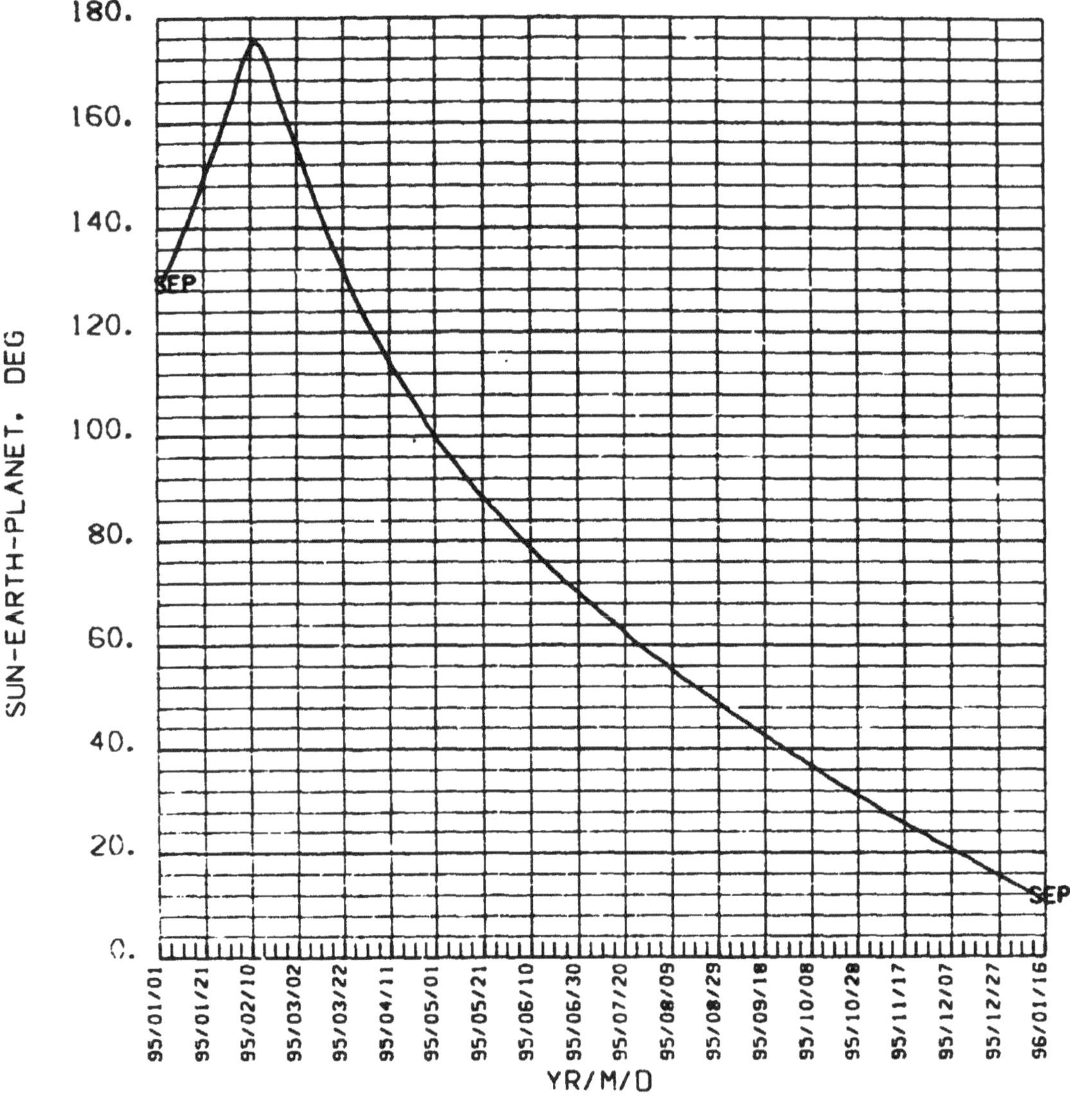

180.
160.
140.
120.
100.
80.
60.
40.
20.
0.
SUN-EARTH-PLANET, DEG
SEP
SEP
95/01/01
95/01/21
95/02/10
95/03/02
95/03/22
95/04/11
95/05/01
95/05/21
95/06/10
95/06/30
95/07/20
95/08/09
95/08/29
95/09/18
95/10/08
95/10/28
95/11/17
95/12/07
95/12/27
96/01/16
YR/M/D

www.ingramcontent.com/pod-product-compliance
Lightning Source LLC
Chambersburg PA
CBHW080240180526
45167CB00006B/2348
9781503148338